High-Temperature Oxidation-Resistant Coatings

COATINGS FOR PROTECTION FROM
OXIDATION OF SUPERALLOYS,
REFRACTORY METALS, AND GRAPHITE

Prepared by the

Committee on Coatings
National Materials Advisory Board
Division of Engineering
National Research Council

NATIONAL ACADEMY OF SCIENCES / NATIONAL ACADEMY OF ENGINEERING
Washington, D.C. 1970

Library of Congress Catalog Card Number 78-606278

Available from

Printing and Publishing Office
National Academy of Sciences
2101 Constitution Avenue
Washington, D.C. 20418

ISBN 0-309-01769-6

Manufactured in the United States of America

Committee on Coatings

PROFESSOR G. M. POUND, *Chairman*, Department of Materials Science, Stanford University, Palo Alto, California 94305.

DR. JOHN GADD, Manager, Technological Development & Control, Jet Ordnance Division, TRW, Inc., Harrisburg Works, 1400 North Cameron Street, Harrisburg, Pennsylvania 17105.

DR. ROBERT I. JAFFEE, Senior Fellow, Department of Physics, Battelle Memorial Institute, 505 King Avenue, Columbus, Ohio 43201.

MR. DONALD L. KUMMER, Chief Ceramics Engineer, Materials and Process Department, McDonnell Douglas Corporation, P.O. Box 516, St. Louis, Missouri 63166.

DR. JAMES R. MYERS, Civil Engineering School, Air Force Institute of Technology, Wright-Patterson Air Force Base, Ohio 45433.

MR. ROGER A. PERKINS, Department 52-30, Palo Alto Research Laboratories, Lockheed Missiles & Space Company, 3251 Hanover Street, Palo Alto, California 94304.

MR. LAWRENCE SAMA, Manager, High Temperature Composites Laboratory, Chemical & Metallurgical Division, Sylvania Electric Products, Inc., Hicksville, New York 11802.

PROFESSOR L. L. SEIGLE, Department of Materials Sciences, College of Engineering, State University of New York, Stony Brook, L.I., New York 11790.

MR. FRANK TALBOOM, Project Metallurgist, Materials Development Laboratory, Pratt & Whitney Aircraft Company, East Hartford, Connecticut 06108.

MR. JAMES M. TAUB, Group Leader—CMB-6, Los Alamos Scientific Laboratory, University of California, P.O. Box 1663, Los Alamos, New Mexico 87544.

DR. HERBERT VOLK, Parma Technical Center, Carbon Products Division, Union Carbide Corporation, P.O. Box 6116, Cleveland, Ohio 44101.

MR. JOHN C. WURST, Department of Ceramic Engineering, University of Illinois, Urbana, Illinois 61801.

Contents

v

General Conclusions and Recommendations

CONCLUSIONS

1a. Based on the performance of simple shapes and test coupons, the current generation of coatings for superalloys and refractory metals meets the basic requirements for *existing* applications. These applications have been designed, however, to operate within the limitations of the available coatings. For example, reusability may not be required, and the operational envelope may therefore be restricted so as not to exceed the coating's temperature capabilities. At the hardware stage, characterization and evaluation of existing refractory-metal coating systems are not adequate. In addition, improvements in the processing of large or complex shapes may be required.

1b. The current generation of coating systems is not adequate for the advanced applications that are predicted to evolve within the next five to fifteen years. Major advances will be required in oxidation behavior, microstructural stability, manufacturing technology, and overall system reliability. The prospect of developing adequate systems is not favorable, and a reappraisal of design concepts and performance requirements is therefore indicated.

2. There is a dearth of sound approaches to the development of new or improved coating systems. Current technology depends solely on the formation of Al_2O_3 or SiO_2 as the protective oxide. Coatings that form complex oxides, spinels, or dense glasses are needed to provide a major advance in performance capabilities. These coatings also must possess a self-healing mechanism for reliability in service. Coatings of the fused-slurry type are regarded as having the best potential for meeting future needs.

3. A fundamental understanding of the factors that govern the performance of existing coatings is needed to provide a sound basis for developing of advanced coating systems. To date, only a minimal understanding of a few very simple systems has been attained, and significant advances are required in this area. It is known that the useful life of most coatings is a small fraction of the maximum attainable under ideal conditions. Performance of most coatings is established and controlled by random defects. In the absence of defects, coating life is limited by diffusional processes at low temperature and by vaporization or melting at high temperature.

4. Specific conclusions related to general areas of application are as follows:

a. Gas Turbines

(1) Coated alloys in advanced engines must have a twofold to threefold increase in useful life and a 200°F increase in maximum temperature capability, i.e., to 2200°F maximum.

(2) The ability to coat small internal passages is not adequate, and manufacturing improvements are required.

(3) Recoating processes result in excessive substrate attack, and improved methods are required.

(4) The current techniques for nondestructive testing are not satisfactory, and improved methods are needed.

b. Chemical Propulsion

(1) Available coating systems are adequate for existing applications.

(2) Currently, there is no need for major advances in this area.

1

c. Hypersonic Vehicles

(1) The current generation of coatings is adequate for existing and near-term future applications.

(2) The reuse capability of coated parts will extend to one or two flights. Current test and performance data are inadequate to permit a better assessment of reuse capability.

(3) Coated columbium alloys are limited to a 2500°F maximum operating temperature (coating lifetime of 150–200 hr) by creep properties. Inadequate creep resistance, not coating technology, limits the use of these materials.

(4) Satisfactory coatings do not exist for use on molybdenum or tantalum alloys above 3000°F and on tungsten above 3300°F. Further work on silicide-base coatings to extend the service temperatures of these alloys is not warranted, and a new approach to coating must be taken.

d. Energy Conversion Systems

(1) Oxidation-resistant coatings generally are not needed. One exception is the platinum coating on radioisotope thermoelectric generating (RTG) systems. Current technology is satisfactory for this application.

e. Industrial Applications

(1) Use of coated metals is limited, and current technology appears to be adequate.

RECOMMENDATIONS

1. The minimum performance capabilities and reliability of existing coatings for superalloys and refractory metals should be upgraded by means of processing improvements. Specifically:

a. The mitigation and control of defects should be studied.

b. New coating deposition processes, particularly suited to large surfaces and complex shapes, should be developed.

c. Field repair techniques should be devised.

d. Recoating processes that reduce substrate loss should be developed.

e. Processes for the coating of small internal passages should be improved.

f. Automated manufacturing processes should be considered.

2. New approaches to the development of oxidation-resistant coatings should be explored. Specifically:

a. Compositions that form complex oxides, perovskites, or spinels on oxidation should be investigated as coating materials.

b. Preference should be given to single-layer coatings applied by slurry processes.

3. A basic understanding of the mechanisms that govern coating behavior should be obtained for current systems. Specifically:

a. The origin of critical defects should be determined. Factors governing the nature, size, and distribution of defects must be defined.

b. Layer growth and diffusion studies should be made with existing systems to establish basic levels of stability and kinetics of compositional changes.

c. Phase diagrams (equilibrium), thermodynamics, and kinetics of oxidation processes should be established for representative coating systems.

d. The physical chemistry of deposition (cementation) processes should be studied for silicide and aluminide coatings.

4. Existing coating systems should be more fully characterized and evaluated at the hardware stage. Specifically:

a. A statistical analysis of performance (and failure) in relation to coating defects should be made.

b. Nondestructive test procedures should be perfected for process control and prediction of coating behavior.

c. Flight environment tests of hardware should be conducted.

d. Accelerated coating-life tests for prediction of long-term performance should be developed.

e. Mechanical and oxidation test methods should be standardized.

f. Comprehensive hardware studies should be made to establish the basic levels of quality, reproducibility, and performance for existing coating systems.

1

Background and Summary of the Program

INTRODUCTION

Coatings to protect structural materials against oxidation at elevated temperature have become important only in recent years. In the past, superalloys based on nickel or cobalt containing about 20 percent chromium were used where both strength and oxidation resistance were required, as in the turbine section of aircraft jet engines. In these alloys, strength falls off at elevated temperatures somewhat more rapidly than does oxidation resistance. Thus, the turbine inlet temperature has been maintained below about 1600°F, where strength begins to fall off. High-temperature oxidation is not a matter of concern at this temperature, and superalloys could be used in gas turbines for thousands of hours without incurring significant oxidation or sulfidation damage.

In recent years, stronger superalloys have been developed based on increasing the aluminum and titanium content needed to form strengthening γ' precipitates at the expense of chromium content. This has resulted in a new generation of strong superalloys containing about 10 percent chromium that can be used at higher temperatures without losing strength. Unfortunately, they have significantly poorer hot-corrosion resistance than the 20 percent chromium alloys. To combat this problem, coatings were developed, largely based on monoaluminides produced by diffusion with the substrate alloys. These coatings provide improved oxidation and sulfidation resistance and have been remarkably successful in some applications, even with the increased turbine-inlet temperatures (up to 2200°F) that have been made

possible through the use of bypass air-cooling of the hot components.

Other developments that have prompted increased consideration of high-temperature coatings are glide re-entry and hypersonic vehicles. These advanced aerospace systems have become of serious interest only in recent years, starting in the early 1960's. Aerodynamic heating of lift and control surfaces require structural materials that can operate for short periods of time (minutes to hours) at 2500-3000°F, depending upon the re-entry corridor of the vehicle. Refractory metals afford an excellent basis for fabrication of such advanced structures. During the early 1960's, a national capability was developed in the United States for producing mill products of strong refractory metal alloys based on columbium, tantalum, molybdenum, and tungsten. In a remarkably short time, mill products of aircraft quality and size were produced. The Materials Advisory Board (MAB)* provided guidance during this effort through the MAB Refractory Metal Sheet Rolling Panel (MAB Report 212M, March, 1966). Coatings based on disilicides and monoaluminides capable of protecting refractory metal alloy structures during typical glide re-entry environments were developed concurrently.

In future aerospace systems, projected requirements for coatings for refractory metals will be much more severe than was the case for the early re-entry vehicles of the Dynasoar and ASSET classes. In 1965, the Materials Advisory Board Aerospace Applications Requirements Panel forecast that

*Since January 1969, the National Materials Advisory Board.

3

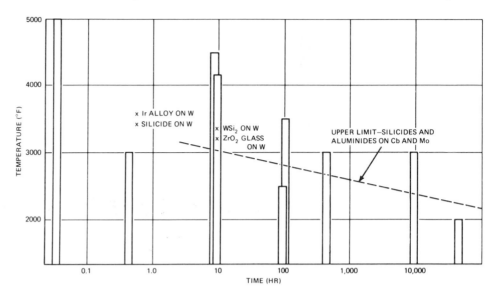

FIGURE 1   Requirements and performance of thin coatings on refractory metals.

future aerospace vehicles would require coatings for refractory metals that would operate in oxidizing environments at the temperatures and times given by the vertical bars in Figure 1. As at present, the future coatings would be thin (<0.005 in. thick) and would have to withstand thermal cycling. On the same figure are plotted the capabilities of oxidation protective coatings as they stood in 1965. The diagonal line represents the best performance of thin disilicide or monoaluminide coatings on columbium or molybdenum tested under laboratory conditions in still air. In addition, a few points on the figure give laboratory performance data for a number of coatings on tungsten. These fall in line with the performance line for coated columbium and molybdenum. Compared with future systems requirements in aerospace vehicles, the performance of state-of-the-art coatings falls considerably short, even under the maximum performance conditions found in the laboratory.

It is apparent that coatings for high-temperature materials will become increasingly important as time passes. The National Aeronautics and Space Administration (NASA), therefore, requested that the Materials Advisory Board convene a committee to review the status of coatings for high-temperature materials, including superalloys, refractory metals, and graphite; to consider the present research and development under way; and to recommend modifications in scope and emphasis.

OBJECTIVES OF THE PROGRAM

The Coating Subpanel of the Refractory Metal Sheet Rolling Panel recommended in 1966 (MAB-212M) that a new Panel on Oxidation-Resistant Coatings be established. In line with their recommendations, amplified by later experience, the following objectives were established to guide this panel study:

1. Define current and future applications for coated superalloys, refractory metals, and graphites.
2. Define the performance requirements for these applications.
3. Review basic coating concepts and current status.
4. Define key technical problem areas in performance, manufacture, and testing in the light of current and future requirements.
5. Define possible approaches to the solution of key problems.
6. Indicate the nature or direction of research and development studies for producing major advances in coating technology.

METHOD OF OPERATION

Work of the Panel on High-Temperature Oxidation-Resistant Coatings was divided into two major phases:

Phase I—Review of the current status and future requirements for coatings on superalloys, refractory metals, and graphite.

Phase II—Detailed analysis of coating concepts, methods of protection and failure, applications technology, manufacturing processes, testing, and inspection.

In Phase I, individual members of the main panel and DOD liaison representatives, in accordance with their special interests or abilities, conducted a cursory review of various aspects of coating technology as shown in Table 1. The results of this study were published in 1968 as a first progress report (MAB-234).

The analysis of Phase I results indicated the areas in which more detailed studies would be required to satisfy program objectives. Task groups were organized, as shown in Table 2, to implement the studies. Leading experts in related fields from industry, government, and education were appointed to the task groups. Each task group met to establish individual assignments for study and analysis. The sections of each subpanel report were submitted to the task group chairmen, who prepared the final input for the main panel report.

The main panel met periodically to review the subpanel reports and conclusions. Each task group was requested to

submit to the main panel for consideration five primary recommendations related to key technical problems in their respective areas, plus any additional secondary recommendations it wished to make. These reports were reviewed and discussed at the last panel meeting. Draft copies of each task group report were reviewed and discussed also by the main panel at its last meeting.

An editorial committee was commissioned to assemble the final report and to prepare the overall conclusions and recommendations; its members were G. M. Pound, R. I. Jaffee, and R. A. Perkins.

The documents that have been generated under this program provide a very broad and comprehensive review of technology for high-temperature oxidation-resistant coatings. The information should be of great value to the National Aeronautics and Space Administration, the Department of Defense (DOD), and industry representatives who are charged with the responsibility of applying coating technology to the design and manufacture of a wide range of devices, structures, power plants, and vehicles. The method of operation selected by the panel to meet its objectives proved to be extremely effective. All members of the panel and all task groups are to be commended for the thoroughness in which they carried out their respective assignments. The wealth of information that has been assembled, analyzed, and correlated attests to the sincerity of their efforts and the overall effectiveness and worth of this study. While the data and conclusions refer to the period 1967-1969, when the Committee was active, there has been relatively little work supported since that time, and the findings remain valid.

TABLE 1 Phase I Study Assignments

Coating concepts	L. L. Seigle
Testing and standards	J. C. Wurst
Coatings for superalloys	J. R. Myers
Coatings for chromium	S. J. Grisaffe
Coatings for columbium	J. D. Gadd
Coatings for molybdenum	R. A. Perkins
Coatings for tantalum and tungsten	L. Sama
Coatings for graphite	H. Volk
Requirements for hypersonic aircraft	E. E. Mathauser
Glide re-entry requirements	L. Sama
Requirements for ramjets	I. Machlin
Requirements in energy conversion systems	R. I. Jaffee
Requirements for rocket nozzles	S. J. Grisaffe

TABLE 2 Phase II Task Groups

Task Group	Chairman
Concepts	L. L. Seigle
Gas-turbine applications	J. R. Myers
Rocket propulsion applications	H. F. Volk
Hypersonic vehicle applications	D. L. Kummer
Energy conversion applications	R. I. Jaffee
Industrial applications	S. J. Grisaffe
Manufacturing techniques	J. D. Gadd
Testing and inspection	J. C. Wurst

TASK GROUP SUMMARIES

FUNDAMENTALS OF COATING SYSTEMS

SUMMARY AND CONCLUSIONS

Coatings developed for protecting refractory metals against oxidation may be classified as (a) intermetallic compounds that form compact or glassy oxide layers, (b) alloys that form compact oxide layers, (c) noble metals that resist oxidation, and (d) stable oxides that provide a physical barrier to the penetration of oxygen. Successful coatings exhibit certain characteristics, among which the following are important:

  1. Low growth rate of the protective oxide layer
  2. Resistance to cracking
  3. Low evaporation rates of the protective oxide and other coating constituents

4. Low rates of coatings–substrate reaction
5. Self-healing ability

It may be stated as a general principle that because of higher activation energies, vaporization processes tend to predominate over diffusion processes at very high temperatures, and, conversely, diffusion processes tend to predominate at lower temperatures. This applies to the silicide and aluminide coatings as well as to ThO_2. For example, at temperatures above approximately 3000°F, the evaporation rate of SiO_2 becomes significant, as do the vapor pressures of SiO in equilibrium with $Si + SiO_2$ and of Si in equilibrium with the higher silicides. With aluminide coatings the vapor pressure of Al is appreciable at higher temperatures. The role of vaporization in coating degradation is also enhanced by low ambient pressures, which not only allow more rapid vaporization but may even lead to circumstances under which a protective oxide layer does not form at all.

The occurrence of defects in coating systems makes lifetimes in practice fall far short of those theoretically achievable with a flawless coating. In silicide and aluminide coatings, cracks seemingly are always introduced during application of the coating or are formed during exposure. Thermal cycling enlarges these cracks until they penetrate through the coating to the substrate, causing localized failures in a fraction of the time required for the coating as a whole to deteriorate. Different susceptibilities to cracking exist in different systems; silicide coatings are prone to the formation of such defects, while aluminide coatings on superalloys are more resistant to cracking. Methods to ameliorate the influence of cracking upon coating lifetimes have been developed, such as the use of a liquid Sn–Al phase to seal cracks, or the use of complex silicides, fused to the surface, that resist or tolerate cracking. In general, however, not enough is known about the nature and distribution of cracks in various coating systems and the mechanisms of crack propagation and crack healing.

From experience gained in practical coating development, as well as from theoretical analysis of coating behavior, it is possible to draw some conclusions about the concepts that underlie or may lead to the development of successful coatings. For an oxide to be protective it must resist transport of both metal and oxygen ions. This implies not only low intrinsic diffusivities of the ions but a composition which remains close to stoichiometric and a minimum of lattice defects. Oxides other than SiO_2 and Al_2O_3 are known to possess these characteristics—e.g., MgO, CaO, and BeO—although they may be unsuitable for other reasons. Complex oxides with a spinel or perovskite structure may exhibit low diffusivities. These are largely unexplored.

The ideal protective oxide would form a sound and coherent layer over the substrate. Preservation of such a layer is aided by the glass-forming ability of the oxide, as with SiO_2, since glassy oxides can flow and accommodate mechanical strain and heal defects. Other glass-forming oxides, such as B_2O_3, may prove useful. The known procedures for enamel-coating development may be of assistance in developing glassy oxide coatings which resist cracking.

Since it is difficult to form a protective oxide that remains absolutely sound under all conditions, the self-healing capacity, i.e., the ability of a coating to regenerate its protective oxide layer in case of surface cracking, has proven to be of great importance. This capability is related to the preservation of an active reservoir of protective-oxide-forming substance (e.g., disilicide or dialuminide) in the coating. Minimizing diffusion into the substrate is an important step in preserving this layer; hence diffusion barriers are a useful concept in coating development. The reservoir layer, however, is often short-circuited by the penetration of cracks formed during processing or exposure; thus the ability of this layer itself to resist cracking is of the greatest importance. Such crack resistance may be obtained by raising the fracture strength of the layer through development of a fine-grained structure or by increasing the plasticity of the layer by the introduction of fluid or plastic constituents.

The concept of employing oxidation-resistant, nonstructural alloys such as Hf–Ta or Nb–Ti–Al–Cr, to coat refractory metals has had some success. Such coatings can deform without cracking, and they afford a means of protection against foreign-object damage.

Intermetallic Compounds

The penetration of cracks is a highly important factor in determining the lifetime of a coating, but even in the absence of cracking, a protective coating would ultimately fail due to the inevitable progress of oxidation, diffusion, and evaporation reactions at high temperatures. Analysis of these mechanisms of deterioration indicates that different processes are important in different coating systems at different temperatures. In simple silicide and aluminide coatings, used at or near 2500°F, it appears that coating–substrate interdiffusion is the important degradation process in air at 1 atm pressure. This occurs more rapidly than either growth or evaporation of the oxide film and leads to a theoretical coating lifetime in the absence of cracking of the order of 1,000 hr, more or less, depending upon the specific system* for a thin (~5-mil) coating. It would take about this long for a 5-mil disilicide or aluminide layer to be converted to lower silicides or aluminides. Observed coating lifetimes under test conditions are, of course, much shorter than the theoretical lifetimes, a result of cracks that develop during processing or testing and penetrate the silicide or aluminide layers. These cracks cause local failures before a large fraction of the layers has reacted with the substrates.

*Theoretical lifetimes are greater for silicide than for aluminum coatings.

Oxidation-Resistant Alloys

Although the data are incomplete, interdiffusion between coatings and substrate is believed to be a major factor in the degradation of Ni- and Co-based superalloy coatings on the refractory metals. In any event, these coatings are limited to temperatures below about 2100°F because of possible occurrence of low-melting constituents in the diffusion zones.

Noble Metals

From the available data, it appears that at temperatures in the vicinity of 2000°C, evaporation of the noble (Pt-group) metals is an important deterioration process since this seems to occur, at least for Ir on W, much more rapidly than interdiffusion with the substrate. A 10-mil coating, for example, would evaporate in about 10 hr in low-pressure air at 2000°C but would require hundreds of hours to diffuse into the substrate.

Stable Oxides

Evaporation also becomes important for stable oxides, such as ThO_2 on W, at the very high use temperatures contemplated for such coatings. Calculations indicate that it would require only minutes for evaporation of a 10-mil ThO_2 layer at 2500°C, while the rate of reaction of ThO_2 with W is negligible at this temperature. On the other hand, at elevated temperatures, the diffusion of oxygen is, surprisingly, many times more rapid in ThO_2 than in SiO_2 or Al_2O_3, and diffusion of oxygen through even a coherent layer of ThO_2 could be a life-limiting factor in a certain temperature range, perhaps just below 2500°C. At higher temperatures, however, it is clear that evaporation theoretically becomes predominant.

RECOMMENDATIONS

The five most important recommendations of this subpanel are:

1. That a thorough study be made of the origin and control of defects during the formation of coatings and during their use in service.
2. That the influence of defects on coating lifetimes be defined quantitatively by the use of statistical analysis.
3. That layer growth studies and diffusion measurements in successful complex coating systems be made in order to identify the reasons for superior performance.
4. That new oxides be sought, around which to design new coatings. Complex oxides with the spinel or perovskite structures are particularly recommended for investigation.
5. That further information be obtained about the phase diagrams, thermodynamic properties, and kinetics of reactions of coating constituents with substrates and oxygen, including the substrate-oxide–coating-oxide system.

A striking fact that emerges from this study is the serious lack of phase diagram, thermodynamic, and/or diffusion data for the systems of interest, necessary for a complete understanding of coating behavior at high temperatures. This is particularly true when the combination oxygen–coating–substrate represents a ternary or higher order system. While sporadic attempts have been made from time to time to obtain portions of this information, it is clear that a fairly large and sustained effort is needed to do the job more rapidly and completely. Since data of this type are often lacking in fields of materials technology, attention should be given by federal agencies to the organization of a sufficiently large and systematic effort to obtain such data on a continuing basis. It is clear that information of this type is not now being obtained quickly enough. It is also clear that there is little hope of changing the situation without an effort such as that recommended.

For protective coatings, even in the absence of complete data on diffusion, etc., it seems clear that much useful information can be obtained by carefully studying the behavior of practical coatings, using metallographic, x-ray diffraction, microprobe, and other techniques, to investigate the reactions occurring and the structures developed in the coating system and to relate these to the success or failure of the coating. For example, it would be particularly valuable to ascertain the reasons for the superior performance of the fused-silicide coatings. Investigations of this type should be included more often as part of the coating evaluation process.

Finally, certain approaches to improving existing coatings, as well as to developing new types of coatings, are indicated. These include the development of means to eliminate or inhibit cracking in silicide and aluminide coatings and of means to prevent diffusion into the substrate. New oxide systems should be sought, around which to design new coatings. The possibilities in complex oxides merit further investigation, as does the development of means to decrease the point defect concentrations in such refractory oxides as ZrO_2 and ThO_2. Glass-forming oxides, such as B_2O_3, may be worth further study. Further consideration of the use of oxidation-resistant alloys as coatings or parts of coatings also seems to be desirable.

APPLICATIONS OF COATING SYSTEMS

GAS TURBINES

Summary and Conclusions

a. *Applications and Environment* Gas-turbine engines are being developed for numerous industrial, marine, and vehicular applications. Industrial engines, which already provide some primary electrical power generation and standby power for larger electric generating facilities, will be used more frequently in future years for a variety of purposes. Marine engines are being, and will continue to be, employed to power high-speed boats and ships, including relatively large vessels in submarine-chaser and transport classifications. Efficient and economical engines will be developed for large commercial and military trucks and military tanks. Gas turbines may also be used to power high-speed ground transportation systems. These newer applications and the increased demand for improved performance in commercial and military aircraft engines will result in unprecedented growth in gas-turbine engine technology and production during the next 15 years.

The requirements for increased performance undoubtedly will result in higher turbine-inlet temperatures. Such temperatures are expected to reach 2400° to $2700^\circ F$, as compared with the 2100° to $2300^\circ F$ maximum currently reached. However, air-cooling schemes will maintain turbine blades in the $1800^\circ F$ regime and turbine vanes at $2000^\circ F$ maximum. This metal temperature level indicates that superalloys will have to be used for some time for turbine hardware.

The phenomena that contribute to the deterioration and failure of hot-section components in gas-turbine engines are:

1. Oxidation and sulfidation (hot corrosion)—Sulfidation is an accelerated form of high-temperature corrosion associated with the presence of Na_2SO_4. Sulfur from the fuel reacts during combustion with ingested sodium chloride from the air to form the Na_2SO_4.

2. Creep—Centrifugal loading at high temperatures can cause extension of rotating blades; gas pressure differentials can cause bowing of stationary vanes.

3. Thermal fatigue—This is a form of low-cycle fatigue; grain boundary oxidation can contribute to the initiation of thermal fatigue cracks.

4. Erosion—Carbon particles, dust, and other ingested particular matter can cause premature removal of engine components.

5. Foreign-object damage (FOD)—Damage from debris ingested by the engine, or failure of upstream components, is a major reason for replacing turbine blades and vanes.

6. Overtemperature—Re-solution of alloy constituents and incipient melting seriously affect the structural integrity of the components.

Cooling techniques are necessary to lower metal temperature of many components, since turbine-inlet temperatures exceed about $1900^\circ F$. Refractory metal alloys of Mo, Cb, W, or Ta probably will not be used to any large extent in near-future engines because sufficiently reliable, oxidation-resistant coatings for these materials are not available. The same would be true of chromium-base alloys in the event a suitable alloy were available. Therefore, for the long life expectancies required for future engines, high-temperature corrosion must be minimized; protective coatings appear to be the most promising approach. In addition to minimizing the hot-corrosion problem, advanced coating systems may also be beneficial in reducing thermal and mechanical fatigue and improving erosion and FOD resistance.

b. *Performance and Reliability* The basic requirements for a coating for superalloy substrates in gas-turbine engines dictate that: it must resist the thermal environment; it must be metallurgically bonded to the substrate; it should be thin, uniform, light in weight, reasonably low in cost, and relatively easy to apply; it should have some self-healing characteristics; it should be ductile; it should not adversely affect the mechanical properties of the substrate; and it should exhibit diffusional stability.

A number of organizations have developed high-temperature protective coatings for superalloys, and many of these coatings have been used effectively in gas-turbine engines to extend the life expectancies of hot-section components. All of the currently important coatings are based on the use of aluminum as the primary coating constituent. Processes used to obtain these aluminum-rich coatings are: pack cementation, hot-dipping, slurry, and electrophoresis. Regardless of processing techniques, however, basic similarities exist among the current coatings in that they are predominantly composed of NiAl (nickel-base alloys) or CoAl (cobalt-base alloys) with minor additions of an alloying element, usually chromium.

Although the current coatings do not satisfy all the basic performance requirements mentioned above, aluminide-type coatings successfully extend the life expectancies of gas-turbine engine blades and vanes. Because the advantages obtained from using coatings outweigh the disadvantages, several hundred thousand superalloy blades and vanes for military and commercial aircraft gas-turbine engines are coated by the engine manufacturers and coating vendors in the United States each month.

However, future coatings for superalloy, hot-section components will have to exhibit performance capabilities superior to those of the currently available aluminide coatings. With improved coatings, blade and vane life expectancies hopefully will exceed 12,000 hr at metal temperatures of $1700^\circ F$ and $1850^\circ F$, respectively. A twofold to threefold improvement in life may be expected in the higher temperature military gas-turbine engines.

For achievement of these goals, future coatings should possess the following characteristics: improved diffusional stability; improved resistance to oxidation, sulfidation, and hot-gas erosion; improved ductility; a $200^\circ F$ increase in operating-temperature capability; improved resistance to thermal and mechanical fatigue; improved resistance to thermal shock; minimum adverse effect of the mechanical properties of the substrate; some self-healing capability; and compatibility with brazing alloys.

c. *Manufacturing Technology* Large-scale coating production for superalloy substrates has been limited primarily to individual turbine blades and vanes, varying in length from about 1.5 to 6 in. A variety of nickel-base and cobalt-base alloys are involved, as well as the various coating processes previously mentioned.

Engine hardware often requires specialized coating techniques, often difficult to apply properly. Blade root sections generally are not coated and so require suitable masking techniques. Coating internal cooling passages is difficult, particularly with certain coating processes. Recoating engine-operated hardware is another requirement. Large hot-section components, up to 48 in. in diameter and 10 in. deep, have been coated experimentally, but processing difficulties were evident.

Future needs indicate that, in addition to an increased production capacity requirement, a capability to coat engine components other than turbine blades and vanes will be needed. Several production problems must be solved, including processing of large and very small components, adaptation of process to more and newer alloys, recoating, coating of internal passage, better process control, and newer coating techniques. Nondestructive test and inspection procedures must also be included.

Recommendations for Advancing the State of the Art

It is believed that a number of research and development programs can be outlined that will greatly assist in the development of advanced coating systems for superalloys to be used in future gas-turbine engines. The primary recommendations of the Subcommittee are as follows:

1. Conduct studies to evaluate thoroughly the potential (i.e., to determine advantages and limitations) of newer coating techniques for depositing advanced coatings on superalloy components used in gas-turbine engines. Processes to be considered should include: cladding, electrodeposition from fused salts, pyrolytic deposition from organometallic vapors or solutions, and physical vapor deposition, especially as applied to alloys. Successful research results in developing new coating processes probably will be necessary in order to permit deposition of advantageous coating alloys.

2. Develop a recoating capability that will minimize (or eliminate) substrate-metal loss.

3. Develop nondestructive testing methods for process control and for predicting the useful remaining life of coated turbine-engine components. Techniques should measure or identify defects, corrosion damage, coating thickness, extent of coating–substrate interdiffusion, fatigue damage, and wall thicknesses of cooled components.

4. Develop manufacturing techniques for the reliable coating of extremely small-diameter internal cooling passages used in convection and transpiration cooling.

5. Develop automated manufacturing processes in order to improve coating reproducibility and reliability.

In addition, a number of secondary recommendations are made to point out areas of study on a more specific basis:

1. Conduct research to define statistically the effect of modifier elements (e.g., rare earths) on resistance of aluminide coatings to oxidation and sulfidation.

2. Determine the effect of alloying elements on the compositional range of NiAl (CoAl). If certain elements can be added that will increase the aluminum content of this intermediate phase, the larger reservoir of available aluminum could increase coating life expectancies.

3. Determine oxidation and sulfidation kinetics (using both kinetic and static tests) for promising binary, ternary, and quaternary alloys in order to find a replacement coating for the basic currently used aluminides. These alloys should have some inherent ductility in order to be considered for use as future coatings.

4. Attempt to raise the melting point of the nickel- or cobalt-poor, aluminum-rich diffusion zone in currently used aluminide coating systems. Modifier elements may prove effective in satisfying this necessity for future aluminide coatings.

5. Conduct additional research to develop a coating for dispersion-strengthened alloys that will function for long periods at temperatures above 2000°F.

6. Continue the preliminary studies of adding discrete particles (e.g., Al_2O_3) to the coating to improve resistance of aluminide coatings to oxidation and sulfidation.

7. Develop a simplified procedure for coating turbine-blade tips to provide for oxidation and sulfidation resistance at this location after minor blade-length modifications have been made to coated blades during engine fit-up.

8. Determine the effect of coating-process variables, coating chemistry, and postcoating heat treatments on (aluminide-type) coating ductility.

9. Establish the effect of long-time service on the mechanical properties of coated components, especially thin sections.

10. Continue research on the use of diffusion barriers (or elements added to the coating to reduce thermodynamic activities of diffusing species) to minimize interdiffusion between the coating and the substrate.

11. Obtain a more comprehensive and fundamental understanding of the degradation phenomena for coated components.

12. Continue research on the development of protective coatings for columbium-base and chromium-base alloys.

CHEMICAL PROPULSION

Summary

a. *Launch Engines*   In almost every case, propulsion rocket engines in present use are regeneratively cooled and uncoated. One exception is the transtage of the Titan II, which employs a thermal insulation coating consisting of a proprietary metal oxide mixture plasma-sprayed onto the nozzle tubes. Another engine employing regenerative cooling supplemented by an insulative coating is the XLR-99, which powered the X-15 experimental aircraft. In this engine the stainless steel nozzle tubes, cooled with ammonia fuel, were further protected by a plasma-sprayed, graded zirconia–nickel chromium alloy deposited on a plasma-sprayed molybdenum undercoating.

Thermal-lag coatings are used in liquid- and solid-fueled rocket engines, providing the firing times are very short (frequently less than 5 sec), so that the heat from combustion does not degrade the mechanical properties of the lightweight structural alloys. The Bullpup missile employs flame-sprayed zirconia on both the aluminum combustion chamber and on the nickel–chromium alloy precoated copper nozzle. The solid-fueled Scout missile uses a flame-sprayed zirconia coating to protect its ZTA graphite nozzle. The Lance surface-to-surface missile uses a metal-graded zirconia thermal-lag coating.

In contrast to current practice, many of the advanced regeneratively cooled rocket engine concepts incorporate thermal insulation coatings. In such concepts, gas temperatures, chamber pressures, and heat-transfer rates are higher than those now encountered. Thus, higher heat fluxes to the structure and the coolant can be expected. These heat fluxes must be reduced if nickel alloys, Type 347 stainless steel, and Inconel-X continue to be used as structural materials. Thermal-barrier coatings may include refractory oxides, refractory metals, or mixtures, or graded layers of both, as well as the refractory metal carbides, nitrides, and borides. These coatings may be applied by plasma-spraying or slurry application. Undercoatings of intermediate melting point may also be used to relieve thermal shock and improve coating adherence.

Examples of regeneratively cooled nozzle conditions for which coatings will be required are hydrogen–oxygen and Flox–propane engines. For nozzles ranging from 3 in. in diameter to very large boosters, H_2-O_2 combustion will develop environments containing H, H_2, O, O_2, and OH with gas temperatures of about $5900°F$ (approximately $3250°C$) and chamber pressures from 1,000 to 2,000 psia. Coating surfaces may reach $4000°F$ ($2200°C$). Coatings having thermal resistances of 120 to 250 in.^{2} sec-$°R$/Btu will be expected to drop heat fluxes to the nozzle from 60 Btu sec-in.^{2} (for uncoated nozzles) to 20 Btu/sec-in.^{2} for a coated nozzle. The latter heat flux will allow the liquid hydrogen to keep the tube surfaces under the coating cooler than $1600°F$ (approximately $870°C$) and thus structurally sound.

For small second-stage engines (3 to 5 in. in diameter), the use of Flox (liquid oxygen containing 76 wt % fluorine) and liquefied gases, such as propane, is attractive. The highly reactive combustion products will include H, HF, F, F_2, and O_2. Gas temperatures will be about $7000°F$ (about $3900°C$), while coating surface temperature will reach $4000°F$ (about $2200°C$). Chamber pressures will be around 100 psia. Here coatings with thermal resistances to 1,400 in.^{2} sec-$°R$/Btu will be needed to decrease heat fluxes from 6 to 3 Btu/sec-in.^{2}

Large solid rocket boosters do not need coatings on either the graphite nozzles or the carbon–phenolic or silica–phenolic exit cones, because any given loss of nozzle material results in a relatively small percentage increase in nozzle diameter. For rockets with smaller nozzle diameters, the same loss of nozzle material has a much larger effect on performance. Here, tungsten or carbide coatings are needed. For example, the Polaris A3 (first stage) has tungsten in the throat and a tungsten liner below.

b. *Engines for Space Maneuver Capability and Attitude Control*   Rocket engines for space propulsion use a wide variety of refractory materials. However, only engines made from molybdenum, columbium alloys, and, in some instances, polycrystalline graphite are customarily coated. Rocket motors of up to 300-lb thrust, incorporating carbon alloy nozzles or nozzle extensions, are now in use or under advanced evaluation. The lunar excursion module (LEM) utilizes a C-103 alloy nozzle with an aluminide coating and ablative throat nozzles. The Apollo service module has ablative throat engines with a C-103 extension nozzle, coated with a modified aluminide slurry. A 100-lb thrust C-103 nozzle, coated with a (Ti, Cr, Si) slurry coating, forms the backup system to a similar molybdenum engine for the Apollo command module. Engines having up to 300 lb thrust, made from C-103, have been tested without failure for 15,000 sec at $2500°F$ (about $1370°C$). Maximum operating temperature was $2800°F$ ($1540°C$), with rapid failure occurring at $3000°F$ ($1650°C$). Recent work has shown the feasibility of protecting columbium alloys to $3600°F$ (approximately $1980°C$), but a substantial amount of fabrication development and testing still remains to be done before these recent advances can be translated into improved coatings for columbium rocket nozzles.

The Apollo program also utilizes molybdenum engines of 50–100-lb thrust for attitude control on the command module and for the LEM module. Similar engines of 100- and 22-lb thrust were envisioned for the Manned Orbiting Laboratory program. The coating for these engines consisted of pure $MoSi_2$ and was applied by pack-cementation. Coating life is limited more by abrasion during handling than by oxidation. Earlier spalling problems have been solved by improvements in coating technique. Molybdenum engines have performed well thus far, and future engines will probably use the same pure $MoSi_2$ coating. The principal reservation to the use of molybdenum nozzles centers on the inherent brittleness of the metal, not on coating performance.

Graphite is generally used uncoated in rocket engines. Exceptions are the Scout and Lance missiles, which use a flame-sprayed zirconia coating on polycrystalline graphite. This coating is thermodynamically unstable and functions only because of the short firing times. Graphite engines with a SiC coating have also been evaluated, but the performance of this coating is limited to approximately $3000°F$ $(1650°C)$. This very limited use of coated graphite results from the fact that, at present, there are no operational coating systems that can protect graphite from oxidation in the $4200°$–$4700°F$ $(2300°$–$2600°C)$ temperature region, where graphite reaches its highest strength. Iridium coatings may potentially protect graphite up to the eutectic temperature of $4140°F$ $(2280°C)$, and an Ir–30 percent Re alloy may offer protection to $4500°F$ $(2480°C)$. Substantial efforts have been made to develop tungsten coatings for graphite for use in solid-fuel engines, but none of these coatings appears to be in use.

Future coating requirements for maneuvering and attitude-control engines are probably somewhat limited. Some use of coatings may be made with hybrid engines and tactical system engines. Attitude control may in the future be achieved by monopropellant hydrazine engines, which operate most efficiently at approximately $1800°F$ $(980°C)$. Uncoated superalloys, e.g., cobalt-based L605, perform successfully in this application. Alternatively, coated molybdenum or columbium engines, or the uncoated beryllium engine, could also satisfy future attitude-control requirements.

Increased maneuverability requirements in the future, i.e., capability to throttle and re-start, may require the use of storable liquids, hybrids, or re-start solids as fuel for tactical missiles. For these advanced requirements, pyrolytic graphite coatings on polycrystalline graphite have shown promise in the laboratory. Some designers favor a "hard" throat for high-performance missiles. Tungsten is a candidate metal, especially if it could form a protective coating with some metal that had been infiltrated into it, possibly zirconium or hafnium. Such "hard" throats should be useful with fluorine or Flox oxidizers, perhaps with hybrid rockets.

Conclusions and Recommendations

Present regeneratively cooled liquid-fueled booster engines do not require coatings. Future booster engines still in the conceptual design stage will operate at much higher exhaust-gas temperatures and will require thermal insulation coatings. Coated hardware requirements for these engines are still several years in the future. NASA has been funding work on oxide based (ZrO_2, ThO_2) thermal-insulation coatings for superalloys. Eventual operational requirements envisioned are a gas temperature of $6000°F$ $(3310°C)$, coating surface temperature of $4000°F$ $(2200°C)$, coating–metal interface temperature of $1800°F$ $(980°C)$, and metal backface temperature between $0°$ and $-200°F$.

A number of liquid-fueled space-maneuver engines utilize coated molybdenum or columbium throats. The $MoSi_2$ coatings for molybdenum engines are adequate, and reservations to the use of these engines are based on the brittleness of the metal, not on coating performance. The modified silicide coatings for columbium also perform well, with a present operational limit at approximately $2800°F$ $(1540°C)$. The development of improved silicide coatings for columbium-alloy engines is largely dependent on the "fallout" from similar coating work for re-entry protection and/or jet engine blades. This is unfortunate because the trade-off considerations are different in these cases.

The Scout and Lance missiles use polycrystalline graphite engines with flame-sprayed zirconia coatings. These coatings appear to perform well, principally because of their low thermal conductivity and short firing times. Other graphite engines are used uncoated, because at present there is no operational coating system that can protect graphite for an adequate time period in the $4200°$–$4700°F$ $(2300$–$2600°C)$ temperature region, where graphite reaches its highest strength. Iridium coatings may eventually protect graphite up to $4140°F$ $(2280°C)$, and iridium–rhenium coatings could extend the protection to $4500°F$ $(2480°C)$. However, the high price and limited availability of these metals will likely restrict their use to the most critical applications.

HYPERSONIC VEHICLES

Summary and Conclusions

The hypersonic vehicles and related propulsion systems that will require the use of high-temperature coatings are:

> Manned hypersonic aircraft
> Lifting re-entry vehicles
> Hypersonic missiles
> Air-breathing propulsion systems

Manned hypersonic vehicles will require modest amounts (relative to superalloys) of coated refractory metal for use in radiative-cooled heat shields and in structures. The coated refractory metals will operate at peak temperatures in the range of $2000°$ to $3000°F$ and must have as long life as possible to be economic. This class of vehicle will be reflown many times. Superalloys will be used for both heat shields and structures. However, the payoff for coating the superalloys is not obvious because present-day coatings are not generally weight-effective. That is, the oxidation rate of the superalloy is sufficiently low that it can be used without a coating. In many cases, the amount of metal that is oxidized will not equal the coating weight until after 50 to 150 flights. The manned hypersonic vehicle is not likely to become a flight system before the year 2000.

Lifting re-entry vehicles are likely to become operational in the near future. Coated refractory metals and superalloys will be used for the same purposes as previously mentioned for the manned hypersonic vehicles. The maximum temperatures are likely to be somewhat higher for lifting re-entry vehicles (3200°F); however, the vehicle life is considerably shorter. Therefore, it is less likely that superalloys will require oxidation-protective coatings. However, the lifting re-entry vehicle life (50 to 100 flights) probably exceeds the capabilities of current coating systems for refractory metals.

Hypersonic missiles are currently under development. Although the environment in which they would operate is more hostile than that encountered by manned hypersonic aircraft, service time is shorter (up to a few minutes) in a single flight. Present-day silicides have potential for service to about 3200°F, and hafnium–tantalum coatings or claddings to 4000°F. Higher temperature capability is desired, particularly for propulsion systems where flame temperatures exceed 6000°F.

Air-breathing propulsion systems (scramjet) generally would be used for the manned hypersonic vehicles. Very high heat rates are attained in the inlet area, so regenerative cooling is required. Thus, the most likely coating need is for thermal insulation to reduce the coolant requirement.

Currently, the superior oxidation-protective coatings for refractory metals for use on hypersonic vehicles are the silicides. This class of coating is most likely to be used in the near future for coating all refractory metal alloys. The aluminides have received considerable use to date because of their processing advantages. However, they are not competitive with the silicides for long life and, in some cases, for maximum temperature resistance. Also, the recently developed fused-slurry silicides have essentially the same processing advantages as the slurry-applied aluminides.

A reasonably good coating capability exists for columbium, molybdenum, and tantalum alloys for use up to about 2800° to 3000°F. There is no doubt about our current ability to satisfactorily and reliably coat most hypersonic-vehicle heat shields and structures for one or two flights. The real questions are how many flights the coating will survive, what the part design constraints are, and what the cost effectiveness of coated refractory metals is, compared to other heat-shielding approaches such as ablators. Answers to these questions are highly dependent, of course, upon mission requirements.

The life of present-day coatings on hypersonic-vehicle components most likely will be determined by the coating defects or thinly coated areas present. The state-of-the-art coatings generally fail because of defects or thin spots, with the time required for the defect or thin spot to induce failure being considerably less than the intrinsic life of the coating. Hypersonic-vehicle components fabricated from refractory metals will tend to be relatively large, of complex geometry, and made from thin-gauge material. Therefore, the probability of a coating defect or thin spot, or of a number of defects or thin spots, is much greater than for the typical test coupon normally used for coating-life determinations. In spite of the probable occurrence of coating defects or thin spots, the chances of a coated refractory metal part surviving one or two flights in a hypersonic airframe application are extremely good, because an early coating defect or thin spot is not likely to cause functional or structural failure of a part until after one or two additional flights. After the local coating failure occurs, it is more readily detected, and repair or replacement may take place as appropriate. The fused-slurry silicide coatings have been prone to thinly coated areas, but have not shown a tendency toward such classical defects as variable composition, holes, or large edge crevices.

The re-use capabilities of the better present-day coatings have not yet been accurately determined. To acquire such data, the coatings must be applied to representative flight hardware and tested in simulated flight conditions where the important environmental factors, including temperature, pressure, stress-strain, and time are simultaneously duplicated. Testing of this kind will also reveal the effects of manufacturing processes and part design on coating and part life. Furthermore, it will establish the defect tolerance of coated refractory metal parts.

Because coating defects are likely to be the factor that controls the life of coated refractory metal parts, coating compositions and processes that are less prone to defects should be emphasized for future development. The fused-slurry silicides are considered the most promising coatings currently available for meeting all the requirements for hypersonic-vehicle applications. This general tendency for defects or thin spots to govern the useful life of state-of-the-art coatings makes nondestructive testing (NDT) very important. Detection of defects and thin spots prior to flight, and of early local failures after flight is necessary for maximum utility of the coatings.

Recommendations

The five major recommendations for research on coatings for hypersonic and re-entry vehicles are as follows:

1. Determine the capabilities of the most promising currently available coating systems for refractory metals by

testing representative hardware under the flight environmental conditions that have the greatest influence on coating performance. Accurate hypersonic-flight simulation is required.

2. Improve the reproducibility and reliability of the most promising coating systems when applied to representative hardware. The deficiencies to be corrected will be identified by the testing recommended above. Improvements should include all influencing aspects of manufacturing technology and processing, as well as design considerations.

3. Improve NDT methods for use in process control, inspection after coating application, and inspection in service.

4. In future development, emphasize simple (nonmultilayer) coatings applied by a fused-slurry method because they are considered to have the greatest potential for meeting the majority of future needs for oxidation-protective coating of refractory metals. Fused-slurry silicide coatings for use to 3200°F should be further developed for tantalum and molybdenum alloys. Exploratory research on new coating systems inherently more reliable than silicides should be encouraged. Desired characteristics include coating ductility at ambient and elevated temperatures and diffusional stability in contact with substrates.

5. In coating large surface areas, weight usually is a critical factor; therefore, thin coatings (~1 mil) that have a long service life (~1,000 hr for superalloys) are needed and should be developed.

The following discussion concerns specific recommendations for selected coating systems:

1. *Diffusion Coatings for Refractory Metals*. Before undertaking any further coating development effort, present coatings should be clearly defined in terms of their ability to protect representative prototype flight hardware reliably and reproducibly. This evaluation must be made in thermal and mechanical environments that correspond to service use. It is only quite recently, after significant progress had been demonstrated by conventional laboratory specimen testing (such as furnace, "slow cyclic," and low-pressure oxidation exposures), that such advanced simulated service environmental testing has become warranted. But intensive testing under mission-oriented, vehicle-design, constrained test parameters is now appropriate. It will permit us to obtain a clear picture of current coating limitations before embarking on ambitious coating or processing improvement programs. This recommendation is generally applicable to all categories of diffusion coatings considered for hypersonic vehicles.

Future developments should primarily advance the practicality and reliability of the better coatings already developed. Specifically, it is recommended that efforts be concentrated on improving the reliable cyclic life of fused-slurry silicide coatings for fabricated columbium components for use at temperatures up to 2500°F. Work to improve higher temperature capabilities of columbium coatings should be deferred until columbium sheet alloy of higher creep strength is available. At temperatures above about 2500°F, creep of the columbium appears more limiting than the coating for long-life, multimission vehicles.

Similarly, more effort is required to establish and improve the reliability of reusable fused-slurry coatings for tantalum and molybdenum alloys for use up to 3200°F. Work should be done to adapt the more refractory silicide compositions as fused coatings for these substrates. Processing times and temperatures need to be minimized for coating molybdenum to avoid ductile–brittle transition problems. Significant improvements in coating emittance compared to available pack silicides appear feasible by slurry composition adjustments for molybdenum.

For vehicle structure and heat-shield applications, the emphasis should continue to be on fused-slurry coatings, because of their inherent advantages for coating large components and complex geometries. To minimize the likelihood of incurring localized coating defects on parts where reusability is critical, pack or sintered coatings should be avoided, if possible; multilayer (multiple-process) coatings should be considered only as a last resort, with simple one-step coatings accorded priority.

In processing, further work is needed to obtain uniform coating thickness on simple and complex parts and assemblies. Process specifications incorporating good quality-assurance provisions are also required. It is essential to define the scope of applicability for available NDT methods and to devise apparatus and techniques amenable to the field inspection of "flight" hardware. Techniques for reliably and quickly detecting small local coating failures after flight exposures are desired. Special attention must be given to the inspection of the interior surfaces of complex fabrications, such as closed corrugations.

Specific effort should be directed to the field repair of coatings. Feasibility for coating repair has been established in the laboratory, but more work is needed to define the limits of repairability and the degree of confidence merited by repaired coatings—i.e., how large and what type of defect can be repaired and how long will the repaired coating be protective? Research is required on quality control and inspection methods for "patch" coatings.

No truly satisfactory diffusion (silicide) coating is available for the protection of molybdenum or tantalum at temperatures much beyond 3000°F, or for tungsten beyond about 3300°F. Moreover, based on considerable past effort, it does not appear that additional work along the silicide coating approach is warranted for these very high temperatures unless the operation time is short. It is recommended, therefore, that no further work be undertaken on diffusion coatings for the protection of refractory alloys at 3500°F or above. Other concepts for very-high-temperature protection are needed.

2. *Superalloys.* There is a definite need for improved protective coatings for superalloys used in gas-turbine engines. Therefore, it is recommended that research and development efforts for satisfying this objective precede work in coating technology for hypersonic flight and re-entry vehicles. This approach is considered logical because much of the information it would provide for developing protective coatings for gas-turbine engines would be directly applicable to the other application if the need should arise.

It is also recommended that improved oxidation-resistant alloys having adequate high-temperature properties be developed. (Such alloys would significantly reduce the need for protective coatings on hypersonic flight and re-entry vehicles.) A promising approach for obtaining these alloys is through minor compositional modifications of existing superalloys and TD–Ni alloys. For example, rare-earth metal additions are known to improve the oxidation resistance of several superalloys, and additional progress can be anticipated if this area of research is more completely exploited.

3. *Platinum-group metal coatings.* Advanced development studies for Pt-group metal protective coatings are recommended only if a specific need occurs, or if no other coating system can be used. If it is found necessary to develop Pt-group metal coatings further, the following considerations should apply:

a. Future research should be limited to Ir and Rh and alloys of these metals.

b. Additional research on diffusion barriers (particularly oxides) is recommended.

c. Research on metallic substrates should be limited to those that will not oxidize catastrophically in the event of localized coating failure.

d. Additional research is recommended on applying Ir- and Rh-base coatings by pyrolytic deposition from organometallic vapors and solutions, or by fugitive vehicle slurries.

e. Continued research on Ir-base coatings for protecting nonmetallic materials (e.g., graphite) is believed to be worthy of additional consideration.

f. Any of the above recommended research efforts should be limited to relatively low monetary expenditures.

4. *Hafnium–tantalum coatings.* Hf-Ta and Hf-Cb alloys should be studied extensively to develop their full potential both for short-time and multihundred-hour service as coatings and claddings, as well as free-standing bodies in oxidizing environments. Detailed understanding should be obtained of: oxidation and degradation mechanisms; alloying for combined oxidation resistance, ductility, and strength; properties of optimum coatings and claddings on Cb alloys,

Ta alloys, and graphite substrates; and component design, fabrication, performance, reliability, and design allowables.

5. *Coatings for graphite.* Practical coatings, having a temperature capability comparable to graphite, are required if graphite materials are to be utilized to any extent for hypersonic vehicles.

6. *Oxide coatings.* New ideas and approaches are required for the reliable utilization of oxide ceramic coatings as thermal insulations for hypersonic vehicle structures.

ENERGY CONVERSION SYSTEMS

Summary

Background Energy conversion systems are used to convert heat or chemical energy into electrical energy. Their efficiency depends on the maximum temperature of the heat source and the lowest temperature at which heat may be rejected. A considerable advantage is associated with the use of high temperatures, which may be translated into weight saving in the energy conversion devices. Efficient energy conversion systems requiring high-temperature materials are needed for auxiliary power units (APU's) for orbital and space vehicles. Nuclear APU's are designated by the acronym SNAP (systems for nuclear auxiliary power) and are of two types: odd number SNAP systems which use radioactive isotopes as the heat source and thermoelectric generators as the energy conversion systems, and are designated RTG systems (see below); and even-number SNAP systems, which employ turbine-alternators and operate as gas, Brayton-cycle or vapor–liquid, Rankine-cycle systems. The coating requirements in these various energy conversion systems are summarized below.

Nuclear Reactors Fuel elements for high-temperature nuclear reactors employ coatings to isolate the fuel elements from the heat-transfer fluid and to retain the gaseous fission products that are released. Coatings for fuel elements have received a great deal of concentrated attention from the Atomic Energy Commission and its contractors. However, they are not considered as coming under the scope of oxidation-resistant coatings for energy conversion systems.

Radioisotope thermal Generators (RTG's) These devices are fueled by radioactive isotopes of such elements as plutonium, polonium, curium. The amount of isotopic material used per capsule is selected to produce a suitable operating temperature for a thermoelectric converter—about 700°C for PbTe and 1000°C for SiGe. The radioisotope containment capsule is generally multilayered for high-temperature applications, with tungsten forming the inside of the capsule, an intermediate layer of high-strength tantalum alloy, and an outer layer of oxidation-protection material, either superalloy or platinum-base alloy. Lower temperature radioisotope sources are contained in a shell of high-strength, oxidation-resistant superalloys. Another RTG requirement is that coatings must enhance thermal emissivity from the nuclear heat source to a dynamic system.

The chief safety hazards in the operational cycle of an RTG occur in ground handling, launch, space operation, and during and after re-entry from orbital flight. It is imperative to provide oxidation protection to maintain fuel container integrity and not disperse fuel into the atmosphere. Requirements include stability against interaction with the fuel during its lifetime, which can extend to hundreds of years.

Thermoelectric Systems Thermoelectric generators have secondary applications for coatings, including insulation and diffusion barrier coatings.

Turbine–Alternator Systems High-temperature turbines intended for space power will not have to be protected against oxidation, since they operate in the vacuum of space. It is considered too risky to use oxidation-resistant coatings for ground testing in air, and such systems are tested in large vacuum chambers.

In turbine–alternator converters, coatings are used chiefly in the radiator section to enhance heat loss by radiation. Thus, they must have high emittance.

Thermionic Diodes Coatings would be needed in thermionic diode systems for improving electronic emission and for insulation purposes. In this case, however, the coating problem is not considered as critical as the many other problems associated with these devices.

Conclusions and Recommendations

Coatings are important to the achievement of long life and high efficiency in energy conversion systems. However, they are used principally for such secondary purposes as maintaining stability of thermoelectric materials, electrical insulation, barriers against diffusion, and thermal emissivity on radiator surfaces. RTG systems have a critical coating requirement for protecting the radioisotope core during launch and space operation and during and after re-entry into the earth's atmosphere. The heavy sheathings of oxidation-resistant alloys probably constitute the best coating system that could be used for this purpose. Coating nuclear fuel elements was considered to be a special case, not within the scope of the Committee.

The task group had no recommendations for future research and development, since the main coating requirements are for secondary purposes, and the only oxidation-resistant coating appeared to be as good as could be developed.

INDUSTRIAL

Coatings for high-temperature service must clearly offer improvements in process or component performance before their additional cost can be justified by an industrial user.

Coatings are most widely used in the metal- and glass-producing industries, where they frequently offer the only way to operate economically and successfully. For example, glass and metal producers must protect their molten product against contamination from the crucible and also protect their control thermocouples from corrosion. In metal forging and extrusion, coatings provide thermal insulation that prevents heat loss from the workpiece and so insures optimum fabricability. Such coatings also protect the hot metal against atmospheric contamination.

Elsewhere, high-temperature coatings are used to extend the lives of critical components of a wide variety of products. Home furnace burners, automotive exhaust-control devices, and parts of petroleum cracking plants are a few of the many applications of these protection systems.

While the industrial application of high-temperature coatings is not extensive at present, their use by selected industries was found to be more widespread than expected. As more industries increase processing temperatures, the use of the coating technology base established by defense and space research can be expected to expand considerably.

MANUFACTURING TECHNOLOGY

SUMMARY AND CONCLUSIONS

A thorough survey was made of all industrial and laboratory organizations in this country that are active in the application and development of high-temperature, corrosion-resistant coatings. Each organization was requested to provide information relative to its coating processes, substrates protected, coating designations and chemistries, coating facility sizes, and process status (production or development). About 80 percent of the organizations responded to the questionnaire. The principal coating organizations did reply, and the information collated represents a good summary of the manufacturing capabilities available for the application of coatings to high-temperature materials.

Nine basic coating processes were defined. Comments pertinent to each of these processing categories are summarized below.

Pack Processing

Pack coating involves immersion of the material to be coated in a granular pack medium, and a thermal cycle which is performed in a nonoxidizing environment. The coating is formed by vapor transport and diffusion. For the refractory metals, several promising coating systems are pack-applied; and, for the limited requirements for coated refractory metals, adequate coating facilities are generally available. Only experimental or low production quantities of coated aerospace and aircraft gas-turbine components have been required; consequently, little impetus has been provided for establishing reproducible and reliable production coating methods. Should demand for production coating of refractory metal components arise, a major effort would be required to advance current processes to the required technological level. Silicides are the principal coatings that are pack-applied to refractory metals. An unresolved problem is reliable repair methods for pack-applied silicide coatings.

Pack-aluminizing of nickel-alloy and cobalt-alloy turbine airfoils is a large-scale manufacturing activity in the gas-turbine industry and is the major pack-coating requirement for high-temperature nonrefractory alloys. At present, adequate facilities are available in the industry for treating the hundreds of thousands of airfoils processed each month. The principal manufacturing problems related to coating nonrefractory metal alloys are (a) coating of component air-cooling passages and internal holes, (b) masking during coating, (c) processing of large-size components such as gas-turbine transition ducts, (d) processing capability for coating dispersion-strengthened alloys, and (e) processes and facilities for aluminide recoating of service-operated gas-turbine components. The industry has directed its efforts primarily to the aluminizing of non-air-cooled blades and vanes; consequently, the problem areas outlined above will require substantial attention in future R&D efforts. A significant need for pack-coating capabilities beyond those required by the aerospace and gas-turbine industries was not identified in this survey.

Slurry Processes

Slurry coating entails the application of a slurry mixture containing the coating elements on a substrate surface, generally by dipping or spraying, followed by a thermal treatment in a protective environment. Coating formation is achieved by vapor transport, liquid reaction, or solid-state sintering. Slurry processes have been widely used for the formation of high-temperature coatings on refractory metals, e.g., aluminides, silicides, and noble metals. A major problem with slurry-applied coatings has been thickness control. From a manufacturing standpoint, the lack of significant requirements for coated refractory metals precludes the need for expanded manufacturing capabilities. Laboratory development of improved coatings and improved slurry application methods is recommended.

Slurry application of aluminide coatings to superalloy blades and vanes is a large-scale production activity in the gas turbine industry. Automation of the slurry approach to aluminizing blades and vanes would be required to make this method economically competitive with pack processing. For the aluminide coating of large nickel alloy AGT components, such as combustors and transition ducts, the slurry approach may offer advantages over pack processing. In this event, development of methods for applying uniform slurry coatings to large complex shapes will be required.

Chemical Vapor Deposition

Chemical vapor deposition (CVD) is an entirely gaseous process in which the coating elements are transported to the substrate surface in vapor form and are deposited on the surface by chemical or pyrolitic reduction of the gaseous compound. This process has been successfully employed to apply many high-temperature protective coatings; however, the available facilities are predominately of laboratory scale. Substantial problems are encountered in depositing uniform coatings on complex shapes by CVD. Until a demand exists, and significant advantages are defined for CVD deposition of certain coatings, a specific effort aimed at developing expanded CVD manufacturing capabilities is not warranted.

An immediate problem in the aerospace field where CVD may be applicable is the deposition of a tungsten barrier layer coating on tantalum alloys. Additional R&D effort in this area is recommended.

Electrolytic Deposition

Electrolytic deposition includes conventional aqueous electroplating, electrolytic deposition from fused-salt baths, and electrophoresis. The last-named technique involves deposition of charged particles from a liquid suspension onto an oppositely charged substrate. Aqueous electroplating is not widely used for forming high-temperature coatings, and specific development efforts in this area are not recommended. Electrolytic deposition from fused salts has been successfully employed on a laboratory scale for depositing many high-temperature coatings. However, owing to the problems associated with scaling-up fused-salt coating methods and the absence of clearly defined advantages for this method, manufacturing development in this area is not recommended.

Electrophoresis provides an excellent means of applying uniform slurry coatings to complex shapes. It is the most promising approach to resolving the problem of uniformly, reproducibly, and economically applying slurry coatings to high-temperature materials. Fused-silicide coatings can be electrophoretically deposited, and consideration should be given to this coating method in future programs.

Fluidized Bed

Fluidized bed processing involves immersion of the material to be coated in a fluidized granular medium of the coating elements or inert materials. Fluidization is accomplished with an inert or reducing gas and/or a gas containing a halide form of the coating species. The process is a variation of chemical vapor deposition. Facility expense and high operating costs are disadvantages of this technique. No immediate need for process development exists for the fluidized bed coating method.

Plasma and Flame Spray Processes

Coating deposition with arc plasma or gaseous fuel–oxygen flame torches has proved extremely successful for the application of hardfacing and wear-resistant materials. These techniques have not been used extensively for applying high-temperature protective coatings, owing to problems related to density, bonding, thickness control, cost, and so on. Although expansion of manufacturing capabilities in these areas is not warranted currently, the use of such techniques for the deposition of coatings not easily formed by diffusion methods should be explored.

Hot-Dipping

Hot-dip coatings are formed by immersion of the part to be coated in a molten bath containing coating elements. Hot-dipping has been widely used to apply coatings to ferrous and nickel-base alloys; however, the family of high-temperature coatings applicable to these materials are better formed by other methods. Adequate hot-dip manufacturing capabilities are available for current high-temperature coating needs.

Cladding

This process entails sandwiching a structural high-temperature material between surface layers of a protective high-temperature alloy. Edge closure is a major problem associated with the cladding application of high-temperature coatings; consequently, little effort has been directed to applying this approach to high-temperature coating development. For the refractory metals, very few protective systems are amenable to cladding application; therefore, manufacturing development in this area is not warranted. In the case of superalloys, heat-resistant coatings not easily formed by diffusion-controlled processes may be applied by cladding, and this approach should be considered.

Vacuum Vapor Deposition

Vacuum vapor deposition encompasses all methods associated with the physical vaporization and deposition of coating materials. This process has been widely used for the deposition of thin-film coatings on such things as mirrors, ornaments, and reflectors. Recently, methods have been developed for the deposition of 3.5-mil-thick high-temperature coatings on superalloys and refractory metal substrates. The flexibility of this method for depositing coatings not amenable to formation by nonconventional diffusion processes makes it attractive for further development. Since relatively expensive facilities are required, laboratory demonstration of the capabilities of this technique must be provided.

PRIMARY RECOMMENDATIONS ON MANUFACTURING TECHNOLOGY

1. Demonstrate the reliability and reproducibility of available protective coating systems on refractory metal hardware.

2. Develop reliable and reproducible methods for refurbishing service-operated aluminized superalloy hardware, and assess the influence of coatings and recoating operations on the mechanical properties of these hardware materials.

3. Develop reliable methods for applying coatings to the internal passageways of air-cooled gas-turbine hardware.

4. Study the physical chemistry of current coating processes to provide a sound basis for process control and improvement.

5. Develop field repair methods for diffusion-coated refractory metals and superalloys.

SECONDARY RECOMMENDATIONS

1. Establish manufacturing methods for applying Cr–Al protective coatings to large dispersion-strengthened (TD–Ni and TD–Ni–C) components.

2. Develop reproducible processing techniques for depositing a tungsten-barrier layer coating on complex tantalum alloy shapes.

3. Establish improved electrophoretic techniques for depositing a wide range of slurry coatings on complex refractory metal and superalloy hardware.

4. Develop vacuum vapor deposition methods for applying complex, heat-resistant alloy coatings to nickel- and cobalt-based superalloys.

5. Introduce automation into high-volume coating activities, such as the aluminizing of superalloy turbine blades and vanes, for cost reduction and improved process reproducibility.

6. Promote communication between hardware designers and coating suppliers to assure the design and fabrication of coatable components.

TESTING AND INSPECTION

SUMMARY AND CONCLUSIONS

The successful development of an effective high-temperature protective coating depends, to a great extent, upon the proper application and interpretation of tests at critical stages during the development phase and in the final evaluation of the finished product. The technology of testing and inspection is broad and interdisciplinary. This discussion is concerned with only one aspect of this technology, that which pertains to the evaluation of high-temperature protective coatings for service above 1800°F. Consideration of only those applications above 1800°F virtually eliminates all industrial coatings. Aerospace and propulsion systems are the primary areas where such high service temperatures are normally encountered. Reviews of current testing procedures for this temperature range show them to be profoundly influenced by the data requirements that lead directly or indirectly to the design of coated components for aerospace and propulsion systems.

Coating evaluation procedures fall into four general classes: standard characterization, nondestructive testing, screening, and environmental simulation. Standard characterization tests are those that measure basic characteristics of a coating and/or the coating–substrate composite. This would include the measurement of such properties as hardness, thickness, density, melting point, emittance, thermal conductivity, tensile and creep strength, elastic modulus, and fatigue strength. Many of these measurements are routine procedures for which there are ASTM standard tests. Nondestructive testing (NDT) includes any inspection or measurement technique that neither disturbs nor alters the physical or chemical characteristics of a coating system. Nondestructive testing is frequently employed in process control to measure coating thickness and to detect nonvisible coating flaws. These techniques employ dye penetrants, ultrasonic transmission or emission, eddy-current emission, radiography, infrared emission, and thermoelective response. The technology of NDT is advancing rapidly and will have a significant influence upon the character of future testing and inspection techniques.

Screening tests are generally designed to provide a preliminary evaluation of the capability and potential of a coating system for a given application. Usually the data derived from such tests are of a comparative or relative nature. The test environment ordinarily includes only one or two independent variables, and the emphasis is on high-volume–high-rate testing rather than on environmental simulation. While screening tests normally involve a minimum of equipment, some facilities, notably those which are used to screen gas-turbine coatings, may represent capital investments exceeding $50,000. Perhaps the most common screening test for high-temperature coatings is the so-called cyclic oxidation test in which small coated tabs are subjected to a series of isothermal air exposures until the original lot of specimens has failed. This test, like other screening tests, can be quite effective in eliminating from further consideration ineffective coatings, thereby minimizing the number of candidate coatings that must be carried into advanced proof testing. Simulated environmental tests are those in which the use environment is simulated in the laboratory. This description is somewhat of a misnomer, since most aerospace and propulsion environments are either too severe or of too long duration to be realistically simulated. Most

"simulated environments" are abbreviated representations of the real environment, and frequently significant parameters must be omitted because it is physically impossible to simulate them.

Laboratories are in general agreement regarding the performance requirements for evaluating coatings for a certain class of applications, for example, gas-turbine hardware. There is almost total disunity, however, with respect to the tests employed and the significance of data obtained from these tests. This lack of agreement stems principally from the independent evolution of testing procedures by different laboratories in an atmosphere of limited communication. It is virtually impossible to compare test results from different laboratories. This has been the cause of substantial duplication of testing efforts. The procedural difficulties of high-temperature testing have contributed to the nonuniformity of evaluation techniques. At elevated temperatures, even simple measurements of temperature and strain are difficult, and none of the several techniques now employed is completely satisfactory. Problems with high-temperature reactions between coatings and specimen-support media and the unavailability of simple inexpensive furnaces for long-term operation in air above 3000°F have further complicated the situation.

Another elusive factor which complicates interlaboratory data comparisons is failure criteria. The appearance of substrate oxide has been generally regarded as evidence of coating failure. However, the amount of oxide that accumulates to constitute a failure is a subjective quantity and will vary from laboratory to laboratory. Moreover, this failure criterion is strictly artificial and may have no relevance to what would constitute a service failure.

The first step toward elimination of the current confusion in evaluating coatings would be establishment of meaningful test standards. An effort of this nature was the subject of an earlier MAB study.* Establishment of test standards for high-temperature protective coatings is currently being investigated by ASTM Committee C-22. The MAB test procedures have received only limited acceptance; the progress of ASTM C-22 has been relatively slow. This is due, in part, to the reluctance of laboratories to discard self-developed test procedures in favor of new and unfamiliar routines. Eventually, tests will be standardized, but not in the near future.

The confusion and inefficiencies caused by the current lack of test uniformity can only worsen as coating requirements become more stringent. Higher operating temperatures, longer service life, and greater reliability will be demanded of future coatings. Improved measurement techniques, refined reliability analyses, and more effective NDT must be developed so that the ability of coated parts to

---

*Procedures for Evaluating Coated Refractory Metal Sheet. Report MAB-201-M, August 3, 1964.

provide satisfactory performance for extended periods in hostile environments can be demonstrated with a high degree of confidence. Unified test standards must be cooperatively evolved by testing laboratories to minimize unnecessary duplication of effort. Furthermore, as service lives increase, meaningful accelerated tests that can be interpreted accurately in terms of service life will be required.

RECOMMENDATIONS ON TESTING AND INSPECTION

In anticipation of these future requirements, the following recommendations are made:

1. Standardize test methods, including standard characterization, nondestructive inspection, screening, and simulated environmental tests. In the interim, cooperation among coating vendors and evaluation groups must be encouraged for the purpose of unifying current testing procedures and eventually standardizing all tests.

2. Develop new and improved NDT methods that yield quantitative data. Exploitation of reliability analyses and nondestructive characterization techniques should proceed as an integral part of new coating development programs.

3. Develop meaningful, accelerated coating-life tests that are thoroughly correlated with service performance.

2

Fundamentals of Coating Systems

INTRODUCTION

Protective coatings for refractory metals have been perfected over the past two decades largely through empirical development. From the body of experience thus obtained, a rationale has evolved concerning the basic concepts underlying the development of successful coatings, and knowledge has been accumulated regarding the mechanisms of coating failure. The purpose of this survey is to review the basic concepts and principles of protection, as well as mechanisms of coating failure, in order to summarize our knowledge of those subjects, in the expectation that this will be of assistance in the development of new coating systems.

This chapter begins with a general discussion of concepts and principles. Next, oxidation, evaporation, and coating substrate reactions are considered as mechanisms of coating degradation. Finally, the predominant role of defects and flaws in the failure of coatings is analyzed.

ANALYSIS OF COATING SYSTEMS

The five types of coatings used to protect metallic systems at elevated temperatures are:

1. Intermetallic compounds that form compact oxide layers.
2. Intermetallic compounds that form glassy oxide layers.
3. Alloy coatings that form compact oxide layers.

4. Noble metals and alloys that either do not react with the environment or react very slowly forming volatile oxides.
5. Stable oxides that provide a physical barrier.

Much of the technology developed in coating high-temperature components has been concerned with refinements and modifications of the basic functions of the coating types listed.

The following paragraphs analyze the successful development of protective coatings and attempt to answer the questions: Why were the coatings successful? What were the predominant basic factors involved? This type of *post hoc* reasoning serves as an introduction to the more traditional scientific approach of predicting from first principles, described in later sections on mechanisms of failure of protective coatings.

INTERMETALLIC COMPOUNDS

CRYSTALLINE OXIDE FORMERS

The most important of the protective coatings in this category are aluminides on superalloy substrates. The aluminides oxidize preferentially to form a layer of crystalline Al_2O_3 because of the high activity of aluminum relative to the more noble constituent. Aluminum oxide is an excellent protective layer because the rate of diffusion of aluminum cations from the substrate aluminide through Al_2O_3 to the surface to react with oxygen is very slow. Transport of oxygen anions through the layer also is very slow. As a result, the oxide film remains thin, with consequent good adherence and mechanical characteristics. Indeed, the most probable failure mechanism is diffusion of aluminum into the metallic substrate. As long as an aluminide layer is present as a reservior of aluminum, mechanical defects in the Al_2O_3 oxide layer are not detrimental. However, mechanical defects in the aluminide layer itself that permit the substrate to be exposed are a source of failure. A special class of aluminide-type coating with a built-in self-healing mechanism is the tin–aluminum coating that is applied to refractory metal substrates. These coatings employ a molten tin vehicle which flows to repair any defect that might form during service. Otherwise, the mechanism of selective oxidation of aluminides formed by reaction with the substrate remains unchanged.

Many other crystalline oxides have the melting points and stability to be potentially useful, including the oxides of beryllium, chromium, thorium and rare earths. In the course of empirical development, these other systems have been tried, and, for one reason or another, have failed. Some of the reasons for failure of systems forming protective oxides other than Al_2O_3 are obvious. For example, the oxides of Group IVa (titanium, zirconium, and hafnium) and Group Va (niobium and tantalum) do not grow as a compact layer during oxidation, thus permiting gaseous transport of oxygen to the substrate. Chromium-rich alloys and compounds produce a Cr_2O_3 layer, but this forms volatile CrO_3 on further oxidation. Reasons for the failure of beryllium or beryllides to form a useful family of coatings is not so clear. It appears that the large mismatch of volume and thermal expansion between beryllides and metallic substrates is the cause of the difficulty. Also, beryllium oxide reacts with water and water vapor. Other deficiencies of beryllide coatings are the occurrence of low-melting eutectics with the substrate and high diffusion rates of beryllium into the substrate. Crystalline oxides based on cobalt and nickel probably are not used because metal cation diffusion through the scale is too rapid. On the other hand, the oxides of Th and Zr exhibit quite high rates of oxygen diffusion, rendering them of questionable value for thin high-temperature coatings.

GLASSY OXIDE FORMERS

Silica is the predominant oxide that can form a glass, although other oxides, such as boron oxide, might be considered as a basis for forming glassy oxides. The silicide coatings have been very successful for protection of refractory metal substrates. It is found experimentally that, of the silicides, only the disilicides are effective in producing continuous glassy oxide layers on the surface. Lower silicides are much less effective, probably because the silicon activity is too low to produce preferential oxidation of

silicon. Silica glass coatings have a tendency to lose silicon through the formation of gaseous SiO at low pressure, either by direct dissociation of SiO_2 or by reaction with silicon in the disilicide.

Glassy oxide coatings can flow and accommodate mechanical strain, healing over the mechanical defects formed in the coating. The art of silicide coating of refractory metals may be based in part on the use of elements that form oxides that modify the glassy range by lowering the flow point, decreasing the viscosity, and enlarging the range over which defects may be healed more effectively. Self-healing is limited at lower temperatures where cracking of the glass may occur.

Diborides form a family of intermetallic compounds that have good oxidation resistance at very elevated temperature. It is possible that the basis for oxidation resistance of the diborides is the formation of B_2O_3 glass modified with other oxides. Thus far, however, diborides have not been successful as coatings, perhaps because of dissolution of boron into the substrate. Another factor is the high volatility of B_2O_3, which would be undesirable in the case of a thin coating. It appears that insufficient effort has been expended in the development of diboride coatings to permit an adequate appraisal of their potential.

When $MoSi_2$ is used as a coating, the molybdenum trioxide that may be formed evaporates, leaving a relatively pure SiO_2 glass. If the silicide coating contains other elements that form nonvolatile oxides (i.e., Cb_2O_3), these may enter and modify the glass. The more effective coatings are balanced in composition to provide glasses that give the best protection and possess self-healing characteristics. The development of fused slurry processes permits a broad range of compositions to be applied as coatings and is considered a major advance in coating technology.

OXIDATION-RESISTANT ALLOYS

Most of the work in this category is based on oxidation-resistant, nickel- and cobalt-base alloys. The basis for oxidation resistance in these alloys is the formation of compact oxides in which the diffusion of metal cations is inhibited. The most desirable oxide layers are those that are thin enough or plastic enough not to fail mechanically during service. Cobalt- and nickel-base alloys for high-temperature application generally contain substantial amounts of chromium. The mechanism for the enhancement of oxidation resistance by chromium appears to be the formation of a dense spinel oxide layer of the type $CoCr_2O_4$ or $NiCr_2O_4$. The rate-controlling process is believed to be chromium ion diffusion through the spinel, which is vastly slower than cobalt or nickel ion diffusion through CoO or NiO scales (e.g., the alloy scale formed without chromium or at low chromium levels).

It is possible to use superalloys as coatings on refractory metals. Here, their maximum use temperature is limited by the possibility of eutectic melting between the superalloy coating and the refractory metal substrate. However, the eutectic temperatures are somewhat higher than the useful temperature range for oxidation resistance and thus are not the ultimate limiting factor in the use of superalloy coatings. A secondary limitation is diffusion of nickel or cobalt into the refractory metal substrate, particularly in the case of the Group VIa metal substrates. Diffusion of these elements reduces the recrystallization temperature and generally causes the base metal to deteriorate. Brittle intermetallic diffusion zones also may be formed. Thermal expansion mismatch is another factor that has limited this approach.

In the case of refractory metals there has been an interesting and moderately successful use of oxidation-resistant nonstructural alloys to coat high-strength substrate alloys. The advantage of this approach is that the concentration gradient for diffusion is much less than in the case of coatings that are chemically very different from the substrate. Some columbium-base alloys containing high concentrations of titanium, aluminum, and chromium have good oxidation resistance in the $800°$–$1200°C$ range, approaching that of the oxidation-resistant superalloys. Their high-temperature strength, however, is inferior to that of superalloys. Hence there has been little incentive to use them. However, such an oxidation-resistant compatible cladding over structural columbium-base alloys could comprise an interesting coating system. Furthermore, in gas-turbine applications, where foreign-object damage is an environmental threat, the ductile columbium-base alloy affords a means of protection through plastic deformation without cracking. The mechanism for protection of columbium by large additions of titanium or zirconium is the formation of complex oxides of the type $Cb_2O_5 \cdot TiO_2$ and $Cb_2O_5 \cdot 6ZrO_2$.

Another example of a compatible oxidation-resistant refractory metal coating is the use of Ta-Hf alloys to clad structural columbium- and tantalum-base alloys. The Ta-Hf alloys oxidize with a parabolic growth rate, indicating diffusion-controlled oxidation. The oxidation rate is rather high so that thick Ta-Hf coatings are necessary for protection.

NOBLE METALS

The noble metals, particularly platinum and iridium, may be used to protect refractory metal substrates and graphite. The maximum temperature and time of usefulness can be estimated from the loss of the noble metal through evolution of volatile oxides; this loss is strongly dependent on the flow rate of oxygen. No mechanism of self-healing is available, so that the coatings must be perfect in order to be protective. A secondary concern is interdiffusion with the substrate. Substrate interdiffusion may be avoided through the use of a relatively thick barrier of nonreactive oxide. For example, platinum-coated molybdenum and tungsten components are used in glass-melting, in which the noble metal is simply a sheath over the refractory metal substrate from which it is separated by an oxide barrier layer, usually Al_2O_3. The noble metal coating merely serves as a barrier to keep out oxygen and does not contribute structurally. Thus, applications for this system are limited to low stresses, where transmission of shear stress across the coating–substrate interface is minimized.

OXIDES

Oxides inert to the substrate may be used in coatings as a physical barrier to oxygen. The mechanism of failure is diffusion of molecular oxygen through pores or cracks in the oxide coating to the substrate. Protection is based on maximizing the diffusion length for molecular oxygen, and such coatings tend to be rather thick. The technology involved in the development of such coatings is based on avoidance of shortcuts to molecular diffusion, such as might occur at large cracks in the oxide coating. A typical example of such a high-temperature protection system is sintered ThO_2 over tungsten. The application technique employs a metallic mesh or network that reduces the area of oxide "tiles" to that which can accommodate thermal stresses without cracking.

It is also effective to add fluxing agents to the oxide to form a frit, or enamel-type coating. There are many examples of successful use of enameled metals, but most of these are for relatively low temperatures, below the solidus temperature of the frit. The mechanism of failure of enamel coatings is the same as that for sintered coatings, namely, molecular diffusion of oxygen through cracks to the exposed substrate.

LIMITING FACTORS IN COATING BEHAVIOR

OXIDATION KINETICS

Metals are thermodynamically unstable with respect to their oxides over such a broad range of temperatures and oxygen pressures that the widespread use of metals (or metallic materials in general) as structural materials would be precluded except for the intervention of chemical kinetics. Fortunately, once an oxide has formed on the surface of a metallic material, it acts as a barrier to further reaction by mechanically separating the two reactants. Oxidation can then proceed only by transport of metal atoms to the oxide–ambient interface, by transport of oxygen atoms to the oxide–metal interface, or by breakage or physical removal of the oxide film. The purpose of this section of the report is to define as precisely as possible the minimum oxidation rates that are theoretically attainable in projected re-entry or engine environments. Consideration of the loss of an oxide barrier by vaporization will be deferred to the next section. The *minimum* rates are generally defined by the rate of transport of reactants through mechanically stable and structurally sound oxide films. In practice, attainment of minimum rates may be hampered by spallation, breakaway, or defective oxide structures. A vital intermediate problem to which some attention must be directed thus involves means for achieving maximally effective oxide barriers. In the following paragraphs a brief summary is presented of the current state of understanding of oxidation mechanisms. The next four sections deal in turn with coatings of silicides, aluminides, refractory oxides, and oxidation-resistant alloys. For each of these types of coating, the ideal long-range potential is assessed and compared with the behavior of current state-of-the-art systems. Finally, pertinent directions for future developmental efforts are considered.

MECHANISMS

In general, a study of the oxidation behavior of any system begins with some measurement of the extent of conversion of the original system to oxide as a function of time, t, under selected conditions of temperature and oxygen pressure. The specific measurement might be weight change, oxygen consumption, thickness of oxide film, recession of the original metallic system, or some combination of these. The data are then analyzed to yield a rate law, which for the sake of clarity can be expressed in the form

$$\frac{dx}{dt} = f(x,t), \qquad (1)$$

where x is the thickness of the oxide film and $f(x,t)$ is some function of film thickness and time. For a large number of systems, the function $f(x,t)$ assumes one of a number of fairly simple forms. If $f(x,t)$ is constant, $f(x,t) = k_l$ and the system is said to oxidize linearly—i.e., the loss of metal proceeds at a constant rate independent of time and independent of the amount of previous oxide buildup. Clearly, when a linear rate law is observed, the oxide is not functioning as an effective barrier. If $f(x,t) = k_p/2x$, the system is said to oxidize parabolically, where k_p is the parabolic rate constant, and the rate of oxidation declines with time as the oxide builds up—i.e., the thicker the oxide film at any point

of time, the slower the rate of subsequent oxidation. If $f(x,t) = (k \log t)$, the system is said to oxidize logarithmically, and the rate of oxidation decreases with time, although at a different rate, of course, than when the parabolic rate law is followed. The linear, parabolic, and logarithmic rate laws are by no means sufficient to describe the experimental rate data for all of the systems that have been studied. However, systems that obey other rate laws are best considered individually, since the rate-controlling processes frequently are quite complex.

For purposes of engineering design, oxidation rate data as summarized in the rate equation are often sufficient. For purposes of rational materials development, it is highly desirable to gain some understanding of the rate equation in terms of a reaction mechanism.

The linear rate law, which implies continuous contact between solid and gaseous reactants, need not concern us. While it is conceivable that, over a finite period, slow linear oxidation may lead to less recession of material than rapid parabolic or even logarithmic oxidation, mechanisms leading to linear oxidation are still essentially nonprotective. When a transition is observed from a parabolic to a linear rate law, as a result of spallation of a thick oxide film or breakaway, then it is important to explore means for retaining the protective oxide intact.

In many ways, the most significant insight into oxidation mechanism has been provided by the Wagner theory of parabolic oxidation. Wagner showed that if the rate-determining step in an oxidation process is volume diffusion of reacting ions (or corresponding defects) through a compact oxide scale, a parabolic rate law results. The derived relationship between the parabolic rate constant k_p and the diffusion coefficients of anions and cations in the oxide is as follows[1]:

$$k_p = \frac{c_o M_{M_a O_b}^2}{2 \widetilde{V} b^3 \rho^2 kT} \int_{\mu_o^i}^{\mu_o^o} \left(\frac{m}{2} D_M + D_O \right) d\mu_O \qquad (2)$$

$$= \frac{c_M M_{M_a O_b}^2}{2 \widetilde{V} b^2 \rho^2 akT} \int_{\mu_m^i}^{\mu_m^o} \left(\frac{2}{m} D_O + D_M \right) d\mu_M, \qquad (2')$$

where

$M_a O_b$ = the chemical formula for the oxide

c_o and c_M = concentrations of oxygen and metal ions as calculated from the density of the oxide

$M_{M_a O_b}$ = the molecular weight of oxide

$\widetilde{V}$ = the equivalent volume of the oxide

k = Boltzmann's constant

T   = absolute temperature
ρ = the density of the oxide
m   = the valence of the metal ions
D_M and D_O = diffusion coefficients for metal and oxygen atoms
μ_O and μ_M = chemical potentials of the respective ions

and the integration is between the metal oxide and the oxide–ambient interfaces. The diffusion coefficients must be relevant to the oxidation experiment in the sense that, for p-type oxides, the diffusion coefficient must be measured for an oxide in equilibrium with oxygen at the same partial pressure used to determine k_p and for n-type oxides, the diffusion coefficient must be measured for the oxide in equilibrium with the metal. It has been pointed out by Brett *et al.*[2] that the expression relating k_p and D_i (the diffusivity of the principal migrating species) can be simplified by making various assumptions about the nature of the defect gradient. The usual assumption[2] for oxides that form n-type metal-excess semiconductors (BeO, ZrO, ZrO_2) is that the concentration of dissolved metal at the oxide–oxygen interface is approximately zero ($C_x = 0$), while the assumption for metal-deficit, p-type semiconductors (Cu_2O, UO_2, NiO) is that the concentration of cation vacancies in equilibrium with the metal is very small compared with that in equilibrium with the oxide–oxygen interface ($C_o = 0$).

These assumptions can be combined with the assumption that the self-diffusion coefficient of the diffusing ion in the oxide is related to the defect diffusivity by $D = D_i\overline{C}$, where $\overline{C}$ is the average defect concentration. Then we may write

$$\frac{x^2}{t} = k_p = 2D_i(C_o - C_x). \quad (3)$$

If the defect concentration is linear, we have $\overline{C} = 1/2(C_o + C_x)$ and

$$k_p = \frac{4D(C_o - C_x)}{(C_o + C_x)} = 4D \text{ when } C_o \text{ or } C_x = 0. \quad (4)$$

Alternatively, if there is a uniform concentration of defects across the oxide and $C_x = 0$, then

$$k_p = \frac{2D}{\overline{C}} \cdot C_o = 2D, \quad (5)$$

since the average concentration is the same as that at the interface where the defects originate. The more exact expression[2] reduces to the simplified expressions in the case of monovalent ions and also yields the relationship $k = 2D(C_o - C_x)$ for divalent ions.

The observation of a parabolic rate equation is not proof that a Wagner mechanism applies unless it can also be shown that the parabolic rate constant is properly related to the diffusion coefficients in both magnitude and temperature dependence. Nonetheless, it seems self-evident that if a transport mechanism controls the effectiveness of an oxide barrier, minimum rates of oxidation should be correlated with minimum diffusion coefficients. This statement assumes, of course, that grain boundary and short-circuit diffusion paths have been eliminated so that only lattice diffusion is possible. For comparison of systems for potential oxidation resistance, diffusion coefficients in the proposed protective oxide should provide a good criterion for identifying the most promising systems.

The logarithmic rate law characterizes the room-temperature oxidation of aluminum, which everyone agrees is the ideal oxidation-resistant material. Unfortunately, it functions over too narrow a temperature range. The extensive work on aluminides is probably largely motivated by a desire to find something that behaves at high temperatures as aluminum does at room temperature. If one compares the integrated forms of the parabolic and logarithmic rate equations as follows:

$$x^2 - x_o^2 = k_p(t - t_o) \text{ (Parabolic)} \quad (6)$$

$$x - x_o = k_{log}(\log t - \log t_o) \text{ (Logarithmic)} \quad (7)$$

where an oxide thickness of x_o is assumed at a time t_o, the logarithmic behavior is clearly advantageous for long-term oxidation resistance. At sufficiently long times, the oxide will grow as $\sqrt{t}$ if the parabolic rate law is followed, but only as $\log t$ if the logarithmic rate law applies. Thus, regardless of the values of the rate constants, k_p and k_{log}, the logarithmic rate law will ultimately lead to slower rates of oxide growth.

The logarithmic room-temperature oxidation of aluminum was associated by Mott[3] with a mechanism involving field induced transport through a thin oxide film. When the oxide is sufficiently thick that the electrostatic field due to adsorbed oxygen ions at the oxide–ambient interface can no longer influence the outward migration of Al^{+3} ions from the metal, oxidation virtually ceases. At higher temperatures, of course, when thermal transport becomes comparable with field-induced transport, a transition to parabolic kinetics is expected. A logarithmic rate law, due to a Mott type of mechanism, is thus highly unlikely at the elevated temperatures within the scope of this report. Fortunately, as is discussed below, normal lattice diffusion through aluminum oxide is also fairly slow, so that while aluminides at high temperature do not behave like aluminum at room temperature, they are still comparatively resistant to oxidation.

An exhaustive treatment is not attempted here of the numerous oxidation mechanisms that have been proposed when, as frequently happens, neither the Wagner mechanism nor the Mott mechanism can account for the experimental results, although other ideas may be introduced in discussing specific systems. It is important to be constantly aware of the fact that the oxidation mechanism may be different in different temperature–pressure regions. It is not possible to extrapolate results obtained in air at atmospheric pressure over a limited temperature range to predict behavior at lower or higher pressures or at widely different temperatures, unless it can be demonstrated that no change in oxidation mechanism occurs.

The silicide, aluminide, and oxidation-resistant alloy coatings all depend for their oxidation resistance on the *in situ* formation of a protective oxide diffusion barrier. The alternative approach of direct application of a protective oxide barrier introduces no new principle. In every case, one tries to design for slow diffusion rates across an oxide barrier.

SILICIDES

The remarkable oxidation resistance of molybdenum disilicide coatings on molybdenum has led to extensive investigation of an enormous number of other binary and more complex silicides as coatings for virtually all of the refractory metals and alloys. When a binary silicide was found to exhibit relatively poor high-temperature oxidation resistance, as in the case of $CbSi_2$, $TaSi_2$, $TiSi_2$, $ZrSi_2$, and $CrSi_2$, improvement was often obtained by the addition of other elements. For all of these systems, however, the mechanism of oxidation both before and after the addition of the "doctoring" elements is almost certainly quite different from that of molybdenum disilicide, and hence the ultimate potential will also be different.

The oxidation resistance of molybdenum disilicide coatings is due to the formation of a surface layer of pure or very nearly pure silica, through which diffusive transport of reacting elements is slow. The silicides that can be discussed as a group, and whose ultimate potential oxidation resistance should be similar, other things being equal, are those that, like $MoSi_2$, owe their oxidation resistance to the formation of a pure silica scale. This group should include WSi_2, $ReSi_2$, and the platinum group disilicides, where either the metallic element is more noble than silicon, so that the silicon is oxidized preferentially, or the oxide of the metallic element is volatile, so that only SiO_2 remains in the condensed phase. A similar oxidation mechanism should also apply to SiC and to pure Si, although in the latter case the low melting point of the element, rather than a deterioration of the diffusive properties of the oxide, limits the overall oxidation resistance of the system.

The sequential steps in the protective oxidation of silicides and the respective steady state fluxes, F_j, when an approximately pure silica film provides the protection are summarized by Krier and Gunderson[4] from the work of Deal and Grove[5] as follows:

1. Transport of oxygen from the gas phase to the surface of the silica scale where either adsorption or reaction occurs:
$F_1 = h(C^* - C_o)$.

2. Transport of the oxygen through the scale from the silica/ambient to the silica/silicide interface:
$F_2 = D_{eff}(C_o - C_i)/x$.

3. Reaction with silicon and/or metal at the silicide/silica interface:
$F_3 = kC_i^a C_{Si,i}^b$.

In the case of silicides for which an intermediate layer of reduced silicon content develops between the original silicide and the SiO_2 scale:

4. Transport of silicon across the intermediate layer to the reaction zone:
$F_4 = D_{Si}(C_{Si,o} - C_{Si,s}/x_s)$

where, with reference to Figure 2

C^*	= the concentration of oxygen in silica in equilibrium with the ambient gas
C_o	= the concentration of oxygen in the outer surface of the silica at any time
C_i	= the concentration of oxygen at the silica-silicide interface
$C_{Si,i}$	= the concentration of silicon at the silica-silicide interface

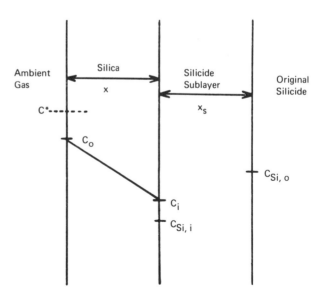

FIGURE 2 Model for the oxidation of silicides.

x	=	the thickness of the silica film
a and b	=	constants that define reaction order
h	=	a gas-phase transport coefficient
k	=	the reaction rate constant
D_{eff}	=	the effective diffusion coefficient for oxygen in the silica scale
x_s	=	the thickness of the intermediate silicide layer
$C_{Si,o}$	=	the concentration of silicon at the interface between the two silicides
D_{Si}	=	the diffusion coefficient of silicon in the intermediate silicide.

In the oxidation of silicon, Deal and Grove[5] found that the interface reaction 3 is not fast compared to the transport reaction, one of the requirements for a Wagner mechanism to apply. The interface reaction should be of at least comparable importance in the oxidation of silicides.

The equilibrium concentration C^* was assumed to be related to the ambient oxygen pressure P_{O_2} by a Henry's law relationship:

$$C^* = k_{P_{O_2}} \qquad (8)$$

where k is the Henry's law constant. The final equation derived for the rate of growth of silica on silicon[5] is

$$x = \frac{A}{2} \left[1 + \frac{t + \tau}{A^2/4B} \right]^{1/2} - 1 \qquad (9)$$

where

$$A = 2 D_{eff} (1/k' + 1/h) \qquad (10)$$

$$B = 2 D_{eff} C^*/N_1 \qquad (11)$$

$$\tau = (x_o^2 + Ax_o)/B \qquad (12)$$

$$F_3 = k' C_i. \qquad (13)$$

The quantity x_o is the thickness of the silica scale at time $t = 0$, after which the equation is valid; N_1 is number of oxygen molecules per cubic centimeter in the oxide as calculated from the density of the scale; and k' is the rate constant for step 3 in the case of pure silicon.

By considering the rate of growth of the silica scale after the formation of some initial oxide thickness x_o, Deal and Grove[5] circumvented the complicated problems involved in the very first stages of growth. This simplification is particularly pertinent to the oxidation of silicides, where generally both the metal and the silicon are oxidized concurrently in

the initial stages and protective oxidation is not established until a continuous, coherent silica scale covers the entire surface of the sample. For assessing the maximum potential effectiveness of silica as a protective barrier, it is convenient to separate out the problem, albeit nontrivial, of establishing such a barrier rapidly. In accordance with the findings of Deal and Grove[5] on the oxidation of silicon in dry oxygen, we assume $x_o = 250$ Å in analyzing the data for silicides.

When an intermediate silicide is formed, and step 4 above assumes significance, the Deal and Grove analysis requires modification. Once steady state conditions have been established, the fluxes permitted by the various rate-determining steps should be equal. Under most flow conditions of practical interest, $h \gg k$, and equilibrium is established at the silica-ambient interface, so that $C_o \simeq C^*$ and $F_1 = 0$. From the conditions $F_2 = F_3$ and $F_3 = F_4$, one can solve for the instantaneous steady-state concentrations C_i and $C_{Si,i}$ in terms of the diffusion coefficients, the fixed concentrations C^* and $C_{Si,o}$, the rate constant k, and the instantaneous thicknesses x and x_s. We will assume that the reaction at the silica-intermediate-silicide interface is first-order in both oxygen and silicon so that a and b in the expression for F_3 (p. 25) are both equal to one. The equation for C_i becomes

$$C_i = \frac{C^*}{1 + \dfrac{kx\, C_{Si,i}}{D_{eff}}}, \qquad (14)$$

and $C_{Si,i}$ is determined as the solution to the quadratic equation

$$\frac{kx}{D_{eff}} C_{Si,i}^2 + C_{Si,i} \left(1 - \frac{kx\, C_{Si,o}}{D_{eff}} + \frac{x_s k}{D_{Si}} C^* \right) - C_{Si,o} = 0. \qquad (15)$$

The general expression for $C_{Si,i}$ derived from Equation (15) is complicated and leads to a differential equation for oxide film thickness, x, which is not readily integrable. Hence, simplified solutions for particular values of the parameters will be considered.

In the case of a material such as SiC where an intermediate silicide layer does not form, $x_s = 0$, and hence $C_{Si,i} = C_{Si,o}$, and from Equation (14) $C_i = C^*/1 + (kx\, C_{Si,o}/D_{eff})$. The differential equation for the rate of growth of the silica scale then becomes

$$\frac{dx}{dt} = \frac{F_2}{N_1} = \frac{F_3}{N_1} = \frac{k}{N_1} \frac{C^* C_{Si,o}}{1 + \dfrac{kx\, C_{Si,o}}{D_{eff}}}. \qquad (16)$$

Equation (16) is the same as that developed by Deal and Grove for the oxidation of silicon, except that the rate constant k' in the Deal and Grove development, with $h \gg k'$, is here replaced by $(kC_{Si,o})$. The solution to Equation (16) is thus the same as that given by Equations (9) through (12) if k' is set equal to $(kC_{Si,o})$ and l/h is considered negligible.

Now let us consider the case of $MoSi_2$ where oxidation results in the formation of Mo_5Si_3 between the silica scale and the disilicide. If a silica scale of thickness x is formed via the reaction

$$MoSi_2 + \frac{7}{5} O_2 \rightarrow \frac{1}{5} Mo_5Si_3 + \frac{7}{5} SiO_2,$$

then the corresponding thickness x_s of the Mo_5Si_3 layer is given by the following relationship:

$$x_s = \frac{(x)(\rho_{SiO_2})(M_{Mo_5Si_3})}{(M_{SiO_2})(7)(\rho_{Mo_5Si_3})} = 0.42x \qquad (17)$$

where the densities are $\rho_{SiO_2} = 2.32$ g/cc and $\rho_{Mo_5Si_3} = 7.4$ g/cc, and M_Y is the molecular weight of Y.

The diffusion coefficient of silicon in Mo_5Si_3 was determined by Bartlett and Gage[6] from experiments on the rate of growth of Mo_5Si_3 between $MoSi_2$ and Mo_3Si. Due to uncertainties in silicon concentrations, the results were expressed in the form

$$D_{Si} = 1.3e^{-\frac{86,000}{RT}} \frac{cm^2}{\Delta C} \frac{}{sec} \qquad (18)$$

where ΔC is the difference in silicon concentration in moles/cc between the $Mo_5Si_3/MoSi_2$ interface and the Mo_5Si_3/Mo_3Si interface. On the basis of equilibrium vapor pressure measurements,[7] the activity of silicon, a_{Si}, in equilibrium with Mo_5Si_3 and $MoSi_2$ is 0.1; the corresponding activity over the two-phase region $Mo_3Si-Mo_5Si_3$ is 0.005. If it is assumed that activity is proportional to concentration, we may estimate $C_{Si} = C_{Si}^o a_{Si}$ where C_{Si}^o, the concentration of silicon in pure solid silicon, would be given by $C_{Si}^o = \rho_{Si}/M_{Si} = 2.4/28$ moles/cc. Hence in Equation (18) $\Delta C = C_{Si}^o (\Delta a) = (2.4/28)(0.1-0.005)$. The final expression for D_{Si} becomes

$$D_{Si} = 57.5 \times 10^8 e^{-86,000/RT} \mu^2/hr. \qquad (19)$$

From the work of Norton,[8]

$$D_{eff} = 10^8 e^{-27,000/RT} \mu^2/hr, \qquad (20)$$

and from the work of Deal and Grove[5] at 1000°C and an oxygen pressure of one atmosphere,

$$C^* = 5.2 \times 10^{16} \text{ molecules/cc.} \qquad (21)$$

In accordance with the above discussion,

$$C_{Si,o} = 5.15 \times 10^9 \text{ atoms/}\mu^3. \qquad (22)$$

To get a feeling for the relative orders of magnitude of the coefficients in Equation (15) as applied to molybdenum disilicide, let us consider a temperature of 1000°C. From Equations (19) and (20), we calculate $D_{eff} = 2.34 \times 10^3$ and $D_{Si} = 1.07 \times 10^{-5} \mu^2/hr$. For the experiments on pure silicon, Deal and Grove defined the flux F_3 by means of the expression $F_3 = k'C_i$. Our expression for F_3 is $F_3 = k C_i C_{Si,i}$. Since in the Deal and Grove experiments $C_{Si,i}$ is a constant, which we may calculate as 5.15×10^{10} atoms/μ^3 from the density of silicon, we may assume that $k' = 5.15 \times 10^{10}$ xk. Deal and Grove measured $k' = 3.6 \times 10^4 \mu/hr$ at 1000°C in oxygen, and hence we may assume $k = 0.7 \times 10^{-6}$. Referring to the parameters in Equation (15), we thus find for $MoSi_2$ at 1000°C and an oxygen pressure of one atmosphere

$$\frac{kx}{D_{eff}} = 2.99 \times 10^{-10} x \qquad (23)$$

$$\frac{kx C_{Si,o}}{D_{eff}} = 1.54 x \qquad (24)$$

$$\frac{kx_s C^*}{D_{Si}} = \frac{(k)(0.42 x) C^*}{D_{Si}} = 1430 x \qquad (25)$$

$$C_{Si,o} = 5.15 \times 10^9 \text{ atoms/}\mu^3. \qquad (26)$$

On the basis of these parameters, the linear term in Equation (15) may be neglected and, to a good approximation, the expression for $C_{Si,i}$ becomes

$$C_{Si,i} \approx \sqrt{\frac{C_{Si,o} D_{eff}}{kx}}. \qquad (27)$$

Substituting for $C_{Si,i}$ in Equation (14), we find

$$C_i = \frac{C^*}{1 + \sqrt{\frac{kx C_{Si,o}}{D_{eff}}}} \qquad (28)$$

Finally, the differential equation for the growth of the silica scale on $MoSi_2$ becomes:

$$\frac{dx}{dt} = \frac{F_2}{N_1} = \frac{F_3}{N_1} = \frac{F_4}{N_1} = \frac{\frac{k}{N_1} C^* \sqrt{C_{Si,o} D_{eff}}}{\sqrt{kx}\left(1 + \sqrt{\frac{kx C_{Si,o}}{D_{eff}}}\right)}, \quad (29)$$

which integrates to

$$x^{3/2} + Ax^2 = B(t + \tau) \quad (30)$$

where

$$A = \frac{3}{4} \sqrt{\frac{k C_{Si,o}}{D_{eff}}} \quad (31)$$

$$B = \frac{3C^*}{2N_1} \sqrt{C_{Si,o} D_{eff} k} \quad (32)$$

$$\tau = \left(x_o^{3/2} + 3 \sqrt{\frac{k C_{Si,o}}{D_{eff}}} \frac{x_o}{4}\right)^2 (1/B). \quad (33)$$

As before, x_o = 250 Å is the thickness of the silica film at a time t = 0, when the mechanism described becomes effective.

At relatively long times, such that $Ax^{1/2} \gg 1$ and $t \gg \tau$, Equation (30) reduces to the form

$$x^2 \simeq \frac{B}{A} t, \quad (34)$$

which is simply a parabolic rate equation with the parabolic rate constant given by $(2C^* D_{eff})N_1$. The Deal and Grove equation, Equation (9), reduces to exactly the same form with the same parabolic rate constant. Note that the final parabolic rate depends only on the properties of the silica scale. At short times, near t = 0, Equation (30) takes the form

$$x = x_o + Bt/\left(\frac{3}{2} x_o^{1/2} + 2Ax_o\right), \quad (35)$$

which, like the Deal and Grove equation for short times, is a linear rate equation. However, the linear rate constants for the two equations are not the same.

As a consequence of the limiting Equation (34), one predicts that once the oxidation of a silicide has become parabolic, the parabolic rate constant should have approximately the same temperature dependence as D_{eff} (ΔE_a = 27 kcal/

mole) and the same pressure dependence as C^* (linear). D_{eff} is at most weakly pressure-dependent and C^* is approximately independent of temperature.

From the data available on the oxidation resistance of silicide-coated refractory metals, it is virtually impossible to draw conclusions about the oxidation mechanisms. Data are reported, for the most part, in terms of "protective lifetime" under given conditions of temperature and pressure. It does seem clear however, from the relatively short lifetimes reported, that oxidation as discussed above is not the limiting factor. Monolithic molybdenum disilicide heating elements, for example, can be used continuously in air at 1650°C for 2000 hr, during which time the oxide scale thickness is of the order of 20 μ. This compares with a thickness of 10 μ calculated from Equation (34) and constitutes remarkable agreement. In contrast, a 1.5-mil-thick layer of Disil-1, a molybdenum-disilicide-based coating, on Mo–0.5 Ti showed edge failure after 15 hr at 1480°C. The possibility of edge failure is essentially excluded from the mechanistic discussion given above, where oxidation is treated as a one-dimensional phenomenon. The special problems of edges and corners are very familiar to coating developers and can sometimes be reduced by building up the coating at the edges and/or by avoiding sharp edges. The problem is not primarily an oxidation problem, but rather a structural problem of obtaining a smooth adherent oxide film that "fits" satisfactorily around the edges without developing undue stresses. Even in the absence of edge failure, however, the limiting factor in coating performance is probably diffusion of silicon into the substrate and attendant coating loss rather than oxidation.

Metcalfe and Stetson studied the oxidation resistance of arc-melted monolithic $MoSi_2$ and WSi_2 in air at temperatures of 815°C and 1315°C. At the lower temperature, WSi_2 failed by "pest," which has recently been shown to be due to enhanced oxidation at the tips of Griffith flaws under the influence of residual stresses.[9] $MoSi_2$ exhibited a fairly low weight gain of 0.03 mg/cm² in both 16 hr and 96 hr, but the observation of a porous black oxide precludes the diffusion-controlled oxidation mechanism proposed above. At 1315°C, the net weight gains for $MoSi_2$ were 0.33 and 0.86 mg/cm² after 16 and 96 hr respectively. However, since the net weight change combines a weight loss due to volatilization of $MoO_3(g)$ and a weight gain due to formation of a silica film, the results are not readily translatable into oxide film thickness. The WSi_2 weight changes were too small to measure. In both cases, however, protective behavior seemed to have been established.

The lower silicides, Mo_5Si_3 and Mo_3Si, shown by Berkowitz-Mattuck and Dils[10] to have excellent oxidation resistance once a continuous silica film is established, have generally been found unsatisfactory for coatings because of the large amount of the silicide which must be consumed prior to establishment of such a continuous film. The pre-oxidizing of silicide coatings to form a silica layer prior to use has never been satisfactory in practice, since the silica glass cracks on cooling. Introduction of a degree of room-temperature ductility into the silica might improve this situation. However, the dream of ductile ceramics has long been with us with no real prospect of being realized.

For those silicides that do form pure SiO_2 scales, the mechanism discussed earlier will probably be proved correct if the right experiments are done and the right parameters measured. Berkowitz-Mattuck and Dils,[10] Glushkv et al.,[11] and Wirkus and Wilder,[12] who studied the oxidation of monolithic molybdenum disilicide, did not continue their experiments for sufficiently long times to insure that they were in the diffusion-controlled region. By fitting their oxygen-consumption or weight-change data to an empirical rate equation, these workers obscured the complexity of the overall oxidation process. If further work is done, the initial stages, when both molybdenum and silicon are oxidized simultaneously, must be clearly separated from the steady-state oxidation process, which should be observed once a compact, adherent silica scale of about 250 Å in thickness has been formed.

Other silicide coatings, such as Cr-Ti-Si, which do not provide protection by formation of a pure silica scale but rather by formation of a much more complex oxide, defy analysis at the present time. Ternary diffusion problems in general require far more study and analysis before they become tractable. Nonetheless, since the complex silicides do provide protection by an essentially different mechanism than that of the molybdenum or tungsten silicides, a fairly basic study of the mechanism of oxidation of some of the empirically developed formulations could result in elucidating new fundamental principles for providing oxidation resistance in coatings.

ALUMINIDES

If the oxidation resistance of aluminides can be ascribed to the formation of a pure or nearly pure Al_2O_3 barrier scale,

and if a Wagner mechanism is valid, then the minimum possible oxidation rates should be calculable for diffusion data on aluminum oxide from Equations (2), (4), or (5). Aluminum oxide has generally been considered an n-type metal-excess semiconductor,[13] but this point of view has been challenged by Pappis and Kingery,[14] who suggest that at $1300°-1750°C$ Al_2O_3 is n-type at low oxygen pressures (less than 10^{-5} atm) and p-type at high oxygen pressures. The self-diffusion data on Al_2O_3 that have been reported in the literature are summarized in Table 3 and plotted in Figure 3.

Between 1600° and 2050°C, the melting point of aluminum oxide, $D_{Al} \gg D_O$, and hence the outward migration of aluminum is expected to determine the parabolic rate constant in diffusion-controlled kinetics. It is interesting to note that Doherty and Davis[15] demonstrated that crystalline Al_2O_3 at 600°C grows beneath an amorphous film by inward diffusion of oxygen. This mechanism is perfectly consistent with the extrapolated relative diffusion rates of Al and O at the lower temperature.

On the basis of the simplified relationships of Equations (4) and (5) between diffusion coefficient and parabolic rate constant, one would predict that the rate of growth of Al_2O_3 on aluminides that tend to form pure Al_2O_3 scales should be given by a rate equation of the form

$$x^2 = 4Dt \text{ (linear defect concentration)} \qquad (36)$$

or

$$x^2 = 2Dt \text{ (uniform defect concentration).} \qquad (37)$$

One of the first and most interesting practical aluminide coatings for refractory metals is the Sn–Al coating developed by Sylcor.[16] In the context of the present discussion, the Sn–Al coating owes its oxidation resistance to the formation of a pure Al_2O_3 scale. In anticipation of some of the problems discussed below that have been encountered with other aluminide coatings, it should be noted that under some conditions the Sn–Al coating is liquid; therefore, the Al_2O_3 scale floats on a liquid reservoir. This introduces difficulties at low oxygen pressures, which are taken up in the next

TABLE 3 Self-Diffusion Coefficients in Alumina

Diffusing Species	D_0, cm²/sec	ΔE_a, cal/mole	Temperature Range (°C)	Notes
Al	28	114,000	1670–1900	Annealed in air
O	1.9×10^3	152,000	>1600	—
O	6.3×10^{-8}	57,600	<1600	Results varied with heat treatment and impurities

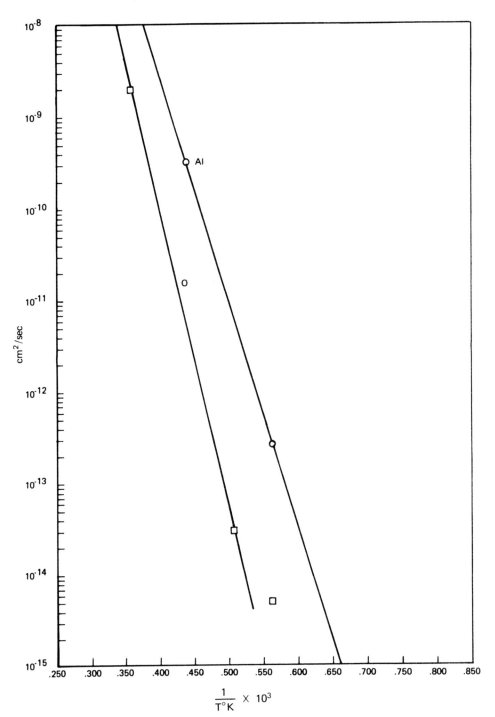

FIGURE 3   Self-diffusion in Al_2O_3.

section, but under ordinary atmospheric conditions the presence of the liquid sublayer serves to eliminate stresses in the outer protective oxide and hence to maintain its structural stability.

Data at 1370° and 1540°C for the Sylcor 50 Sn–50 Al

coating 34S on Ta-10W are presented in Table 4, and the results are compared with the prediction of Equation (4). At the higher temperature, agreement between the experimental and predicted results is quite good. At the lower temperature, the experimental results are higher than those predicted by a factor of about ten. The slight decrease in weight at the lower temperature between 70 and 90 hr also indicates that something other than a simple diffusion mechanism is operative. It is important to note that the data given in Table 4 were taken under isothermal conditions. Under cyclic exposure at the same two temperatures, where the oxide probably fractures on cooling, weight changes are very much greater.

Kofstad and Espevik studied the oxidation behavior of a $TaAl_3$–Al coating on tantalum at an oxygen pressure of 0.1 Torr.[17] Preferential oxidation to Al_2O_3 was observed, and protective behavior was established between 1000° and 1600°C. Below 1000°C, breakaway was observed, so that short-circuit diffusion paths were opened up, and there is no reason to expect oxidation rates to be related to diffusion coefficients. In Table 5, data at 1400°, 1500°, and 1600°C are compared with the theoretical diffusion-controlled oxide thickness calculated from Equation (4) and (5). We note first of all that the observed oxide thicknesses are greater than the predicted values by a factor of about five, so that the potential of Al_2O_3 as a protective barrier does not seem to have been realized. Of course, the diffusion data were obtained for Al_2O_3 annealed in air, while the correct data to be applied to aluminide oxidation would be for Al_2O_3 in equilibrium with aluminum. This might account for the discrepancy. Unfortunately, there are no data to prove this conjecture. Alternatively, since breakaway is known to occur at lower temperatures, short-circuit diffusion paths may open up at the higher temperatures too, although to a lesser extent. To test this possibility, it would be very worthwhile to develop methods for studying the structure of the growing oxide film. The data of Kofstad and Espevik are approximately parabolic, as shown by the relative constancy of the ratio of oxide thickness to square root of time at each temperature, Table 5, column 6.

The average parabolic rate constants calculated from the data are plotted in Figure 4 and lead to an activation energy of 83.2 kcal/mole. This is to be compared with a value of 114 kcal/mole for the self-diffusion of aluminum in aluminum oxide. The agreement is reasonable and the discrepancy might be ascribable to one or another of the factors said to explain differences in absolute rates.

OXIDES

It is generally conceded that the coating systems of greatest promise are those of the reservoir type with a built-in mechanism for repair or self-healing. The aluminides and silicides discussed above are examples. Nonetheless, serious attention has been paid to spraying or otherwise depositing protective oxide barriers onto refractory substrates. Naturally, the oxides of greatest interest are those that exhibit the lowest oxygen-diffusion rates. This assumes of course that oxidation resistance will result if the oxygen activity at the substrate–oxide interface is sufficiently low. If by chance the substrate atoms should diffuse rapidly through the outer oxide, so that they become subject to attack at the oxide–ambient interface, the efficacy of the entire concept of an oxide barrier could be lost. If the problem of metal atom diffusion through the oxide is ignored for the moment, the best possible protection that an oxide barrier can give is determined by the oxygen diffusion coefficient through a structurally perfect oxide. Values for oxygen diffusion in a number of refractory oxides are plotted in Figures 5 and 6. The numerical values must be applied with some caution in attempting to predict oxidation kinetics. First of all, as pointed out above in the discussion of the Wagner mechanism, the diffusion coefficient pertinent to oxidation kinetics must be obtained for the oxide in equilibrium with the ambient oxygen pressure of the kinetics experiments. Secondly, oxygen that does diffuse through the oxide can react with the substrate to form a new oxide, which may in turn react with or modify the properties of the original barrier. Finally, if short-circuit diffusion paths are present in the original oxide barrier, or if they develop during exposure, the effectiveness of the oxide as a diffusion barrier will be greatly reduced.

Among the single oxides, minimum diffusion rates are found for Al_2O_3, MgO, and CaO. These oxides exhibit only small deviations from stoichiometry and have low point-defect concentrations. Zirconia, a very stable refractory oxide, generally has a large concentration of oxygen point defects as it is ordinarily prepared and hence it must be considered unsatisfactory as a diffusive barrier. The situation could be very different if the defect concentration could be reduced by some mechanism. Diffusion in complex oxides is largely unexplored, although the few data available suggest that oxides with a spinel or perovskite structure should

TABLE 4 Oxidation Behavior of 50 Sn–50 Al (Sylcor 34S) on Ta-10W

Temperature, °K	Time, hr	Weight Gain mg/cm^2	Oxide Thickness, μ	$\sqrt{4Dt}, \mu$
1643	10	1.3	3.28	0.40
	30	3.0	7.55	0.69
	50	4.0	10.04	0.89
	70	4.1	10.32	1.05
	90	4.0	10.04	1.19
1811	10	2.0	5.05	2.66
	30	4.7	11.83	4.66
	50	6.0	15.10	6.0
	70	7.0	17.62	7.14
	90	7.7	19.40	8.05

TABLE 5   Oxidation of Ta Coated with $TaAl_3$–Al

Temperature, °K	Time, hr	Oxide Thickness, μ	$\mu/\sqrt{t}$	$\sqrt{4Dt}$	$\sqrt{2Dt}$	$10^3/T$	k_p, av., μ^2/hr
1673	5	3.7	1.65	0.51	0.36	0.598	2.34
	15	5.9	1.52	0.89	0.62	–	–
	25	7.5	1.50	1.14	0.80	–	–
	35	8.8	1.49	1.35	0.95	–	–
	45	9.85	1.47	1.53	1.08	–	–
1773	5	6.6	2.94	1.4	0.97	0.564	10.5
	10	10.0	3.16	1.95	1.37	–	–
	15	13.2	3.40	2.38	1.68	–	–
	20	15.6	3.48	2.76	1.94	–	–
1873	1	5	5.0	1.44	1.02	0.534	33.6
	2	7.8	5.5	2.04	1.44	–	–
	3	10.6	6.1	2.50	1.76	–	–
	4	13.2	6.6	2.89	2.04	–	–

exhibit low diffusion rates. In any case, a real question exists about the practical utility of a single-layer oxide barrier. Oxides are generally brittle, and a single-layer barrier has no mechanism for repair or self-healing in the case of rupture or failure. Furthermore, the oxide would have to be applied as a compact, pore-free, and coherent layer, which is extremely difficult within the framework of current technology.

Since single-oxide barriers are much less attractive than reservoir systems, very few are being seriously considered for application. One exception is ThO_2 sintered onto tungsten. While ThO_2 is one of the most refractory oxides (thermodynamically stable, with a high melting point and a low vaporization rate), and while it is chemically compatible with tungsten, the diffusion coefficient for oxygen, as shown in Figure 6, is not particularly low. It is orders of magnitude higher than diffusion through Al_2O_3, for example. The real question, however, is what happens to the oxygen when it arrives at the W–ThO_2 interface. Presumably, it reacts rapidly with tungsten. Nonetheless, since the oxygen activity will be greatly reduced compared with that of the ambient gas, it is not clear that such oxidation will be immediately deleterious to the properties of the entire system. Thus WO_3 (g), for example, would probably not build up to a level sufficient to rupture the outer layer of thoria. Furthermore, with both thoria and zirconia, the intrinsic diffusion that might be obtained at lower point-defect concentrations should be very much less than that indicated in Figure 6. Test results on ThO_2–W systems have been reported.[18]

OXIDATION-RESISTANT ALLOYS

Intermetallics and oxides share one unfortunate property: They tend to be brittle at room temperature. For this reason, if for no other, metal alloys as coatings or monolithic bodies would be vastly more attractive than the materials considered thus far, provided that comparable oxidation resistance could be assured. Certainly, the idea of developing oxidation-resistant alloys based on the refractory metals that might be used in the uncoated state was abandoned long ago, after many years of optimistic experimentation. Nevertheless, the superalloys and the Hf–Ta alloys remain as highly exciting developments over the range of conditions where they provide satisfactory oxidation resistance.

The Hf–Ta system displays a rather complex oxidation behavior. Three regions can be distinguished in oxidized specimens: a dense, adherent, fully oxidized outer layer that appears to be tantalum-stabilized HfO_2; a subscale consisting of stringers of HfO_2 interspersed with stringers of oxygen-saturated β-tantalum; and a two-phase oxygen contamination zone consisting of oxygen-saturated β-Ta and a hafnium-rich phase. Thus, the factors that must be considered in attempting to analyze oxidation behavior in the Hf–Ta system are diffusion through the external oxide scale, solution of oxygen in the tantalum phase, and internal precipitation of HfO_2. A theoretical analysis of a general model of oxidation processes for internal oxidation in combination with a parabolically thickening external scale has been given by Maak,[19] but it has not been applied to the Hf–Ta case. There is experimental evidence that the temper-

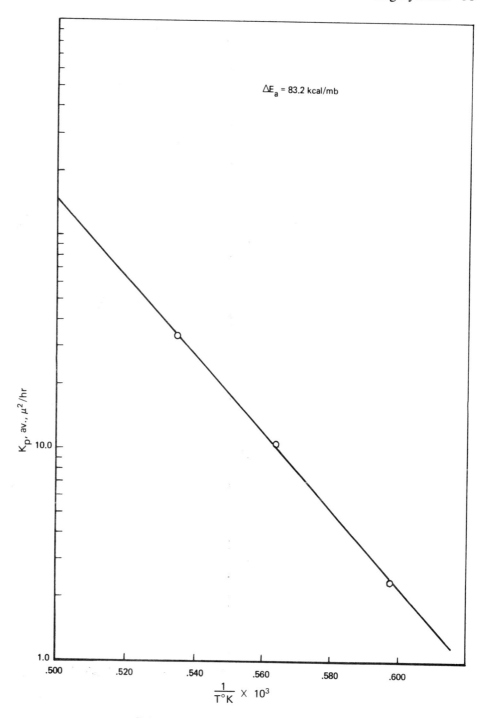

FIGURE 4   Oxidation of Ta coated with TaAl₃–Al.

ature of the alloy controls the rate. Also, if a gradient exists within the oxide so that the external surface is hotter than the interior, abnormally low overall oxidation rates are observed.[20] The general problem of internal oxidation and the effect of internal oxide precipitates on oxidation behavior would seem to merit further investigation.

The superalloys are multicomponent systems, generally developed with specific properties for defined structural applications. Within the design specifications, they usually function extremely well. The large number of components involved in each system and the significant differences between alloys makes a general discussion of oxidation behav-

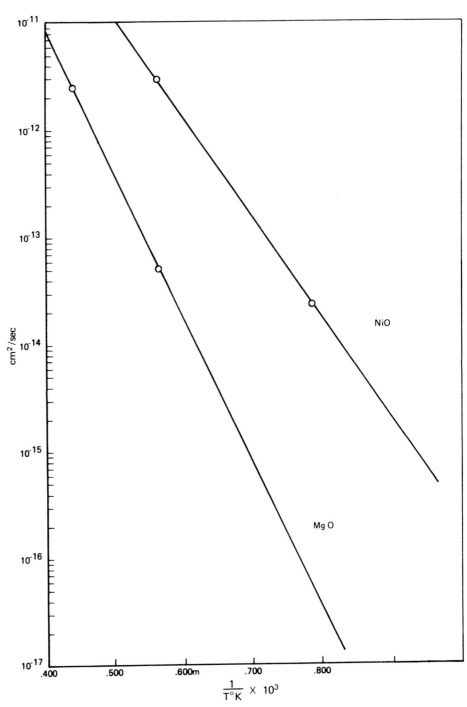

FIGURE 5 Oxygen diffusion in refractory oxides.

ior impossible. Each system must be analyzed separately. The work of Wlodek on René 41 and Udimet 700[21] is an example of what should be done to understand complex systems. There is no question that some understanding of the rate-controlling oxidation mechanisms can be helpful in suggesting a direction for the development of improved alloys.

VAPORIZATION

The alloys of interest as structural materials for advanced aerospace applications are, in order of increasing temperature capability, those based on columbium (B-66, D-36, D-43), molybdenum (TZM), tantalum (Ta-10W), and tungsten. If these materials are to be used in an oxygen-contain-

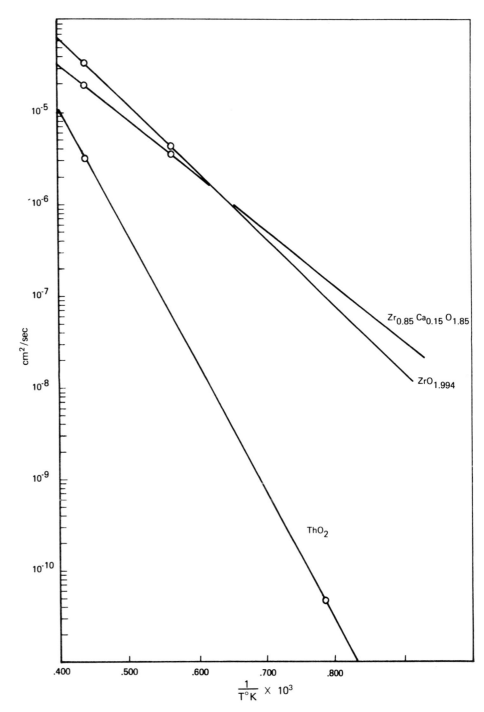

FIGURE 6   Oxygen diffusion in refractory oxides.

ing atmosphere, all require a protective coating—the columbium and tantalum alloys because they dissolve large
amounts of oxygen with consequent embrittlement, and
the molybdenum and tungsten alloys because they are
rapidly vaporized by the formation of volatile oxides. The
primary function of the protective coating is, of course, to
prevent, or at least retard, oxygen access to the substrate.

A consideration of vaporization problems associated with
coatings under development gives an insight into the kind of
vaporization data required for evaluating the potential of
new materials.

Five distinct vaporization processes can influence the
behavior of protective coatings:

Simple evaporation of the original coating, either through the protective oxide or prior to the establishment of a protective oxide

Simple evaporation of the outer protective oxide

Formation of high-temperature gaseous species at internal interfaces within the coating system as a result of chemical interaction between adjacent phases

Formation of an aerodynamic boundary layer at the coating surface that limits oxygen access to the surface

Formation of volatile products by reaction between ambient gases and the outer surface of the coating system

In the following paragraphs, the influence of each of these factors on coatings currently under development is discussed. It is shown that the data required to evaluate each of these factors are (a) equilibrium rates of vaporization of each pure component of a coating system into vacuum, (b) activities in any solid solutions or compounds that form, (c) stoichiometry of important reactions, and (d) geometric configurations, flow velocities, and gaseous diffusion coefficients.

SILICIDES

The pack-silicide coating for molybdenum is one of the oldest and best-studied coating systems. The coating is primarily molybdenum disilicide, $MoSi_2$, which oxidizes with the formation of silica glass. The SiO_2, as discussed above, acts as a diffusion barrier to oxygen. The $MoSi_2$ is mechanically compatible with both molybdenum and SiO_2 and oxidizes very rapidly to form the protective silica layer. In oxygen or air at one atmosphere and at temperatures below 1700°C, the protective coating is stable and effective for hours. At low oxygen pressure and at temperatures above 1700°C vaporization processes occur that can lead to coating failure. These processes, which characterize silicide coatings in general, are:

1. Evaporation of the silicide itself during exposure in vacuum or inert atmosphere, prior to the establishment of an oxide barrier

2. Simple evaporation of the protective SiO_2 layer

3. Reaction between the silicide and the SiO_2 to form gaseous $SiO(g)$ at the silicide-oxide interface

4. Failure to form a protective $SiO_2(g)$ layer at all due to an insufficient oxygen supply at the silicide surface

Each of these processes will be discussed in relation to $MoSi_2$, in particular, and will then be generalized to other silicide systems.

Direct Evaporation of Silicide Coatings

Silicide coatings are normally applied in such a way that the silicide of highest silicon content is formed. In the Mo–Si system, this is $MoSi_2$.* If the $MoSi_2$ surface is heated in vacuum, silicon tends to evaporate with the formation of a surface layer of Mo_5Si_3. The rate of evaporation can be estimated by means of the Langmuir equation[22] from the equilibrium vapor pressure data of Searcy and Tharp[7] on $MoSi_2$ and Mo_5Si_3. The Langmuir equation, based on the kinetic theory of gases, relates Z, the rate of vaporization from a solid surface at temperature T, to p, the equilibrium vapor pressure over the solid at the same temperature, as follows:

$$Z(moles/cm^2/sec) = \frac{44.4\, p\,(atm)a}{\sqrt{MT(°K)}}, \qquad (38)$$

where M is the molecular weight of the vaporizing species, a is the evaporization coefficient, and the units consistent with the coefficient of 44.4 are given in parentheses. The evaporation coefficient a equals one when molecules evaporate at the equilibrium rate and is less than one when there are kinetic barriers to evaporation. Of course, the maximum evaporation rate for any system is obtained for the condition $a = 1$. Maximum rates of evaporation of silicon over the two-phase regions $MoSi_2$–Si, Mo_5Si_3–$MoSi_2$, and Mo_3Si-Mo_5Si_3 are plotted in Figure 7. In practice, the initial rate of evaporation of an $MoSi_2$ coating may be governed by the pressure of silicon over $MoSi_2$–Si. As the surface becomes converted to Mo_5Si_3 via the reaction $MoSi_2 \rightarrow 1/5\, Mo_5Si_3 + 7/5\, Si(g)$, the rate of silicon loss will diminish until it becomes controlled by the silicon pressure over Mo_5Si_3–$MoSi_2$. In an inert atmosphere, silicon will also evaporate from $MoSi_2$ and leave behind a surface layer of Mo_5Si_3, but the rate of conversion will be less than in vacuum[23] by from one to several orders of magnitude.

If exposure conditions prior to oxidation have resulted in extensive conversion of $MoSi_2$ to Mo_5Si_3, the silicide coating will not be nearly as effective in providing oxidation resistance. Eventually, a protective layer of silica does form over Mo_5Si_3, but the volume of Mo_5Si_3 oxidized prior to formation of the protective layer is about six times that of $MoSi_2$.[10] Expressed in another way, an Mo_5Si_3 layer would have to be about six times thicker than an $MoSi_2$ layer to afford equal protection.

To evaluate the effect of coating loss by vaporization in

---

*Since $MoSi_2$ is thermodynamically unstable in contact with molybdenum, one normally finds layers of Mo_5Si_3, Mo_3Si, and the terminal solid solution of silicon in molybdenum between the outer layer of $MoSi_2$ and the molybdenum substrate. For kinetic reasons, one or another of these layers may be missing.

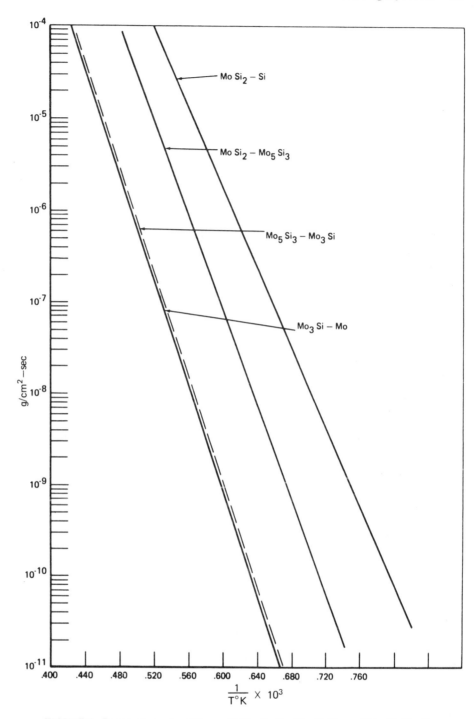

FIGURE 7   Evaporation rate of Si over $MoSi_2$–Si, Mo_5Si_3–$MoSi_2$, and Mo_3Si–Mo_5Si_3.

vacuum on oxidation protection, one must know both the equilibrium vaporization behavior of the coating material and the oxidation behavior of the depleted coating. For the Mo–Si system, fairly complete data are available.[10] For the W–Si system, both vaporization and oxidation data are sparse. The dissociation pressures of silicon at 1765°K were measured over the two-phase regions, WSi_2–Si(W) solution, WSi_2–W_5Si_3, and W_5Si_3–W(Si) solution.[24] Although a high uncertainty for temperatures was reported, a comparison of tungsten and molybdenum results suggests lower silicon vapor pressures in the W–Si system than in the Mo–Si system at comparable temperatures and compositions. In other words, a WSi_2 coating would be more stable in vacuum with respect to conversion to the next lower silicide (W_5Si_3) than an $MoSi_2$ coating would be under the same conditions. Oxidation of WSi_2 and W_5Si_3 was studied briefly by Berkowitz.[25] She found that both materials are superior to $MoSi_2$ in oxidation resistance. A coating of W_5Si_3 would have to be about three times thicker than one of WSi_2 to provide the same protection for the substrate.

For the Cb–Si and Ta–Si systems, vaporization data are completely lacking, as is comparative oxidation data for the various silicides.

The possible deleterious effects of inert gas or vacuum exposures on ultimate performance of coating systems in oxidizing environments has not received sufficient attention. The performance characteristics of W/WSi_2 coatings on Ta–10W, for example, were shown to be seriously degraded by an hour's preheating in helium, even at temperatures as low as 1000°C.[26]

Simple Evaporation of the SiO_2 Layer

The loss of a silicide coating by evaporation, as discussed above, is important only if the coated metal is exposed to vacuum without a protective layer of oxide. If, for example, $MoSi_2$ is exposed to oxygen to form a protective layer of SiO_2, and the oxidized sample is then subjected to vacuum at the same temperature, vaporization of silicon from $MoSi_2$ is retarded. However, loss of the surface *oxide* by evaporation could be very significant. As shown in Figure 8, the rate of vaporization of SiO_2(c) is similar in vacuum and in pure oxygen at 0.2 atm, although the vapor species are mainly SiO(g) and O_2(g) in vacuum, and SiO_2(g) in oxygen. In oxygen, oxidation to form additional solid SiO_2 can partially balance evaporation losses. In vacuum, depending upon the temperature, pressure, and time of exposure, the SiO_2 layer could be lost altogether, and the silicide evaporation discussed above could assume importance. In any case, upon re-exposure to oxygen, additional oxidation of the silicide coating would be necessary to re-establish the protective layer of silica. Thus, if a system coated with $MoSi_2$ were to be cycled between vacuum and atmospheric oxygen, a thicker coating would be required than if the system were simply held in oxygen.

The protective oxide formed in the high-temperature oxidation of $MoSi_2$ and WSi_2 is pure SiO_2(c). Hence the loss of oxide in vacuum would be the same for both materials at the same temperature. In the case of $CbSi_2$ and $TaSi_2$, it is likely that the protective oxide contains both Cb (or Ta) and Si. Since the vaporization behavior of these mixed oxides is unknown, their rates of loss cannot be calculated at present.

Formation of Gaseous SiO(g) at the Silicide–Oxide Interface

Metallic silicon can react with solid silicon dioxide to form gaseous silicon monoxide. In the case of a coating system, for the phase boundary between SiO_2 and a silicide of silicon activity a_{Si}, this reaction can be written in the form

$$Si(a_{Si}) + SiO_2(c) \rightarrow 2SiO(g). \qquad (39)$$

The pressure of SiO(g) due to Equation (39) can be calculated from the standard free energy of the reaction and the silicon activity in the silicide:

$$\ln p_{SiO} = -\frac{\Delta F°}{2RT} + \frac{1}{2} \ln a_{Si}. \qquad (40)$$

If the SiO(g) pressure at the silicide–oxide interface becomes comparable with the ambient pressure, disruption of the protective SiO_2(c) layer can be expected. For silicon at unit activity, the equilibrium pressures of SiO(g) calculated for reaction Equation (39) are plotted in Figure 9. In $MoSi_2$ the silicon activity is between 0.1 and 1; in Mo_5Si_3 it lies between 0.1 and 0.005; in Mo_3Si it is about 0.004. The effect of the activity change in the pressure of SiO(g) is also shown in Figure 9. At atmospheric pressure an $MoSi_2$ coating is expected to fail by evolution of gaseous SiO(g) at temperatures around 2000°K when the pressure of SiO(g) is of the order of 1 atm. At lower pressures, failure would occur at lower temperatures. This is in accord with experimental observations.[10,27] In like manner, a protective layer of SiO_2(c) over Mo_5Si_3 would be stable to higher temperatures than a similar layer over $MoSi_2$ because of the lower silicon activity in the alloy. This too has been verified experimentally.[10,27]

The few thermodynamic data available[2] for WSi_2 suggest that the silicon activity in WSi_2 is less than that in $MoSi_2$ by perhaps a factor of two. Hence, a WSi_2 coating should be useful at lower pressures and higher temperatures than a $MoSi_2$ coating. Recent findings seem to confirm this conclusion.[28]

Silicon activities or silicon vapor pressures have not been measured in the columbium and tantalum silicides. Furthermore, Equation (40) may not be strictly applicable because the protective oxide probably contains Cb or Ta. Thus, currently available thermodynamic data do not permit an analysis of the low-pressure failure of Cb or Ta silicides due to gaseous SiO(g) formation at the silicide–oxide interface.

Failure to Form a Protective Oxide Due to Insufficient Oxygen Supply

The importance of the vaporization processes discussed so far can be assessed from equilibrium thermodynamic data. In silicide coatings, however, it has been found that a pro-

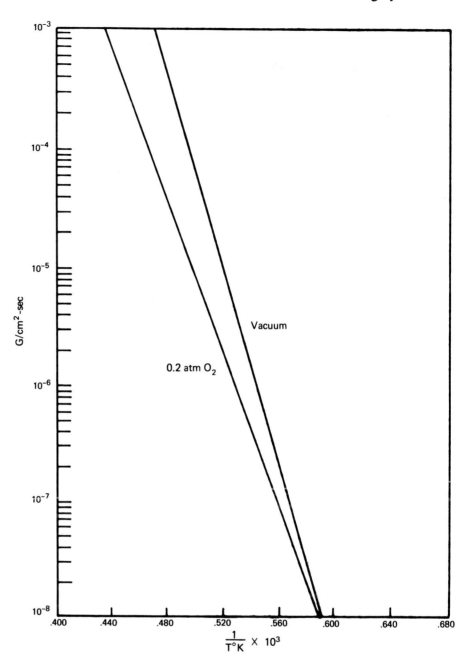

FIGURE 8   Rate of vaporization of SiO_2 in vacuum and in oxygen at 0.2 atm.

tective layer of SiO_2(c) fails to form at all, even at ambient oxygen pressures for which the oxidation reaction is strongly favored on thermodynamic grounds. Similar behavior has been observed in the oxidation of silicon.[29] In the case of $MoSi_2$, an oxygen pressure in excess of 5 Torr is required to form a protective SiO_2(c) layer. At lower pressures, the oxidation is linear, and no condensed phase oxide is observed. In the analogous case of pure silicon, it was shown by Wagner[30] that aerodynamic considerations be-

come significant. At any temperature, there is a finite pressure p_{SiO}(eq) of SiO(g) in equilibrium with silicon and SiO_2, given by Equation (40). Unless that pressure is exceeded, SiO_2 cannot condense on a silicon or silicide surface. Wagner showed that the formation of a condensed layer of SiO_2 is dependent on the rate of transport of oxygen across a boundary layer of gaseous SiO(g) formed via reaction Equation (39).

If p_{SiO}^*, the pressure of SiO(g) at the surface of a silicide

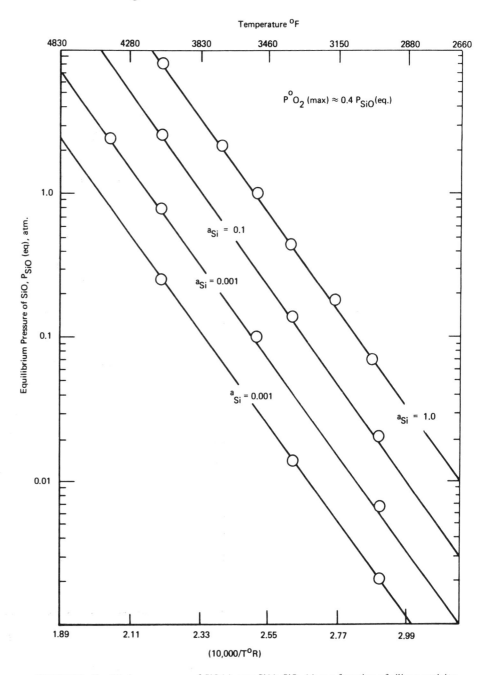

FIGURE 9   Equilibrium pressures of SiO(g) over Si(s)–SiO$_2$(s) as a function of silicon activity.

of activity a_{Si}, is less than the equilibrium partial pressure p_{SiO} (eq) for reaction Equation (39), a condensed oxide will not form. (In general, p_x represents the pressure of x at the solid surface.) The condition for a protective SiO$_2$(s) film to form, or in Wagner's terminology, the point of transition between active and passive oxidation is

$$p_{SiO}^* = p_{SiO(eq)}.$$

(41)

Since oxygen should react readily at a bare silicide surface, oxidation in the active region will be characterized by $p_{O_2}^* \ll p_{O_2}^o$, where $p_{O_2}^o$ is the oxygen partial pressure in the ambient gas stream. From boundary layer theory, the rate of transport of oxygen toward the surface will be given by

$$\dot{m}_O^- \ (\text{atoms/cm}^2\text{-sec}) = -\frac{2D_{O_2} \ C_{O_2}^o}{\delta_{O_2}} , \qquad (42)$$

where the negative sign denotes motion towards the surface, D_{O_2} is the diffusion coefficient for O_2 molecules, $C_{O_2}^o = p_{O_2}^o/RT$ and δ_{O_2} is the boundary layer thickness for mass transport of O_2. Similarly, the rate of transport of oxygen away from the surface as SiO(g) is given by

$$\dot{m}_O^+ = \frac{D_{SiO} \ C_{SiO}^*}{\delta_{SiO}} . \qquad (43)$$

In the steady state, there will be no net oxygen transport, and therefore

$$C_{SiO}^* = 2(\delta_{SiO}/\delta_{O_2}) \ (D_{O_2}/D_{SiO}) \ C_{O_2}^o. \qquad (44)$$

Wagner approximates the ratio of the boundary layer thicknesses by $(\delta_{SiO}/\delta_{O_2}) = (D_{SiO}/D_{O_2})^{1/2}$. Then, setting $p_{SiO}^* = p_{SiO(eq)}$, he finally solves for the maximum ambient oxygen pressure compatible with a bare silicide surface:

$$p_{O_2}^o \ (\text{max}) \simeq 1/2 \left(\frac{D_{SiO}}{D_{O_2}}\right)^{1/2} p_{SiO(eq)} \approx 0.4 \ p_{SiO(eq)}. \qquad (45)$$

The ratio of diffusion coefficients is estimated by Wagner to be $(D_{SiO}/D_{O_2}) = 0.64$. The equilibrium pressure, $p_{SiO(eq)}$, for reaction Equation (39) is plotted in Figure 9 as a function of temperature for various activities of silicon. Calculations were based on the data given in the JANAF Tables.[31]

In the range where the rate of gaseous diffusion is slow compared to the rate of surface oxidation, the oxygen partial pressure at the solid surface is very much less than that in the bulk gas. Thus, at 1960°K, according to the Wagner theory, an oxygen pressure of 20 Torr is required in the bulk gas to form a layer of $SiO_2(c)$ on a silicide with a silicon activity of 0.1. This is the approximate silicon activity over the two-phase region $MoSi_2$–Mo_5Si_3. Experimentally, it was found that at 1960°K and 2 Torr, the oxidation of $MoSi_2$ is linear and leads to the formation of gaseous products only—$MoO_3(g)$ and SiO(g). At 1960°K, a condensed film of SiO_2 did form on an $MoSi_2$ surface at an oxygen pressure of 13 Torr, in reasonable agreement with theoretical prediction.[10]

There is a hysteresis in the transition point between active and passive oxidation. The above discussion was concerned with the failure to form a protective silica film as a function of ambient oxygen pressure. Once an $SiO_2(s)$ film has been formed, the ambient oxygen pressure may be permitted to fall well below $p_{O_2}^o$ (max) before oxidative protection is lost. Wagner's analysis[30] gives the lowest ambient oxygen pressure $p_{O_2}^o$ (min) at which a layer of SiO_2 formed at higher pressure is stable. The final equation is

$$p_{O_2}^o \ (\text{min}) = 1/2 \ [(1/4)^{2/3} + (1/2)^{2/3}] \ K^{2/3} \ \frac{D_{SiO}}{D_{O_2}} , \qquad (46)$$

where K is the equilibrium constant for the reaction

$$SiO_2(s) = SiO(g) + 1/2 \ O_2(g). \qquad (47)$$

Values of $p_{O_2}^o$ (min) based on JANAF thermodynamic data are plotted against $10^3/T$ in Figure 10. According to Figure 10 even at 4000°F, a silica film once formed should be stable down to oxygen pressures of 4.7 Torr.

The influence of an aerodynamic boundary layer on effective oxygen pressure must be considered for all silicides. In general, at any given temperature, the ambient oxygen pressure required to form a condensed-phase silicon dioxide on a silicide surface increases with increasing silicon activity of the alloy. For a given silicide, the required oxygen pressure increases with temperature.

ALUMINIDES

Sn–Al Coatings

In the Sn–Al coatings, a protective layer of $Al_2O_3(c)$ is formed on a liquid Sn–Al substrate. The vaporization processes that can have a deleterious effect on coating performance are analogous to those discussed above for silicides:

Evaporation of the Sn–Al substrate
Evaporation of the Al_2O_3 protective barrier
Aerodynamic effects

In the case of the silicides, the equilibrium vaporization of silicon from the alloys is not sufficient to rupture a film of SiO_2 under any projected service conditions. In the case of the Al–Sn coatings, the vapor pressures of both Sn and Al are of the order of 0.1 atm at 1800°C. At ambient pressures in the millimeter range, the protective Al_2O_3 layer could be destroyed by vaporization of Sn and Al at temperatures in the vicinity of 1300°C. As in the silicide case, the data required to identify possible conditions of failure are activities of Sn and Al over molten Sn–Al alloys as a function of temperature.

Evaporation of the protective layer of Al_2O_3 would not shorten coating lifetime significantly at temperatures below about 2200°C.

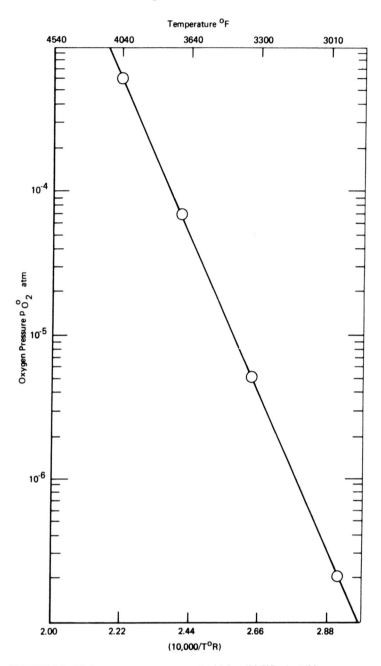

FIGURE 10   Minimum oxygen pressure at which solid SiO_2 is stable as a function of temperature.

A possible aerodynamic effect that might prevent formation of condensed Al_2O_3 on the liquid substrate at low oxygen pressures is analogous to the effect observed in the oxidation of zinc. At temperatures where zinc has an appreciable vapor pressure, oxidation to form solid $ZnO(c)$ occurs primarily in the gas phase. In other words, oxygen reacts with zinc vapor at some distance from the metal surface to form a cloud of condensed particles of $ZnO(c)$, and the surface of the metal remains bare of condensed oxide. This phenomenon has never been observed experimentally with Sn–Al coatings. However, following Turkdogan *et al.*,[32] we find that in a stream flowing at 80 cm/sec, an ambient or bulk oxygen pressure of 100 Torr might be required in order to condense $Al_2O_3(c)$ on an Sn–Al substrate held at as low a temperature as $1100°C$. At lower flow velocities and the same substrate temperature, lower oxygen pressures would suffice. At higher temperatures, for a given configuration and flow velocity, higher bulk oxygen pressures might be required.

TABLE 6   Vaporization of Simple Refractory Oxides

Refractory Oxide	Melting Point (°C)	Temperature (°C) for Vaporization in Vacuum at		Vaporization Reaction	Temperature (°C) for Vaporization in 0.2 atm of O_2 at 10 mils/hr
		1 mil/hr	10 mils/hr		
HfO_2	2900	2394	2601	1. $HfO_2(c) \rightarrow HfO(g) + O(g)$ 2. $HfO_2(c) \rightarrow HfO_2(g)$	2720
ThO_2	3220	2165	2345	1. $ThO_2(c) \rightarrow ThO_2(g)$ 2. $ThO_2(c) \rightarrow ThO(g) + O(g)$	2387
Y_2O_3	2410	2120	2290	1. $Y_2O_3(c) \rightarrow 2YO(g) + O(g)$ 2. $Y_2O_3(c) \rightarrow 2Y + 3O(g)$	2570
ZrO_2	2710	2110	2290	1. $ZrO_2(c) \rightarrow ZrO_2(g)$ 2. $ZrO_2(c) \rightarrow ZrO(g) + O(g)$	2325
Al_2O_3	2034	1925	2080	1. $Al_2O_3(c) \rightarrow 2Al(g) + 3O(g)$ 2. $Al_2O_3(c) \rightarrow Al_2O(g) + 2O(g)$ 3. $Al_2O_3(c) \rightarrow 2AlO(g) + O(g)$	see text
BeO	2520	1910	2070	1. $BeO(c) \rightarrow 1/3 \, (BeO)_3(g)$ 2. $BeO(c) \rightarrow Be(g) + O(g)$	2200
La_2O_3	2305	–	1990	1. $La_2O_3(c) \rightarrow 2LaO(g) + O(g)$ 2. $La_2O_3(c) \rightarrow 2La(g) + 3O(g)$	2230
UO_2	2880	1815	1960	1. $UO_2(c) \rightarrow UO_2(g)$ 2. $UO_2(c) \rightarrow UO(g) + O(g)$	see text
CaO	2570	–	(1845)	1. $CaO(c) \rightarrow Ca(g) + O(g)$ 2. $CaO(c) \rightarrow CaO(g)$	(1855)
Cr_2O_3	2430	–	1815	1. $Cr_2O_3(c) \rightarrow 2Cr(g) + 3O(g)$ 2. $Cr_2O_3(c) \rightarrow CrO(g) + CrO_2(g)$	see text
MgO	2800	–	1760	$MgO(c) \rightarrow Mg(g) + 1/2O_2(g)$	2055
SrO	2435	–	1654	1. $SrO(c) \rightarrow SrO(g)$ 2. $SrO(c) \rightarrow Sr(g) + O(g)$	1654

For Sn-Al coatings, the reaction analogous to reaction Equation 39 for silicides would be formation of gaseous $Al_2O(g)$ or $AlO(g)$ at the alloy–oxide interface. However, the pressures due to this reaction are not expected to be of practical importance.

OXIDES

Simple Oxides

The oxides known to have melting points in excess of 2000°C are listed in Table 6 in order of increasing volatility. Column 3 defines the temperature at which the oxide vaporization rate in vacuum is equal to 1 mil/hr. Column 4 gives maximum service temperatures in vacuum for each oxide if a loss of 10 mils/hr is tolerable. Column 5 gives mode of evaporation in cases where it is known. For oxides which vaporize primarily to the elements, evaporation will usually be suppressed by the presence of oxygen. Column 6 gives the temperature at which a maximum loss of 10 mils/hr can be expected at an ambient oxygen pressure of 0.2 atm.

In the case of $UO_2(c)$, $Cr_2O_3(c)$, and probably $Al_2O_3(c)$, the oxygen enhances, rather than suppresses, the rate of vaporization of the oxide. In the case of $UO_2(c)$, a reaction

with oxygen occurs to form the gaseous species $UO_3(g)$. In the case of $Cr_2O_3(c)$, $CrO_3(g)$ is formed. In the case of $Al_2O_3(c)$, it has been postulated that other vaporization reactions become dominant at high oxygen pressures.

The presence of water vapor in the environment has been shown to increase the rate of vaporization of BeO[33] and MgO[34] by the formation of gaseous hydroxides. Most of the other oxides have not been investigated.

It should be pointed out that we are discussing here only the effect of oxygen and water vapor on the position of thermodynamic equilibrium. From an aerodynamic point of view, vaporization rates are generally suppressed by the presence of a finite environmental pressure.[23] On the basis of the results presented in Table 6 there are six simple oxides that have vaporization rates of less than 10 mils/hr at 2000°C: HfO_2, ThO_2, Y_2O_3, ZrO_2, Al_2O_3, and BeO. Of these, only Al_2O_3 and Y_2O_3 appear promising as diffusion barrier materials.[2] We are led, therefore, to consider vaporization behavior in complex oxides.

Complex Oxides

Although there is considerable uncertainty about the vaporization rates of simple refractory oxides, some experimental data are available for most of them, and recession rates can probably be estimated to at least an order of magnitude. For complex oxides, there are virtually no vaporization data. Frequently, the complex oxides cannot be purchased from a commercial supplier. For the most part, they have been discovered as intermediate compounds in phase studies of the binary oxide systems. A complete study of vaporization behavior in even a single mixed-oxide system at temperatures close to 2000°C is a costly and time-consuming job. It is extremely valuable, therefore, to be able to estimate rates of vaporization of mixed oxides from the phase diagrams and the available data for simple oxides. Such estimates can serve as guides to the selection of useful experiments and can also aid in interpreting experimental data.

Figure 11 shows the liquidus curve of a typical binary system AB having an intermediate line compound of molar composition $A_{1-X}B_X$.[39] According to the Wagner-Hauffe development,[35-38] the activities of A and B at the melting point of the intermediate compound $A_{1-X}B_X$ are related to the activities at any point along the liquidus in the neighborhood of the compound by the equations

$$\ln a_B^\ell (x, \theta) = \ln a_B^\ell (X, \theta) -$$

$$\frac{\Delta H_f}{RT_{mX}\theta} \left[\frac{1-x}{X-x} (T_{mX}-\theta) - (1-X) \int_X^x \frac{(T_{mX}-\theta)}{(X-x)^2} dx \right] \quad (48)$$

$$\ln a_A^\ell (x, \theta) = \ln a_A^\ell (X, \theta) -$$

$$\frac{\Delta H_f}{RT_{mX}\theta} \left\{ \frac{x}{x-X} (T_{mX}-\theta) + X \int_X^x \frac{(T_{mX}-\theta)}{(X-x)^2} dx \right\}, \quad (49)$$

where $a_B^\ell (x, \theta)$ is the activity of B in a liquid of mole fraction x at temperature θ; θ, the temperature along the liquidus; x, the mole fraction of B; T_{mX}, the congruent melting point of the intermediate phase; ΔH_f, the heat of fusion of the intermediate compound; R, the gas constant; and X, the deviation from stoichiometry. The development of Equations (48) and (49) is based almost entirely upon equilibrium thermodynamic arguments. The additional assumptions are:

1. The heat and entropy of fusion of the intermediate phase, ΔH_f and ΔS_f, are independent of temperature over the temperature range of interest.

2. The solid that separates out along the liquidus is the intermediate phase of interest, $A_{1-X}B_X$, i.e., there is no appreciable solid solubility.

3. The activities in a melt of composition $A_{1-X}B_X$ are independent of temperature.

Equations (48) and (49) define the activities in the mixed oxide relative to the activities $a(x, \theta)$ along the liquidus. In order to obtain absolute values for the mixed oxide, single values of a_B^ℓ and a_A^ℓ must be known at one (x, θ) point (not necessarily the same point, of course) for both A and B. For a simple system like that of Figure 11, we know that at (x_1, T_{E1}), the eutectic between pure A and $A_{1-X}B_X$, the activity of A is unity; i.e., $a_A^\ell (x_1, T_{E1}) = 1$. At (x_2, T_{E2}), the eutectic between pure B and the compound, we know that $a_B^\ell (X_2, T_{E2}) = 1$. Thus, by substituting $x = x_1$, $T = T_{E1}$ in Equation (49), we can calculate the activity of A in the intermediate phase; substituting $x = x_2$, $T = T_{E2}$, in Equation (48), we get an equation for computing the activity of B in the intermediate phase. The integrals in Equations (48) and (49) may be evaluated graphically. In many practical cases, the liquidus curve in the neighborhood of its maximum may be fitted to a good approximation by a parabola of the form $(T_{mX}-\theta) = k(X-x)^2$, where k is a positive constant.

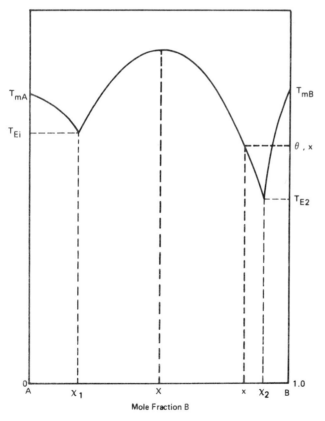

FIGURE 11   Liquidus curve for a two-component system.

On the basis of Equations (48) and (49), the activities will be small in systems where (a) the melting point of the intermediate is very much higher than that of the eutectic mixture, i.e., $T_{mX} \ggg T_{E1}, T_{E2}$; and (b) the eutectic compositions are close to the composition of the intermediate phase, i.e., $x_1, x_2 = X$. Conditions (a) and (b) imply a large value of k. In short, the most stable intermediate compounds in a binary system are those with the sharpest peaks in the liquidus curve in the neighborhood of the compound.

Unless the pure oxides vaporize without decomposition of any kind, it is invalid simply to multiply the activity of a component in a mixed oxide by the total rate of evaporation of the pure oxide to get the total rate of evaporation of that component from the mixed oxide. Assuming congruent and stoichiometric vaporization of the oxides and an evaporation coefficient of unity, it may be shown[40] that the following relationships describe the rates of loss of XO and YO in moles/cm^2/sec due to vaporization of the atoms X, Y, and O:

$$\frac{Z_X}{Z_X^o} = \frac{p_X^o \sqrt{M_Y} \, a_{XO}}{\sqrt{(p_X^o)^2 M_Y \, a_{XO} + (p_Y^o)^2 \, a_{YO} M_X}} \tag{50}$$

and

$$\frac{Z_Y}{Z_Y^o} = \frac{p_Y^o \sqrt{M_X} \, a_{YO}}{\sqrt{(p_X^o)^2 M_Y \, a_{XO} + (p_Y^o)^2 \, a_{YO} M_X}}, \tag{51}$$

where the superscripts refer to standard states, p is vapor pressure, M is molecular weight, and a is activity.

It might be interesting to consider a number of special cases of Equations (50) and (51). If for one of the elements, say X, we do have $Z_X/Z_X^o \approx a_{XO}$, then for Y we must have $Z_Y/Z_Y^o = \sqrt{(1 - a_{XO})} \sqrt{a_{YO}}$. The latter is greater than a_{YO} if $(a_{YO} + a_{XO}) < 1$, and less than a_{YO} if $(a_{YO} + a_{XO}) > 1$. If one of the pure components vaporizes to the elements to a greater extent than the other, say $p_X^o \ggg p_Y^o$, then we will have $Z_X/Z_X^o = \sqrt{a_{XO}}$ and $Z_Y/Z_Y^o = (p_Y^o/p_X^o)$ $(\sqrt{M_X}/\sqrt{M_Y}) (a_{YO}/\sqrt{a_{XO}})$. Since $a_{XO} \leqslant 1$, we will have, in this case, $Z_X/Z_X^o \geqslant a_{XO}$, and unless a_{XO} is very small, $Z_Y/Z_Y^o < a_{YO}$. In other words, compound formation will reduce the rate of vaporization of XO to molecular XO(g) by a fraction a_{XO}, but will reduce the rate of vaporization of XO to the elements X(g) and O(g) by a fraction $\sqrt{a_{XO}}$. The rate of vaporization of YO from the complex oxide as molecular YO(g) will be reduced by a fraction a_{YO}. The rate of vaporization of YO to the elements will generally be reduced by a factor less than a_{YO}. Physically, the high oxygen pressure due to vaporization of XO (as the elements) suppresses the vaporization of YO to the elements.

As discussed above, minimization of activity in a binary system minimizes the partial pressures of the component molecules in the vapor but does not necessarily minimize total vapor pressure. To calculate total vapor pressures, one must know in addition the concentration of any mixed species in the vapor and the dissociation energies of the components of the binary system in the vapor. It should be pointed out that the phase diagrams for binary oxide systems are usually obtained in air or oxygen at one atmosphere. If the phase relations change with pressure, activities calculated from Equations (48) and (49) may be slightly different from experimental partial pressures obtained in vacuum.

The foregoing considerations have been applied to the known complex oxides with melting points above 2200°C and appear to give reasonable predictions.[2]

Thus far, only simple oxides and binary mixed oxides have been considered. Ternary, or more complex systems, have not been discussed at all, but may yet be very important. Dudley,[41] for example, reports that at 1100°C, the binary oxide composition 4BaO:Al$_2$O$_3$ vaporizes at a rate of 290 μg/cm^2/sec, while the ternary 4BaO:2CaO:Al$_2$O$_3$ vaporizes almost ten times more slowly, at a rate of 35 μg/cm^2/hr. Far more work is required on the potential of complex oxide systems for providing oxidation protection.

NOBLE METALS

The oxidation of the platinum metals at elevated temperatures, like the oxidation of bare Mo or W, leads almost exclusively to the formation of volatile oxide products. Alcock and Hooper studied rates of loss of iridium and rhodium into oxygen by means of transpiration measurements in the range 1000°-1600°C.[42] For iridium at 1400°C, the transpiration experiment yielded a metal weight loss of 6.3×10^{-4} mg/cm^2/hr. In contrast, the oxidation kinetics data of Krier and Jaffee[43] indicated a weight loss of 5.3×10^{-2} mg/cm^2/hr at the same temperature. The apparent usefulness of iridium coatings in practical aerospace environments, in spite of the high volatility of the oxide, is almost certainly due to boundary layer effects, which prevent the establishment of true chemically controlled surface reaction rates.

Criscione et al.[44] calculated diffusion-controlled rates of oxidation of iridium for an altitude of 66,000 feet and temperatures of 1400°-2000°C from the boundary layer equations developed by Scala and Vidale.[45] Predicted recessions are less than 10^{-3} in/hr. Kaufman and Clougherty performed similar calculations,[46] based on a simplified boundary layer treatment, and found that the rate of weight loss $\dot{m}$ from a nose radius R_B at a total stagnation pressure P_e is given by an equation of the form

$$\frac{\dot{m}\, R_B^{1/2}}{P_e} = 10^{-\left(3.9 + \frac{900}{T}\right)}.\qquad(52)$$

The experimentally observed recession rates of iridium are consistent with the diffusion-controlled mechanism.

The extent to which vaporization will be a limiting factor controlling coating lifetime can be predicted quite satisfactorily in most cases, if a modicum of thermodynamic and diffusion data is available. On the other hand, meaningful screening tests, involving perhaps a series of low-pressure exposures for specific times at defined temperatures, are difficult to devise, and the exposure conditions selected can easily fail to include the range where problems might be encountered.

The questions that must be asked with respect to the vaporization behavior of a coating system include the following:

1. Can the coating be heated in an inert or vacuum environment without deterioration of its properties due to evaporation losses?

2. Can the vapor pressure of the original coating exceed the ambient pressure under any of the projected service conditions?

3. When the oxidation resistance of a coating system depends upon the formation of an outer protective oxide barrier, how rapidly will such an oxide evaporate at the temperatures and oxygen pressures anticipated in service?

4. Do any of the possible interface reactions lead to the generation of gases at pressures in excess of the ambient pressure?

5. What is the nature of the aerodynamic boundary layer under conditions of coating use, and how does it influence coating loss by evaporation?

If no compounds are present in the vapor phase, the first and second questions can be answered from a knowledge of the activities of the coating components and of the vapor pressures of the pure elements. The vapor pressures are fairly well established. The activities can be determined experimentally. A check should be made by high-temperature mass spectrometry for the presence of compounds in the vapor phase. If they are present, some measure of their thermodynamic stability must be obtained to fully establish the vaporization behavior of the coating.

The third question, pertaining to the vaporization of oxide barriers, can be answered at present for pure oxides vaporizing into vacuum. More work would be useful on the possible stabilization of new oxide vapor species at high oxygen pressures and on the decrease of evaporation rates in the presence of inert gases. The prediction of vaporization rates of complex oxides from phase diagrams merits further exploration.

The fourth question requires both a knowledge of the possible chemical reactions and information about the activities of components at interfaces within the coating system. Reasonable guesses with respect to the stoichiometry of important chemical reactions can probably be made in most cases. The activities are largely unknown, but perfectly feasible to obtain within the framework of present technology.

Finally, the coupling of aerodynamic and chemical factors can be quite subtle, and there is no question that more fundamental understanding in this area would be of great assistance both in the interpretation of laboratory evaluation tests and in the design of new tests of greater relevance to particular end uses.

COATING–SUBSTRATE INTERACTIONS

INTERMETALLIC COMPOUNDS

With intermetallic coatings in general, corrosion failures based mainly on coating–substrate diffusion reactions are difficult to single out. Other limiting factors, perhaps more important, may be melting, evaporation, crack propagation, and oxidation. If these other factors could be isolated and minimized, then substrate reactions might be found to be limiting. It appears that diffusion provides secondary detrimental effects such as interstitial embrittlement or reduction of strength in thin sheet. Further limiting effects are the dilution of the protective surface compounds by interdiffusion with the substrate, e.g., $MSi_2 + M \rightarrow M_x Si_y$ and $MAl_3 + M \rightarrow M_x Al_y$. In general, these lower silicon and aluminum compounds are less oxidation-resistant, particularly in the simple binary systems. To illustrate the life potential of various coatings, the diffusional development of lower compounds is plotted in Figures 12 and 13 and discussed in the following paragraphs.

Silicides

In the case of pure disilicides, extrapolations can be made to show when the MSi_2 layer will be completely consumed. The assumption that this occurrence will lead to loss of protection due to $M_5 Si_3$ formation is probably true for most cases. Figure 12 suggests theoretical order-of-magnitude limitations on coating life as affected by diffusion. These are: (a) for Mo, about 100 hr at 2800°F and 500 hr at 2500°F; (b) for Cb, about 5,000 hr at 2500°F; and (c) for Ta, about 2,000 hr at 2800°F and 10,000 hr at 2500°F. The estimates for Mo appear to be in line with practical coating behavior, probably because with successful systems such as W-3 neither the coatings nor the alloys are heavily modified. In practice, lives far shorter than theoretical are

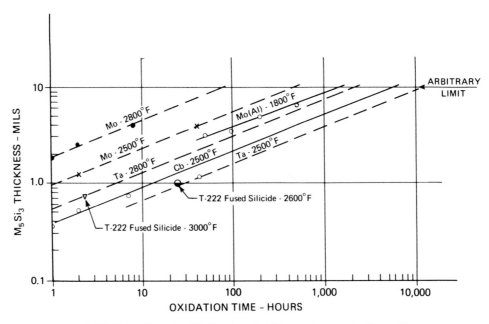

FIGURE 12 Growth of M_5Si_3 phase in MSi_2 coatings on Mo, Cb, and Ta.

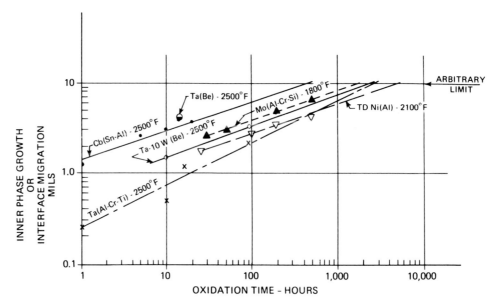

FIGURE 13 Growth (interface migration) of aluminide and beryllide coatings.

obtained with Cb and Ta alloys and coatings, which are more heavily modified. This fact shows that diffusion is not life-limiting for Cb and Ta coatings and that other mechanisms are major factors. Practical coatings for W are based primarily on fairly pure or lightly modified disilicides. Accordingly, diffusion behavior should be close to that of pure silicides on tungsten or molybdenum.

A more complex situation exists with Cb and Ta alloys because their silicide coatings are more heavily modified

than silicide coatings on W and Mo. From analytical investigations of several silicide coating systems, it appears that diffusion effects are considerably changed within the coating and within the substrate to the extent that its surface is modified, particularly within zones that have been titanized. Examples are the Ti-W-Si coating for Ta alloys and the Ti-Cr-Si coating for Cb alloys. The entire character of the silicides is changed because the substrate-element concentration in the silicide is lower than the modifiers. The coating becomes very complex, consisting of mixed MSi_2 phases plus varying amounts of M_5Si_3. Interdiffusion rates with the substrates are increased because of the presence of a highly alloyed titanium-rich layer; there is also some increase in solubility for silicon.

Although diffusion rates are higher in the heavily modified silicides, their overall protectiveness is increased over the simple disilicides of Ta and Cb because of their superior oxidation behavior over a wide temperature range and their resistance to thermal cycling. However, superior coating protection does not necessarily equate with very low oxidation rate. The interrelation between protectiveness and oxidation rates is complex because of (a) the discontinuities in oxidation rates with temperature; (b) the development of more than one oxide; and most significantly, (c) the protection afforded in cracks in the silicide layer. The past-named effect is the most important and the least understood.

For heavily modified silicides it is very difficult to assess theoretical coating-life limitation due to diffusion effects because of the poorly defined coating–substrate interface and the complexity of composition. It cannot be assumed that the consumption of disilicide phases will readily lead to failure.

The fused silicides[47–49] represent a class of highly modified silicides that illustrate complexities in defining limitations due to coating-substrate interactions. These coatings consist of several layers, with the outer layers varying in mixed MSi_2 plus M_5Si_3 zones and inner layers alternating between Cb or Ta and modifier-rich M_5Si_3 zones. The inner layers are sufficiently well defined to permit measurement of their growth during high-temperature exposure. It cannot be assumed, however, that protectivity declines with the consumption of the disilicide phase, because some of these coatings are primarily M_5Si_3 to begin with. Limited data indicate that the inner Cb_5Si_3 and Ta_5Si_3 phases grow more slowly than in the pure silicide, as shown in Figure 12.

A promising new coating system for tantalum alloys[50–52] is difficult to assess from a diffusion standpoint. This coating consists of a porous tungsten alloy layer impregnated with silicon to form a disilicide coating. It has proven difficult to reproduce, but it appears to have no discontinuities in oxidation behavior as a function of temperature, has good life at high temperatures, and may prove to be limited in life by diffusion. The existence of porosity at the silicide-substrate interface may act as a barrier. This system suggests that other dissimilar combinations of coating compound and substrate should be sought that have, or lack, unique interactions. It also suggests that a completely dense coating may not always be desirable.

Aluminides

Less is known about refractory metal aluminide compounds and their diffusional behavior than about the silicides. The compounds WAl_2, $MoAl_3$, Mo_3Al, $CbAl_3$, Cb_2Al, and Cb_3Al have been identified. $TaAl_3$, Ta_2Al, and Ta_3Al appear to be identified, but Kofstad[1] indicates that a compound between $TaAl_3$ and Ta_2Al exists.

The major difference between silicides and aluminides in regard to diffusional behavior appears to be related to the larger concentration gradients that exist in the aluminides and to their lower melting points. The higher aluminum compounds are of the MAl_3 type and are generally followed by M_2Al or M_3Al. Accordingly, it would be expected, and is indeed found, that more rapid diffusion occurs than with silicides. Also, substrate metal losses and grain-boundary attack are more common than with silicides. The formation rates of lower aluminide and beryllide compounds are illustrated in Figure 13.

Tungsten   Although the compound WAl_2 has been reported, no W-WAl_2 diffusion data are available. WAl_2 appears to form slowly even at 1650°C, i.e., at a rate of about 0.25 mils in 5 min.[2] However, aluminum sometimes attacks tungsten intergranularly. Sn–Al coatings behave similarly, but they can be protective against oxidation almost to the melting point of Al_2O_3.[53] The texture of the substrate, e.g., wire or sheet, appears to affect grain boundary attack.

Brett *et al.*[2] found with Mo-Mo_3Al-W couples that little or no Al diffuses into the W side at 1950°C in 1 hr. This suggests that W may be a good barrier and Mo_3Al a suitable overlay coating. It is not clear from the literature, however, whether Mo_3Al has suitable oxidation resistance to act as a coating at temperatures above 1800°F.

Molybdenum   Although considerable work was done some years ago in applying and evaluating Al and Al–Cr–Si coatings, most of the evaluation was done at 1800° and 2000°F,[54] relatively low temperatures compared with those required today. Examination of micrographs where $(MoCr)_3$ $(AlSi)$ was identified as the inner layer of the coating showed the growth of this layer to be quite rapid at 1800°F, as shown in Figure 13 and superimposed in Figure 12 for comparison with Mo_5Si_3 growth. It is seen that at 1800°F, Mo_3Al modified by Cr and Si grows rapidly enough that its life is limited to about 1000 hr, if 10 mils of Mo_3Al is assumed to be a limiting value in the consumption of the $MoAl_3$. A weakness in this assumption is that it is not clear whether Mo_3Al, modified or otherwise, is oxidation-resistant at 1800°F. It is significant that the growth rate of Mo_3Al at 1800°F is of the same magnitude as that of Mo_5Si_3 at 2500°F.

Tantalum   Data for an Al–5Cr–5Ti coating[55] were used as a basis for approximating the diffusion behavior shown in Figure 13. The original $TaAl_3$ layer converts completely to a lower aluminide of unidentified composition in 1 hr at 2500°F. If an oxide layer is formed concurrently, the coating continues to protect during subsequent exposure at 2500°F. Otherwise, this compound is not protective in itself. The next compound, $Ta_2Al(?)$, grows more slowly. The data taken from micrographs are somewhat scattered but when extrapolated indicate that protection might theoretically extend to several thousand hours at 2500°F. This is at least several times less than with silicides of Ta. At higher temperatures, Perkins and Packer find with Sn–Al coatings that intergranular attack may be a limiting factor.[56]

Columbium The growth of Cb_2Al, which has no oxidation resistance by itself, is approximated from substrate interface migration measurements taken from Sn–Al-coated samples. Diffusion is relatively rapid, compared with silicides, and a limiting life of 500–1000 hr at 2500°F appears likely. This suspicion was verified in practice to some extent at General Electric ANP.[57] The silicides, by comparison, have a theoretical multithousand-hour life. General Electric also found that Cb–Ti–Al-base alloys with approximately 50 a/o Al had excellent oxidation resistance but when clad on Cb interdiffused very rapidly at 2500°F.

Nickel and Cobalt The situation with regard to Ni- and Co-base superalloys is significantly different, compared with the refractory metals. Aluminides are the major coating systems, and coated gas-turbine hardware, for example, has demonstrated a practical life of thousands of hours under service conditions. Therefore, neither diffusion nor oxidation limitations have prevented the utilization of these coated materials. However, maximum continuous operating temperatures have been around 1500°–1900°F. With the projected need for higher-temperature duty and the availability of stronger cast nickel-base alloys to suit these needs, the questions regarding diffusion and oxidation stability now become important.

Most superalloy aluminide coatings are of the NiAl or CoAl type, in all cases modified deliberately or by the presence of Cr and other additives in the substrate. The substrates are complex, consisting of nickel or cobalt and chromium together with solid-solution strengtheners and carbide formers such as the refractory metals W, Mo, Ta, or Cb (and in the case of nickel-base alloys, precipitation strengtheners such as Ti and Al). Therefore, coating-substrate reactions can be rather complicated.

A great deal of testing has been done, but diffusion data are not readily available from these tests. The tests are accelerators and promote failure by corrosion attack; the coatings are compared on a relative order-of-merit basis. A recent evaluation program at General Electric,[62] however, presented electron-probe analyses, as-coated and after 50 hr of hot corrosion at 1800°F, for four different coatings and four different substrates (two cobalt-base and two nickel-base). All coatings were of the approximate composition MeAl but contained small amounts of a second phase; in addition, there was a two-phase diffusion zone which appeared to contain appreciable Me_3Al. Slow growth of this zone can be detected at 1650°F in 150 hr, and the zone be-

comes significantly large in the 1800°–2000°F range in 50–150 hr. However, pore formation at the coating interface also becomes appreciable in the Ni alloys. This porosity leads to spalling on severe thermal cycling.

Since considerable hot-salt corrosion takes place at the higher temperatures, even with MeAl present, it is obvious that the lower aluminide Me_3Al presents a real "no go" composition level. More significantly, there appears to be a gradual increase in concentration of refractory metal elements in the diffusion zone that present a less protective corrosion and oxidation barrier in the event of coating penetration and/or cracking due to thermal cycling.

In general, from practical coating evaluation work, it can be concluded in the case of superalloys that diffusion and oxidation may not be limiting for multithousand-hour operation at temperatures up to approximately 1600°F. However, both hot corrosion (sulfidation) and diffusion become limiting factors at somewhat higher temperatures, 1800°–2000°F. Therefore, the simple theoretical analogies used for the refractory metals appear to be somewhat optimistic when applied to the superalloys. For example, if the simple system of NiAl on TD Nickel is used for comparison purposes, the growth of Ni_3Al, when extrapolated in the plot of Figure 13, indicates a maximum potential life of 5,000 hr at 2100°F. In some cases, however, interface porosity assumes greater importance and may be life-limiting. Obviously, this is a highly optimistic extrapolation and represents an unlikely goal for the superalloys as presently constituted. Higher temperature performance by superalloys, in the 2000°–2200°F range, becomes even more impractical, not only because of strength limitations but also because of the formation of eutectics between these substrates and aluminides (particularly with the nickel-base alloys).

Beryllides

These compounds have much more rapid initial diffusion rates than aluminides on Ta.[55] The data are penetration measurements since the system is quite complex. There are six compounds: $TaBe_{13}$, Ta_2Be_{17}, $TaBe_4$, $TaBe_3$, $TaBe_2$, and Ta_3Be_2, the two higher ones being considered oxidation-resistant. Accordingly, it is difficult to make assumptions for calculating theoretical lives based on diffusion behavior. It is significant that the Ta-10W alloy substrate has considerably lower diffusion rates. No data are available for Mo, W, and Cb alloys.

Interstitial Diffusion

There is little evidence that contaminant gases such as nitrogen and oxygen diffuse at significant rates through aluminides and silicides. Usually, interstitial access is through flaws or cracks, or it occurs after coating failure. In the case of chromium alloys, nitrogen diffusion may be considered life-limiting because of its embrittling effects. These effects occur in a relatively short time at 2100°-2400°F. Aluminide coatings eliminate this behavior but substitute embrittlement with interdiffusion with the substrate.[58] Accordingly, some sort of diffusion barrier is required to tie up the aluminum.

An interesting aspect of interstitial diffusion effects is the reverse diffusion of carbon from Cb alloy substrates into coatings containing titanium-enriched sublayers,[59] i.e., the "interstitial sink" effect which lowers high-temperature strength by eliminating carbides. Strength losses occur at 2500°F in a matter of hours. It may be speculated that this general phenomenon can be put to good use through the *in situ* formation of a compound to act as a diffusion barrier for other elements at the coating-substrate interface. The coating would provide one element and the substrate the other to form the compound.

OXIDES

Ceramic coatings are understood to be layers of refractory oxides, applied by troweling and sintering, flame-spraying, or other deposition methods, that rely for their protectiveness upon their inertness to both the oxidizing atmosphere and the underlying substrate. Examples are ThO_2 on W, zircon-glass on W and alumina-glass on Cb. The only such coating that appears to be actively considered at the present time is ThO_2 for the protection of W at very high temperatures (> 3500°F). It is obvious that for inertness at the coating-substrate interface, the substrate metal should not reduce the oxide coating. The free energy of formation of the oxide coating should therefore be much more negative than that of the oxide of the base metal, a condition that is clearly fulfilled by ThO_2 or ZrO_2 on all of the refractory metals. Even in the absence of a simple replacement reaction, however, a degree of interaction between coating and substrate can occur, depending upon the specific thermodynamic and kinetic properties of the oxide-metal system under consideration.[72]

For example, at the very least, the coating must dissolve to a small extent in the substrate, at a rate dependent upon the solubility relationships and diffusion coefficients of the coating components in the base metal. A more drastic interaction would be formation of a compound phase such as a complex oxide or an intermetallic compound involving both metals at the interface. Such an occurrence could seriously influence the course and extent of the coating-substrate interaction, particularly if the compound were liquid at operating temperature.

In principle, it is possible to predict the nature and extent of the coating-substrate interactions from a knowledge of the phase diagrams, thermodynamic properties, and diffusion constants of the systems involved.[73] Since, even in the simplest cases, these are ternary systems (oxygen plus two

metals), the amount of basic information required for a complete description is quite large. A cursory survey reveals that only a fraction of this information is at hand, and it is necessary to resort to empirical approaches in order to define the coating-substrate interactions in most systems.

A number of studies of the behavior of various combinations of metals and oxides heated in contact have been made[63] on ThO_2 and ZrO_2 and Mo, W, Nb, and Ta. It has been reported, mainly on the basis of microscopic observation, that "little" or no reaction occurs between these oxides and Mo and W after times varying from 1,000 hr at 1540°C to a few minutes at 3000°C. No intermediate phases are reported to form at the metal-oxide interface in these systems, and it appears that the solubility of the oxide in the metal is not extensive at high temperatures.

The "little" reaction that occurs in the above systems is presumably a slow dissolution of the oxide into the metal. Insufficient data are reported to determine the observed rate of this reaction, but the problem can be approached theoretically, as shown by the following illustration using the W-ThO system as an example: On the assumption, for purposes of simplification, that the amount of W dissolved in the ThO_2 is negligible, an equilibrium exists between ThO_2(s), Th dissolved in W, Th (a), and O dissolved in W, O (a) at the metal-oxide interface. This equilibrium is defined by ThO_2(s) = Th (a) + 2O (a), for which the equilibrium constant

$$K = \frac{a_{Th} \cdot a_o^2}{a_{ThO_2}} = a_{Th} \cdot a_o^2. \qquad (53)$$

Letting $a_{Th} = \gamma_{Th}\chi_{Th}$ and $a_o = \gamma_o\chi_o$, where a, γ, and χ are the activity, the activity coefficient, and the mole fraction, respectively,

$$\chi_{Th} \cdot \chi_o^2 = \frac{K}{\gamma_{Th} \cdot \gamma_o^2}. \qquad (54)$$

As shown by Kubaschewski[64], the values of K, γ_{Th}, and γ_o can be obtained from a knowledge of the standard free energy of the formation of ThO_2, the free energy of dissociation of O_2, and free energies of mixing of thorium and oxygen in W. If the latter are not known exactly, they may be estimated from solubility or other data; therefore, a relationship can be obtained between the O and Th concentrations in W at the ThO_2-W interface at a given temperature.

$$\chi_{Th} \cdot \chi_o^2 = const. \qquad (55)$$

The rate of diffusion of O and Th, and therefore of ThO_2, into W depends on the diffusion coefficients D_o and D_{Th} as well as on the interface concentrations χ_{Th} and χ_o. A second relationship is required to define both χ_{Th} and χ_o, and this can be obtained from the fact that the ThO_2 layer is assumed to remain essentially pure. Thus, the number of moles of O, N_o dissolved in the W substrate at any time τ is just twice the number of moles of Th, N_{Th}. Using standard solutions of the one-dimensional diffusion equation for the boundary conditions of constant interface concentration and semi-infinite space,

$$2\,N_{Th} = 4\chi_{Th}d \left(\frac{D_{Th}\tau}{\pi}\right)^{1/2} = N_o = 2\chi_o d \left(\frac{D_o\tau}{\pi}\right)^{1/2} \quad (56)$$

or

$$\frac{\chi_{Th}}{\chi_o} = \left(\frac{D_o}{4D_{Th}}\right)^{1/2} \quad (57)$$

In the above equations, d is the density of the substrate in moles/cm^3, and D_o and D_{Th} are the diffusion coefficients in W of O and Th, respectively. Equations (55) and (56) define the interface concentrations; with these known, the amount of ThO_2 dissolved at any time can be obtained from Equation (56). Application of these equations requires that the free energies, activities, and diffusivities for the compound and its elements be known. From expressions for $\Delta G^{\circ}_{ThO_2}$ and ΔG°_o given by Kubaschewski,[64] the equilibrium constant K in Equation (53) is obtained as

$$RT \ln K = -412{,}400 + 43.6\,T. \quad (58)$$

The solubility of Th in W is not known exactly but is reported to be "very slight," suggesting a value for $\gamma_{Th} \gg 1$. The solubility of oxygen in W is reported as about 0.06 at. % at 1700°C.[62] Using this piece of information in conjunction with the known value of the free energy of formation of WO_2 and assuming the same behavior of oxygen as in Mo, we may estimate that $\gamma_o \cong 1$ at 2500°C. In the neighborhood of 2000°C, $D_{Th} = 1.0 \exp[(-120{,}000)/RT]$.[74] D_o is not known, but the diffusion coefficient of carbon is given as $D_c = 0.31 \exp[(-59{,}000)/RT]$, and D_o should be of similar order of magnitude. If these data are used, values of the constants at 2500°C are estimated to be as follows:

$$K \cong 10^{-28} \qquad D_{Th} = 5 \times 10^{-10}\ \text{cm}^2\ \text{sec}^{-1}$$

$$\gamma_{Th} \cong 100 \qquad D_o = 1 \times 10^{-4}\ \text{cm}^2\ \text{sec}^{-1}$$

$$\gamma_o \cong 1$$

With these values, the constant in Equation (55) becomes about $\cong 10^{-30}$.

From Equations (55) and (57), the values of mole fractions of Th and O dissolved in W in equilibrium with ThO_2 at the ThO_2–W interface are calculated to be $\chi_{Th} \cong 2 \times 10^{-9}$ and $\chi_o \cong 10^{-11}$. The number of moles of Th, and hence of ThO_2, dissolved at any time can now be obtained from Equation (56). From this, one may compute the rate of decrease in thickness of the ThO_2 layer at 2500°C:

$$\Delta W \cong 5 \times 10^{-14}\,t^{1/2}\ \text{cm/sec}^{1/2}. \quad (59)$$

According to these calculations, it would require about 10^{14} hr at 2500°C for the thickness of the ThO_2 layer to decrease by one mil as a result of dissolution into the W. These rough calculations confirm our belief in the great inertness of ThO_2 with respect to W and demonstrate that reaction of the oxide with the substrate is not a factor in the deterioration of this coating system.

The behavior of ThO_2 and ZrO_2 on Ta and Cb has been less completely defined; the available data suggest, however, that there is more rapid reaction of the oxides on these metals than on Mo and W. While no reaction between ThO_2 and Cb or Ta was reported after 1,000 hr at 1540°C,[65] a "slight" reaction between ThO_2 and Cb was reported after 15 min at 1800°C.[66] On the other hand, ZrO_2 reacted "violently" with Cb at 1800°C,[66] and HfO_2 reacted strongly with Ta at 2200°C with the formation of an intermediate phase.[67] More extensive and complex interaction of the oxides with Cb and Ta might be expected for several reasons. First, the free energies of formation of Cb and Ta oxides are considerably lower than those of Mo and W[64] and therefore are nearer those of ThO_2 and ZrO_2. Second, oxygen is much more soluble in Cb and Ta than in Mo and W.[67] Third, both Ta_2O_5 and Nb_2O_5 are appreciably soluble in ZrO_2, and complex oxides with relatively low melting points occur in these systems. It should be possible to use thermodynamic analysis to show how the above factors influence the reactivity of the oxides with the metals. However, in the presence of extensive solid solubility and intermediate phase formation, the calculations become more complex than those for the previous example of ThO_2 on W. Methods for solving this problem have not yet been worked out; they would be a good subject for future investigation.

NOBLE METALS

The platinum-group metals have been considered as protective coatings, particularly Ir. Since a combination like Ir on a refractory metal represents a binary system, a knowledge of the reactions that tend to occur at the coating-substrate interfaces can be obtained directly from the equilibrium diagrams. These are known for at least the systems Mo-Pd,

Pt, Rh; W–Ir, Pd, Pt, Rh; Cb-Pd, Pt; and Ta–Ir, Pt, Rh.[68] It can be seen from the phase diagrams that all of the Pt-metal-refractory-metal systems contain intermediate phases that form eutectics or peritectics with each other or with the terminal phases. The equilibrium diagrams indicate immediately at which temperatures liquid phases can be expected to form at the coating-substrate interface and thus indicate where a catastrophic rate of reaction is anticipated. These temperatures are listed in Table 7.

At lower temperatures diffusion zones are formed containing layers of the various intermediate compounds found in the respective systems. The rate of formation of these layers is governed by the diffusion coefficients and equilibrium interface concentrations for each phase. These quantities have been well defined by microprobe studies for the interdiffusion of Ir, Rh, and Pt into W at 1300°–2000°C,[69] and the data are therefore available for an exact calculation of the rates of reaction at the interfaces in these systems.

Much poorer information is found for the other combinations of interest. The width of the diffusion zone after one hour at 1700°C was estimated by microscopic and hardness studies for the systems W–Ir, W–Rh, W–Pt, Ta–Ir, Ta–Rh, Ta–Pt, Mo–Ir, Mo–Rh, Cb–Rh, and Cb–Pt.[70] The results are given in Table 8. From the data of Rapperport *et al.*,[69] it can be simply calculated that the width of the diffusion zone (layer of intermediate compounds) between Ir and W should be about 0.0005 in. after 1 hr at 2025°C. A 10-mil layer of Ir would last for about 400 hr at this temperature. However, Ir can take up to about 20 percent W into solid solution, and the diffusion of W through the Ir layer may be an important factor in this system. The limitation imposed by this process could, in principle, be calculated from Rapperport's data.

OXIDATION-RESISTANT ALLOYS

These may be coatings of superalloys and also Hf–Ta. Phase diagrams are available for the systems Mo–Cr, Mo–Co, Mo–Ni, W–Cr, W–Co, W–Ni, Cb–Cr, Cb–Co, Cb–Ni, Ta–Cr,

TABLE 7 Minimum Melting Points (°C) in Pt-Metal-Refractory-Metal Systems

| Pt Metal | Refractory Metal | | | |
	Mo	W	Cb	Ta
Ir	–	2300	–	1950
Rh	1940	1970	–	1750
Pt	1750	1750	1700	1600
Pd	1550?	1550	1550	–

TABLE 8 Width of Diffusion Zone (inches) after 1 hour at 1700°C[a]

System	Intermediate Phases	Total Diffusion Zone
Mo–Ir	.0005	.002
Mo–Rh	–	.008
W–Ir	.0006	.002
W–Rh	.0008	.004
W–Pt	.0003	.008
Cb–Rh	.0002	.004
Cb–Pt	–	–
Ta–Ir	.0005	.002
Ta–Rh	.0004	.005
Ta–Pt	.0004	.004

[a]Derived from Passmore.[70]
Further results for the interdiffusion of Ir with Mo, W, and Cb may also be found in a report by Rexer.[71]

Ta–Co, and Ta–Ni, as well as for all four refractory metals with Hf.[68] In addition, portions of the ternary systems Mo–Cr–Ni, Mo–Cr–Co, W–Cr–Ni, W–Cr–Co, Cb–Cr–Ni, and Ta–Cr–Ni are available.[68]

It is reasonable to expect that the presence of an extensive region of liquid phase at low temperatures in the equilibrium diagram is a dangerous situation. A eutectic occurs at 1165°C in the Ta–Cr–Ni system. Extensive regions of liquid phase appear in the W–Cr–Ni system at 1400°C. Furthermore, eutectics are found in the various binary systems at temperatures as follows: Cb–Co, 1235°C; Cb–Ni, 1175°C; Mo–Co, 1340°C; Mo–Ni, 1350°C; Ta–Co, 1275°C; Ta–Ni, 1360°C; W–Co, 1465°C; W–Ni, 1495°C; Ni–Cr, 1345°C. It would be dangerous to use superalloy coatings above these temperatures.

The phase diagrams of the refractory metals with Ni, Co, and Cr show the existence of a variety of intermediate phases in these systems. It is expected, therefore, that intermetallic compounds will form at the interface between a superalloy coating and a refractory metal. While considerable information is available concerning diffusion rates in the terminal solid solutions, virtually nothing is known about diffusion in the intermediate phases. Therefore, a complete theoretical analysis of interdiffusion rates in these systems is not possible. It might be expected that, due to the low melting points of many of the constituents in the phase diagrams, fairly rapid interdiffusion between coating and substrate would occur, and this has been mentioned as one of the observed major shortcomings of Ni- and Co-base coatings.[63]

Some references to available diffusion data are as follows: Diffusion of a variety of elements in W was surveyed by Lement.[74] D (1100°–1300°C) in 95–100 percent Ni has been measured after 1 hr at 1700°C by Passmore.[70] D in

95-100 percent Co has been measured by Darin *et al.* (see Smithells[75]). At the W-rich end, the self-diffusion coefficient of W has been measured (reference 75, p. 644) as has the diffusion coefficient of Fe in W at 1927°-2527°C. According to Smithells,[75] the self-diffusion coefficient of Mo from 1850°-2350°C and the tracer diffusion coefficient of Co in Mo at 1850°-2330°C, D in 90 percent Co at 1000°-1300°C, and D in 90 percent Ni at 1000°-1300°C are known. In addition, Fugardi *et al.*, have measured diffusion zone thicknesses in Ni-Mo couples after 1-10 days at 1000°C (reference 63, p. 53). Sefranek has measured diffusion zone thicknesses in Cr-Mo couples after interdiffusion for 1-1000 hr at 870°-1320°C, in Ni-Mo couples from 1-1000 hr at 870°-1200°C, and in Co-Mo from 1-100 hr at 900°-1275°C (see reference 63, p. 54). Composition-distance curves for simultaneous interdiffusion of Ni and Cr into Mo after 600 hr at 1100°C are given by Couch *et al.* (reference 63, p. 55). Byron reports D for Co-Mo at 900°-1700°C.[76]

In the system Cr-Cb, D values for all phases from 1100°-1600°C are given by Rapperport *et al.*[69] An electron-probe study of Cr-Cb and Ni-Cb at 1100°C is also given by Birks *et al.*[77] Fugardi *et al.* report thicknesses of diffusion zones from 1-10 days at 1000°C in Ni-Cb couples (reference 63, p. 53). Smithells (reference 75, p. 645) gives the self-diffusion coefficient for Cb from 900°-2400°C; D_{Co} in Cb, 1550°-2030°C (p. 650); and D for Cr-Cb, all phases at 1100°-1700°C (p. 664).

Smithells[75] gives the self-diffusion coefficient for Ta at 1200°-2300°C (p. 645) and D_{Fe} in Ta at 930°-1240°C (p. 650), Cr-Ta diffusion couples were studied by Passmore *et al.* for 1 hr at 1200°C.[70] The diffusion of Ni in Ta at 1100°C was studied by Birks and Seebold.[78]

Although data are insufficient for a rigorous treatment of the multiphase-multicomponent diffusion problem presented by the coating-substrate reaction in these systems, some information might be gleaned from the binary diffusion data, as well as from various empirical studies of layer growth rates in certain systems. Table 9 gives the approximate widths of several diffusion zones after 100 hr at 1200°C.

The data suggest that thin superalloy coatings might protect for times of the order of 100 hr at 1200°C. At this temperature, a close competition between oxidation and coating-substrate reactions as degradation processes might be expected, necessitating a closer comparison to decide which is most important.

With regard to Hf-Ta alloys as coatings, phase diagrams are available for the Cb-Ta, Ta-Hf, Mo-Hf, Mo-Ta, W-Hf, W-Ta, and portions of the W-Ta-Hf systems.[68] While the minimum melting point in the Ta-Hf system is 2100°C, eutectics occur in the Mo-Hf and W-Hf systems at 1950° and 1930°C.

In the system of greatest interest, Ta-Hf, complete solid solubility exists at high temperatures. Interdiffusion rates between Ta and Hf have been approximately defined by Hill *et al.*[79] over the temperature range 1300°-1800°C. These can be expressed in the form

$$\log \frac{x^2}{2t} \; (cm^2 \; sec^{-1}) = -\frac{2 \times 10^4}{T(°K)} + 3.5, \qquad (60)$$

where x is the width of the diffusion zone observed. At T = 2000°K, $x^2/2t = 3 \times 10^{-7}$ cm² sec⁻¹, and after 100 hours x $\cong$ 0.10 in. Hill's results suggest that diffusion between Hf-Ta alloys and Ta proceeds a little more slowly than between Hf and Ta, due to an expected increase in D with Ta content.

For a complete analysis of coating-substrate reactions, information is required about phase relationships, thermodynamic properties and diffusion coefficients in the systems of concern. Although sufficient data often exist for binary systems, this is generally not true for ternary and higher order systems. Even when fundamental data are lacking, however, much useful information can be derived from studies of layer growth at the coating-substrate interface, which are not too difficult to perform. Examples of the latter are Perkins' study of the reaction of $MoSi_2$ on Mo[61] and the investigation of Passmore *et al.* of the behavior of ThO_2 on W.[70] Such studies are recommended, therefore, as a normal part of the coating evaluation process.

From the data on hand it appears that coating-substrate reactions may be life-limiting with beryllide coatings and possibly with aluminides, but not necessarily with silicide coatings. It is expected, on theoretical grounds, that the rapidity of layer growth should increase as the melting point of the intermediate phases decreases. In the refractory metal systems, silicides have the highest melting point, followed by the aluminides and beryllides, which may explain the observed rates of reaction. It may be stated as a general principle that high intermediate-phase melting points (low diffusion coefficients), as well as restricted solubility in such phases, are desirable to minimize coating-substrate reactions. It appears that interdiffusion between coating and substrate is also an important factor in the degradation of superalloy coatings on the refractory metals. Here again, fairly rapid

TABLE 9 Width of Diffusion Zones in Various Systems at 1200°C

System	Width of Diffusion Zone (in.)	Reference
Mo–Ni	0.015	70
Mo–Co	0.003	70
Mo–Cr	0.002	70
Cb–Ni	0.003	70
Cb–Cr	> 0.005	69

diffusion rates at the coating–substrate interface may be associated with the presence of relatively low-melting phases in the pertinent systems.

In contrast to the above, Ir seems to react with W quite slowly, and the oxides ThO_2 and ZrO_2 very slowly, so that coating–substrate reactions do not appear to be life-limiting factors in these systems. At the temperature of interest, other processes, such as evaporation or permeation of oxygen through the protection layer, seem to be much more critical.

DEFECTS

Coatings for high-temperature metals act as a barrier to exclude reactive gases such as oxygen or nitrogen from the surface. The mode of protection with few exceptions is based on the selective oxidation of an element in the coating (i.e., Al or Si) to form a protective oxide film on the surface. Theoretically, coating life should be governed by the kinetics of this oxidation process. That is, life at any temperature should depend upon the rate of oxidation of the coating and coating thickness.

As previously discussed, the useful life of coating systems is usually only a small fraction of the theoretical life based on oxidation kinetics of the coating. One of the reasons for this behavior is the fact that the composition of a coating changes with use. Diffusion of elements in the coating outward to the oxide scale or atmosphere and inward to the substrate, as well as diffusion of elements from the substrate into the coating, occurs at high temperature. This results in dilution of the coating and a shift to less oxidation-resistant forms. The kinetics of the diffusion reaction tend to be far more rapid than those of the surface oxidation process. The latter is governed by diffusion through an oxide scale, which is slow compared to atom diffusion in metal matrices. Thus it is not surprising to find in many cases that coating life depends on interdiffusion of coating and substrate rather than on surface oxidation process.

Even in this case, theoretical studies reveal that actual coating life is again only a fraction of that predicted from considerations of interdiffusion. This fact is illustrated for $MoSi_2$ coatings on Mo in air at 2500°F in Figure 14. The life of this system is known to depend on the time required to convert the coating to Mo_5Si_3 by diffusion of silicon into the substrate.[61,80] The calculated life for a 2-mil-thick coating based on this process is 40 hr at 2500°F. As shown in Figure 14, the actual life of a 2-mil-thick coating is only 18 hr.[81] The coating in fact behaves as if it were 1.25 mils thick rather than 2.0 mils thick. The experimental and calculated curves have the same shape, but the experimental performance at all thicknesses is less than predicted. The basic failure mode is the same as that assumed in the calculation; it is the depletion of silicon by diffusion to the substrate that causes ultimate failure. The coating fails when Mo_5Si_3 replaces $MoSi_2$ at the oxidizing surface.

The difference between calculated and experimental results in this case is due to random defects (in the form of cracks and fissures) that reduce the coating thickness in local regions. The minimum or effective coating thickness based on random defects that initiate failure is only 60 to 65 percent of the actual thickness. The defects penetrate the outer third of the coating and only the inner two-thirds of the coating actually provides protection. This fact was demonstrated experimentally by removing a third to a half of a coating with such defects by polishing without effecting any change in coating life.[82] For this particular coating and environment, the life is governed by a specific type of random defect, and the life is said to be random-defect-controlled.

All coatings contain microscopic defects; an ideal defect-free coating for high-temperature metals has yet to be developed. Typical defects that are to be found in oxidation resistant coatings are illustrated in Figure 15. Coating defects can be separated into two broad classifications: defects introduced during application of the coating, and defects introduced during manufacture, assembly, and use of the structure or part. All of the coating defects illustrated in Figure 15 can be introduced during application of the coating. However, the only defects that are introduced during

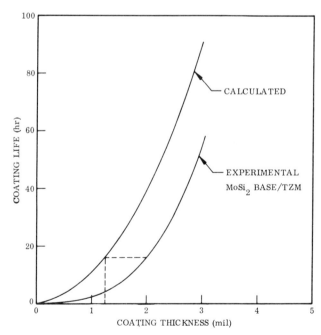

FIGURE 14   Effect of coating thickness on the life of $MoSi_2$–Mo systems at 2500°F.

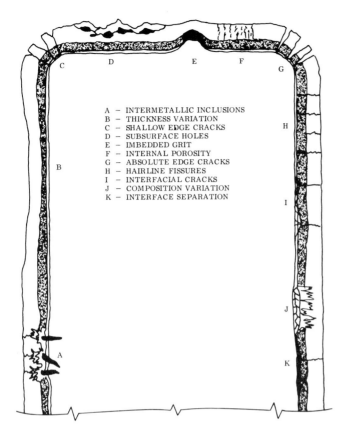

A — INTERMETALLIC INCLUSIONS
B — THICKNESS VARIATION
C — SHALLOW EDGE CRACKS
D — SUBSURFACE HOLES
E — IMBEDDED GRIT
F — INTERNAL POROSITY
G — ABSOLUTE EDGE CRACKS
H — HAIRLINE FISSURES
I — INTERFACIAL CRACKS
J — COMPOSITION VARIATION
K — INTERFACE SEPARATION

FIGURE 15 Representative random defects in coatings.

environment, it may have little or no effect in another. For example, hairline cracks and fissures in silicide coatings on Mo and Cb tend to govern life at atmospheric pressure, while at low pressure such defects tend to have little effect.[82] Wide V-type edge defects in these coatings can be healed at atmospheric pressure but, on the other hand, are a predominant source of failure at low pressure. Variations in coating thickness tend to have a far greater effect on the performance of silicide coatings at low pressure than at high pressure.[82]

The temperature of service also can have a pronounced effect on the contributions of coating defects to performance. For example, as shown in Figure 14, the life of $MoSi_2$/Mo coatings at 2500°F is less than that predicted by diffusion data. This is due to the fact that crack-type defects reduce effective coating thickness. However, as shown in Figure 16, the life of these coatings at 3000°F exceeds that predicted from diffusion data. In this case, a viscous SiO_2 film effectively heals coating defects. Also, the Mo_5Si_3 phase that causes failure at 2500°F is oxidation-resistant at 3000°F. Mo_5Si_3 performs as well as $MoSi_2$ as a coating material at 2850°F or above.[61,80]

Considerations such as these make any assessment of the role of coating defects a difficult if not impossible task. As a general rule, coating thickness has the greatest effect on performance, and any factor that tends to reduce effective thickness will reduce life. This effect is shown graphically for a diffusion-controlled system ($MoSi_2$–Mo at 1 atm) in

manufacture and use are cracks that result from thermal expansion mismatch, excessive mechanical strain, or impact. In many cases, the basic crack patterns are introduced during application of the coating, and the service environments or thermal cycling merely serve to increase the number or severity of these defects.

The existence of coating defects has been recognized from the earliest days of coating development work. In the early 1950's, Blanchard[54,83] observed that imbedded grit and large carbide particles produce defects in Al–Cr–Si coatings on Mo. He also noted that sharp corners increase the probability of failure due to formation of edge cracks. The presence of crack-type defects has been reported by all investigators of silicide and aluminide coatings. In 1963 Jefferys and Gadd[84] stated that cracks and pinholes or porosity often are the origin of coating failure. Most investigators during this period noted that defects exist in coatings and generally have a detrimental effect on performance.

The effect of coating defects on performance cannot be clearly defined. It is a function of the specific coating-substrate combination and the environment of use.[82,86] Each coating system has its own specific set of variables that will influence coating life. Thus, whereas a given type and severity of defect will reduce the performance in one

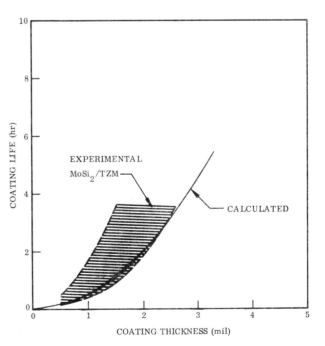

FIGURE 16 Effect of coating thickness on the life of $MoSi_2$–Mo systems at 3000°F. Solid curve, calculated $MoSi_2$–Mo; dashed curve, experimental W-2–Mo-5Ti.

Figure 17. Increasing the thickness from 1 to 3 mils produces a nine-fold increase in life at 2500°F. A 2-mil-thick coating with a 1-mil-deep defect behaves in effect like a 1-mil-thick coating (dashed curve). However, broad generalizations of this type tend to be misleading and may be erroneous. The defect tolerance of a coating is variable within broad limits and must be defined in terms of a specific application. Failure modes are governed both by the basic coating system and by the environment of use.

In the past few years, attention has been given to the development of nondestructive testing methods to identify and characterize coating defects.[86-88] Thermoelectric and eddy current techniques can be used effectively to measure variations in coating thickness. Eddy currents also can be used to monitor the growth or change of coatings by diffusion.[86] This method is particularly suited to inspection of large, flat areas. The thermoelectric process, on the other hand, is well suited to the detection of thin regions at edges.[48] Radiography will reveal internal porosity, while dye penetrants can delineate basic crack patterns.

The available nondestructive testing methods can indicate the nature, size, and distribution of most critical defects found in coating systems. By and large, however, industry is not ready to apply many of these techniques to process or product control. This reluctance results from the fact that the defect-tolerance limits for various coating systems have not been determined. Thus, although defects can be detected and measured, a basis for acceptance or rejection has not been established. Basic and applied research on the role of defects in the performance of coating systems in representative environments must be undertaken.

Statistical analysis of performance provides an effective tool for studies of this type. Wurst and Cherry[85] have pioneered in the use of statistical methods to evaluate coating behavior. Using the Weibull function, these investigators have been able to determine under what conditions defect control of performance will occur for various coating systems. The next step is to establish which defects in the system govern observed behavior. Basic studies of this type are yet to be made. It should be noted that even where a burnout (nonrandom defect) mode of failure is indicated by statistical data, defects still may be the basic cause of failure. For example, a uniform distribution of cracks and fissures produced by thermal cycling can result in an apparent failure of the entire coating. Each individual failure, however, still occurs at a defect site. Defects need not be random to degrade coating performance.

The majority of defects in coatings are introduced in the manufacture of coated parts. Accordingly, in any program to reduce or control coating defects, attention must be given to the basic coating processes. Many defects are merely the result of carelessness or inadequate process controls. Thus, imbedded grit, subsurface porosity, large edge fissures, interfacial separation, and coating composition variations normally can be controlled or eliminated by adjustments in processing. For example, edge cracks can be mitigated by proper rounding of edges and by controlling coating thickness and the rate of coating deposition. In general, defects such as these do not present major problems with the advanced coating systems in use today.

A major problem with coating defects arises from the hairline cracks or fissures that develop during the thermal cycling of a coating system. A thermal expansion mismatch between a brittle coating on a ductile substrate results in a craze-type cracking of the coating. The cracks tend to form on cooling when tensile stresses are developed in the coating in a temperature range where ductility is low. On heating, the coating is placed in compression and the craze cracks tend to be closed.

The susceptibility to and tolerance for this type of defect varies widely between different coating systems. Silicide coatings are very prone to the formation of such defects. The basic crack pattern in these coating systems is established during the application process when coated parts are cooled to room temperature. Subsequent thermal cycling in service tends to enlarge the fissures and generate additional cracks. The aluminide coatings on Ni- and Co-base superalloys, on the other hand, are very resistant to the generation of hairline cracks or fissures. These coatings, as pro-

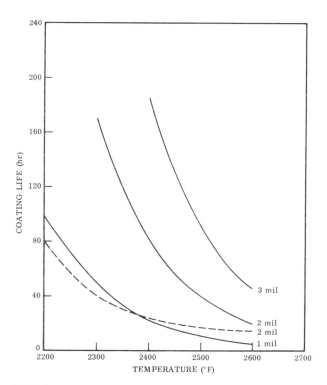

FIGURE 17   Experimental versus calculated life of MoSi$_2$–Mo coating systems.

duced, do not contain cracks. After extended thermal cycling in service, however, cracks do develop, and defects that reduce coating life are generated.

Most investigators recognize the fact that thermal cycling reduces useful coating life and that the formation of cracks due to thermal-expansion mismatch is the primary cause of this behavior. Cox and Brown,[84] for example, demonstrated that hairline fissures introduced during thermal cycling reduce the life of a $MoSi_2$–Mo coating system at 1832°F from 6 months under steady-state to 10 hr under thermal-cycling conditions. Blumenthal and Rothman[90] studied the growth of hairline fissures in $MoSi_2$ coatings on Mo during cyclic oxidation. Fissures initially 0.03 to 0.1 mil wide grew to 0.1 to 0.6 mil wide on cyclic operation.

These defects probably have the greatest effect on coating performance. They are not random but are found over the entire surface. Their effect is often masked by the fact that the coating appears to fail all at once instead of at random sites. Although coatings may appear to fail by wear-out mode, the failure is usually due to the generation of a coating defect that reduces useful life. This type of defect is without a doubt the most difficult to control or eliminate. The mitigation of such defects is one of the most challenging technical problems in the field today.

Surprisingly little work has been undertaken on the elimination or control of hairline cracks or fissures in coatings. Two promising approaches to the solution of this problem have appeared in the past several years. One used by investigators at Battelle Memorial Institute in Geneva[91,92] is a novel method in which a liquid Sn–Al phase is used to seal all cracks and fissures; the porous silicide coating is impregnated with an Sn–Al phase. The approach has been used successfully on Mo and Cb alloys. The basic limitations of this approach are related to compatibility (liquid phase) problems and the high vapor pressure of tin. A more promising and practical approach has to do with alloying of a basic silicide coating to produce a highly complex structure that has improved resistance to the formation of such defects and improved tolerance for defects once formed.[48,93] The exact mechanism by which these coatings provide enhanced defect tolerances is not clear at this time. The coatings are applied as slurries and fused onto the surface. The fact that a truly alloyed coating is produced is probably a contributing factor. These coatings also offer greater protection to faying surfaces, where coating defects often arise with pack-cementation-type processes.[87] Far more detailed and basic studies are needed to fully develop the potential of systems such as these. Results to date indicate good promise for effecting basic changes in defect patterns and for increasing the tolerance for random defects.

REFERENCES

1. Kofstad, P. 1967. High temperature oxidation of metals. John Wiley and Sons, New York.
2. Brett, J., L. L. Seigle, J. Berkowitz-Mattuck, B. Lement, D. Phalen, *et al.* August 1965. Experimental study of factors controlling the effectiveness of high-temperature protective coatings for tungsten. AFML-TR-64-392.
3. Mott, N. F. 1947. Trans. Faraday Soc. 43:429.
4. Krier, C. A., and J. M. Gunderson. 1966. Silicides as oxidation resistant materials. Structures and Materials Technology, Aerospace Group, The Boeing Company, Seattle, Washington. Presented at the Symposium on Oxidation in the Re-Entry Environment, Fall Meeting of the Electrochemical Society, Philadelphia, Pa., October 9–13.
5. Deal, B. E., and A. S. Grove. 1965. J. Appl. Phys. 36:3770.
6. Bartlett, R. W., and P. R. Gage. July 1964. Investigation of mechanisms for oxidation protection and failure of intermetallic coatings for refractory metals. ASD-TDR-63-753, Part II.
7. Searcy, A. W., and A. G. Tharp. 1960. J. Phys. Chem. 64:1539.
8. Norton, F. J. 1961. Nature 191:701.
9. Berkowitz-Mattuck, J., M. Rossetti, and D. W. Lee. Enhanced oxidation under tensile stress—pest in molybdenum disilicide. Submitted for publication, Trans. AIME.
10. Berkowitz-Mattuck, J. B., and R. R. Dils. 1965. J. Electrochem. Soc. 112:583.
11. Glushkv, P. I., V. I. Dorokho, and Ye. P. Nechiporenko. 1962. Phys. Met. Metallurg. (English translation) 13:111.
12. Wirkus, C. D., and D. R. Wilder. 1966. J. Amer. Ceram. Soc. 49:173.
13. Oishi, Y., and W. D. Kingery. 1960. J. Chem. Phys. 33:480.
14. Pappis, J., and W. D. Kingery. 1959. MIT Thesis.
15. Doherty, P. E., and R. S. Davis. 1963. J. Appl. Phys. 34:619.
16. Sama, L., and D. D. Lawthers. June 1962. Aluminide and beryllide protective coatings for tantalum. Presented at Technical Conference on High Temperature Materials, Cleveland, Ohio.
17. Kofstad, P., and S. Espevik. 1967. J. Less-Common Metals 12:117.
18. Licciardiello. April 1967. Development of composite structures for high temperature operation. AFDL-TR-67-54.
19. Maak, F. 1961. Z. Metallkunde 52:545.
20. Kaufman, L. April 1968. Stability characterization of refractory materials under high velocity atmospheric flight conditions. Progress Report No. 4, AF33(615)-3859.
21. Wlodek, S. T. 1964. Trans. AIME 230:1078.
22. Quill, L. L., ed. Sept. 1955. The chemistry and metallurgy of miscellaneous materials. TID Report No. 5212.
23. Blackburn, P. E., K. F. Andrew, E. A. Gulbransen, and F. A. Brassart. 1960. Oxidation of tungsten and tungsten based alloys. WADC-TR-59-575, Part III.
24. WADD-TR-60-463, Part II. October 1962. The vaporization and physical properties of certain refractories. AVCO Co., Wilmington, Mass. AF33(616)-6940.
25. Berkowitz, J. B. ASD-TDR-62-203, Part II. Kinetics of oxidation of refractory metals and alloys at 1000–2000°C. Arthur D. Little, Inc., Cambridge, Mass. AF33(616)-6154.
26. Berkowitz-Mattuck, J. B. Mechanisms of oxidation of Ta-10W alloy coated with tungsten disilicide. Submitted for publication in J. Electrochem. Soc.
27. Perkins, R. A., L. A. Biedinger, and S. Sokolsky. 1963. Space and Aeronautics 39:107.

28. Regan, R. E., W. A. Buginski, and C. A. Krier. 1966. AEC Accession No. 41424. Report No. AED-CONF-66-103-14, Boeing Co., Seattle, Washington.
29. Kaiser, W., and J. J. Breslin. 1958. J. Appl. Phys. 29:1292.
30. Wagner, C. 1958. J. Appl. Phys. 29:1295.
31. JANAF thermochemical tables. Dow Chemical Co., Midland, Michigan.
32. Turkdogan, E. T., P. Grieveson, and L. S. Darken. 1963. J. Phys. Chem. 67:1647.
33. Grossiveiner, L. I., and R. L. Seifert. 1952. J. Amer. Chem. Soc. 74:2701.
34. Alexander, C. A., J. S. Ogden, and A. Levy. 1963. J. Chem. Phys. 39:3057.
35. Hauffe, K., and C. Wagner. 1940. Z. Electrochem. 46:160.
36. Wagner, C. 1958. Acta Met. 6:309.
37. Wagner. C. 1952. Thermodynamics of alloys. Addison-Wesley, Reading, Mass.
38. Chip, J. 1948. Disc. Faraday Soc. 4:23.
39. Levin, E. M., H. F. McMurdie, and F. P. Hall. 1956. Phase Diagrams for ceramists. Amer. Ceram. Soc.
40. Buchler, A., and J. B. Berkowitz-Mattuck. 1963. J. Chem. Phys. 39:286.
41. Dudley, K. 1961. Vacuum 11:84.
42. Alcock, C. B., and G. W. Hooper. 1960. Proc. Roy. Soc. A254: 551.
43. Krier, C. A., and R. I. Jaffee. 1963. J. Less-Common Metals 5:411.
44. Criscione, J. M., et al. Oct. 1964. High temperature protective coatings for graphite. ML-TDR-64-173, Part II.
45. Scala, S. M., and G. L. Vidale. 1960. Int. J. Heat Mass Transfer 1:4.
46. Kaufman, L., and E. V. Clougherty. March 1966. Investigation of boride compounds for very high temperature applications. RTD-TDR-63-4096, Part III.
47. Bracco, D. J., P. Lublin, and L. Sama. May 1966. Identification of microstructural constituents and chemical concentration profiles in coated refractory metal systems. AFML-TR-66-126. GT&E Labs to AFML on AF33(615)-1685.
48. Priceman, S., and L. Sama. Dec. 1966. Development of fused slurry silicide coatings for the elevated-temperature oxidation protection of columbium and tantalum alloys. Sylvania Electric Products Annual Summary Report STR-66-5501.14 to AFML on AF33(615)-3272.
49. Priceman, S., and L. Sama. Sept. 1968. Development of fused slurry silicide coatings for the elevated-temperature oxidation protection of columbium and tantalum alloys. Final Report from Sylvania Electric Products to AFML.
50. Winber, R. T., and A. R. Stetson. July 1967. Development of coatings for tantalum alloy nozzle vanes. NASA-Cr-54529. Solar to NASA on NAS-3-7276.
51. Jackson, R. E. May 1968. Tantalum system evaluation. Interim Technical Reports. MAC to AFFDL on AF33(615)-3935.
52. Berkowitz, J. B. May 1968. Comparison of two commercial W-Si based coatings for TA-10W. Presented at the 14th Refractory Composites Working Group Meeting at Wright Field in Dayton, Ohio.
53. Sama, L., et al. Jan. 1964. Development of oxidation resistant coatings for refractory metals. TR-64-484.1. Presented at the 8th Meeting of the Refractory Composites Working Group in Fort Worth, Texas.
54. Blanchard, J. R. June 1955. Oxidation resistant coatings for molybdenum. WADC TR-54-492, Part II. Climax Molybdenum Company of Michigan to WADC.
55. Lawthers, D. D., and L. Sama. Oct. 1961. High temperature oxidation resistant coatings for tantalum-base alloys. ASD TR-61-233. Sylvania Electric Products to ASD on AF33(616)-7462.
56. Perkins, R., and C. Packer. Sept. 1965. Coatings for refractory metals in aerospace environments. AFML-TR-65-351. Lockheed to AFML on AF33(657)-11150.
57. Lever, R. C. June 20, 1962. Metallic fuel element materials. APEX-913. Flight Propulsion Lab. Dep. of GE to USAEC and USAF on AF33(600)-38062 and AT(11-1)-171.
58. Negrin, M., and A. Block. July 1967. Protective coatings for chromium alloys. NASA Cr-54538. Chromalloy to NASA on NAS-3-7273.
59. Brentnall, W. D., and A. G. Metcalfe. April 1968. Interstitial sink effects in columbium alloys. AFML-TR-68-82. Solar to AFML on AF33(615)-5233.
60. Wagner, C. 1960. Pp. 68–75 in W. Jost. Diffusion. Academic Press, New York.
61. Perkins, R. A. 1965. Status of coated refractory metals. J. Spacecraft 2:520.
62. Allen, B. C., and W. M. Albright. 1960. DMIC Memo no. 50, p. 10.
63. For reviews of this work see DMIC Report no. 162, "Coatings for the Protection of Refractory Metals," Krier, November 1961, and ASD-TDR-62-205, "Analysis of Basic Factors in Protection of Tungsten Against Oxidation," Promatis et al., December 1961.
64. Kubaschewski, O. 1967. Metallurgical thermochemistry. 4th ed. Pergamon Press, New York (p. 259).
65. Farnwolt, D. E., et al. Nov. 1964. Compatability of refractory metals with various ceramic insulation materials. Report no. CIVLM-5942, AEC Contract AT(30-1)-2789.
66. Economas, G. 1953. Behavior of refractory oxides in contact with metals at high temperatures. Ind. Eng. Chem. February 1953: 458. Also J. Amer. Ceram. Soc. 36:403.
67. Hahn, G. T., et al. 1963. Refractory metals and alloys–II, Vol. 17 of AIME Met. Soc. Conf. p. 24. Interscience, New York.
68. DMIC 152 and 183. April 1961, Feb. 1963. Binary and ternary phase diagrams of Cb, Mo, Ta and W.
69. Rapperport, E. J., et al. March 1964. Diffusion in refractory metals, ML-TDR-64-61.
70. Passmore, E. M., et al. Aug. 1960. Investigation of diffusion barriers for refractory metals. WADD-TR-60-343.
71. Rexer, J. Jan. 1967. Diffusion of Ir in W, Mo, and Cb. PR3-NASA-CR77634.
72. Dickinson, C. D., et al. July 1963. Factors controlling the effectiveness of high temperature protective coatings for W. ASD-TDR-63-744.
73. Clark, J. B., et al. 1959. Trans. ASM 51:199.
74. Lement, B. Aug. 1965. Diffusion studies of W and selected materials. p. 179 in AFML-TR-64-392.
75. Smithells, C. J. 1967. Metals reference book, Vol. II. 4th ed. Plenum Press, New York.
76. See p. 206 of "Refractory metals and alloys–II" (ref. 67).
77. Birks, L. S., et al. March 1960. Diffusion of Nb with Cr, Mo, Ni, Fe, and stainless steel. NRL 5461.
78. Birks, L. S., and R. E. Seebold. Sept. 1960. Diffusion in Nb-Pt, Se, Zn, Co, Ta-Ni, and Mo-Fe. NRL 5520.
79. Hill, V. L., et al. Sept. 1966. Protective coatings for Ta-base alloys, AFML-TR-64-354, Part III.
80. Perkins, R. A. 1964. Coatings for refractory metals: Environmental and reliability problems. Materials Sciences and Technology for Advanced Applications, Vol. II, ASM Golden Gate Conference, San Francisco.

81. Mathauser, E. E., B. A. Stein, and D. R. Rummler. Oct. 1960. Investigations of problems associated with the use of alloyed molybdenum sheet in structures at elevated temperatures. NASA TN DO447.

82. Perkins, R. A., and C. M. Packer. Sept. 1965. Coatings for refractory metals in aerospace environments. AFMC TR-65-351.

83. Blanchard, J. R. Dec. 1954. Oxidation resistant coatings for molybdenum. WADC TR-54-492, Part I.

84. Jefferys, R. A., and J. D. Gadd. 1963. High temperature protective coatings for *Columbium Alloys*, *High Temperature Materials II*, AIME Metallurgical Society Conference, Vol. 18. Interscience, New York.

85. Wurst, J. C., and J. A. Cherry. Sept. 1964. The evaluation of high temperature materials. AFML-TDR-64-62, Vol. II.

86. Stinebring, R. C., and T. Struiale. Nov. 1966. Development of nondestructive methods for evaluating diffusion-formed coatings on metallic substrates. AFML-TR-66-221.

87. Karplus, H. B., R. A. Semmler, and B. E. Arneson. May 1968. Evaluation of nondestructive testing techniques of diffusion coatings. AFML-TR-67-358.

88. Lawrie, W. E., and R. A. Semmler. June 1966. Nondestructive methods for the evaluation of ceramic coatings. WADD-TR-61-91, Part VI.

89. Cox, A. R., and R. Brown. 1964. Protection of molybdenum from oxidation by molybdenum disilicide based coatings. J. Less-Common Metals 6:51–69.

90. Blumenthal, H., and N. Rothman. June 1964. Development of a powder and/or gas cementation process for coating molybdenum alloys for high temperature protection. AFML-TR-67-74.

91. Stecher, P., and B. Lux. 1968. Oxidationsschutz von Molybdän Durch Eine MoSi$_2$/Sn-Al Schutzschicht. J. Less-Common Metals 14.

92. Stecher, P., B. Lux, and R. Funk. "Protection contre l'oxydation d'alliages de niobium," Entropie 21 (Mai–Juin).

93. Priceman, S., and L. Sama. 1968. Protectiveness of fused silicide coatings in simulated re-entry environments. Summary of the 13th Refractory Composites Working Group Meeting. AFLM-TR-68-84.

3

State of the Art of
Coating–Substrate Systems

INTRODUCTION

To form a basis for consideration of future research and development, the Committee on Coatings reviewed the state of the art of various coating-substrate systems. Results are presented in the present chapter organized under the various substrates, including superalloys, chromium, columbium, molybdenum, tantalum and tungsten, and graphite.

SUPERALLOYS*

COATING SYSTEMS

Coated superalloys have been successfully used in a wide variety of turbine engine applications for many years.[1-3] Originally developed to obtain improved high-temperature oxidation and erosion resistance, coatings are now used to achieve a number of product improvements, including (a) resistance to corrosion (sulfidation), (b) wear-resistance, (c) improved creep resistance (e.g., reduced stretch in turbine blades), (d) reduction of warpage and distortion (e.g., minimized bowing in turbine vanes), and (e) reduction of thermal and mechanical fatigue. Numerous coating systems have been developed that satisfy many of these objectives, but the increasing turbine gas inlet temperatures and the lower chromium contents in modern superalloys (to obtain the required strength or metallurgical stability at high tem-

*Prepared by J. R. Myers and N. M. Geyer.

peratures) necessitate significant improvements in coating technology.

The basic requirements of a coating for superalloy materials dictate that (a) it must resist the thermal environment, or meet one or more of the special features described above; (b) it must adhere (be bonded) to the substrate; (c) it should be thin, uniform, light in weight, reasonably low in cost, and relatively easy to apply; (d) it should have self-healing characteristics; (e) it should be ductile; and (f) it should not adversely affect the mechanical properties of the substrate. Since no single coating system can possibly satisfy all of these requirements, a large number of coatings have been developed in recent years.

Coatings that are mechanically bonded to the superalloy substrate generally do not provide the required adherence. Therefore, diffusion-type coatings that produce a metallurgical bond at the coating-substrate interface are much more desirable. Mechanically bonded coatings produced by such techniques as electroplating, spraying, or hand-troweling are not reviewed in this report. Rather, attention is directed to those coatings that are metallurgically bonded to the substrate.

COATING TECHNIQUES

The important currently available processes for applying metallurgically bonded coatings to superalloy substrates are (a) pack-cementation (including "halogen-streaming" and "slip-pack" techniques), (b) slurry, (c) hot-dipping, (d) chemical vapor deposition (CVD), (e) fused-salt (both electrolytic

60

and electroless), (f) vitreous or glassy-bonded refractory, (g) physical vapor deposition, and (h) electrophoresis. The basic features of each of these coating processes, including the coating metals frequently applied, are summarized in Table 10. Of these six techniques, probably the most important is pack cementation. However, a number of commercial coatings are applied using slurry, hot-dipping, vitreous or glassy-ceramic, and electrophoresis processes.

PACK-CEMENTATION PROCESSES[4–17]

The basic pack-cementation process is accomplished by packing the object to be coated in a powdered mixture con-

sisting of (a) the coating metal(s) in elemental or combined form, (b) a halide compound (activator), and (c) an inert filler material to prevent sintering of the powders. In most cases, the reaction chamber (retort) containing the coating ingredients is evacuated or filled with an inert atmosphere to prevent oxidation of the object to be coated and the metal powders. Some processes also use an additional compound (e.g., urea) in the powder mix to control the retort atmosphere. The sealed pack system is heated to an elevated temperature where the halide gas, generated by the decomposing halide compound, reacts with the coating metal(s) to form a metal–halide gas. At temperature, this gas either reacts with, or is chemically reduced at, the surface of the object to be

TABLE 10 Summary of Important Current Coating Processes for Superalloys

Process	Steps	Atmosphere	Remarks
Pack cementation (includes "halogen-streaming" and "slip-pack" processes) (Al, Cr, Al–Cr, Be, Al–Cr–X)	Frequently one, but could involve two (or more)	Generally inert at start; then supply halide (usually Cl) at temperature. At least one process uses vacuum at start	Pack supports component being coated and eliminates need for special holding fixtures.[a] Disadvantage of process is long heat-up and cool-down times. Component must be supported in slip-pack process
Slurry (Al, Cr, Al–Cr, Al–Cr–X)	Multiple	Both inert and vacuum are used in heating cycle. May want to apply slurry in vacuum to obtain good coverage	Process avoids long heat-up and cool-down times. Needs special holding fixtures
Hot-dipping (Al, Al–X)	Generally multiple (2)	Flux or inert	Short reaction times. Process basically is limited to Al- or Al–X-type coatings
Chemical vapor deposition (CVD) (Mostly Cr and Al, but no real limitations)	Generally 2	Vacuum or inert	Somewhat similar to pack-cementation, except halide carrier is generated externally. Special holding fixtures needed. Can get undesirable directional depositions
Fused-salt (Includes electrolytic and electroless processes) (Numerous metals, including rare earths)	Could be one or two. Additional steps could be needed to deposit multi-component systems	Inert	Scale-up could be a problem. Technique is not limited by coating material
Vitreous or glassy-bonded refractory	Usually multiple	Air or inert	Brittle, relatively thick coatings which are limited to about 1800°F in service
Physical vapor deposition (PVD) (Al, Cr, Si, various alloys)	Generally 2	Vacuum	Poor throwing power and expensive scale-up. Not readily repairable
Electrophoresis (Al-rich)	Multiple	Organic dielectric solvent	Limited to relatively small components

[a]Special holding fixtures can cause "blind spots" in coating.

coated. During the coating process, the deposited metal(s) diffuses into the substrate. (This frequently requires a long hold-time at a high temperature, that can adversely affect the mechanical properties of the substrate.) The coatings are solid solutions or intermetallic compounds (or both) composed of alloys of the coating metal(s) and the substrate. They are an integral part of the base material and have excellent bonding strength. The properties of these coatings can be controlled by a careful choice of coating metal(s) and proper control of the time–temperature cycle during coating.

A modification of the pack-cementation technique, as described above, is a method known as "halogen streaming." In halogen-streaming, the halogen gas is introduced into the retort from an external source. Otherwise, the two processes are identical. Some of the advantages claimed for the halogen-streaming technique over the use of a halide compound added to the powder mix are[17] (a) regulation of the gas-flow rate and halogen content in the retort, (b) timely introduction (and discontinuation) of the halogen into the retort, (c) elimination of extraneous elements associated with the halide compound, (d) possibility of using different halogen gases during one coating cycle (at any desired temperature),* and (e) provision for purging the retort with an inert gas at any time during processing.

Still another modification of the basic pack-cementation process is especially advantageous for the economical coating of large objects. This process, known as the "slip-pack" technique, involves the use of a slurry consisting of the coating metal(s) (with or without a halide compound and an inert material) suspended in some vehicle such as water or lacquer. The slurry is applied to the substrate by dipping, brushing, or spraying and then allowed to dry. When the slurry contains the halide compound, the component to be coated is heated, usually in an inert atmosphere, to the elevated temperature where the coating process occurs. If the slurry does not contain the halide compound, the slurry-coated object is placed in a retort and supported by an inert refractory oxide powder (e.g., Al_2O_3). A quantity of halide compound placed on the bottom of the retort provides the halide gas required for the coating process.

Unfortunately, processing details on the various pack-cementation coatings for superalloys are considered proprietary information by the coating vendors. Since there is no free exchange of process information by the developers of pack-cementation coatings, it follows that progress in producing optimum coating systems has been seriously impeded.

Some basic information on the pack-cementation chromizing of nickel and nickel-base alloys is given by Samuel and Lockington.[15,16] They report that nickel and nickel-base alloys (Inconel, Hastelloy B, and the Nimonic series)

can be chromized by using modifications of the methods developed for ferrous materials. One pack composition described by these authors[15] consists of 2 parts chromium metal powder, 1 part unvitrified kaolin (filler) and $0.5^{w}/o$ ammonium iodide. Using this pack and a circulating stream of purified hydrogen gas, chromium reportedly can be deposited on nickel at coating temperatures of 1050°C or above. (Kelley[5] cautions that nickel can be chromized providing the deposition temperature does not exceed about 1300°C, where Ni–Cr eutectic melting occurs.) Samuel[16] proposed, from thermodynamic considerations, that it is unlikely that the coating process occurs by interchange at the substrate surface (e.g., $Ni + CrAl_2 \rightarrow NiCl_2 + Cr$). Rather, he believes the relevant reactions are $CrX_2 + H_2 = 2HX + Cr$ (depositing on Ni)—a reduction process; and $CrX_2 = X_2 + Cr$ (depositing on Ni)—a thermal decomposition process, where X is some halide. These conclusions apparently were confirmed by experimental results. The same authors[16] also report that most of the features discussed for nickel apply equally to the chromizing of cobalt and cobalt alloys; however, the diffusion rate in cobalt is lower than for nickel, and thinner coatings are obtained.

Some of the coating systems that can be deposited on superalloy substrates using pack-cementation processes are Cr, Al, Al–Cr, Al–Cr–X, and Be, where X is some additive metal. A typical modern-day pack-cementation process for applying these coatings is outlined below.

1. Deburr (round corners).
2. Grit blast with 100-mesh Al_2O_3.
3. Glass-bead hone.
4. Degrease in trichloroethylene.
5. Pack component in retort containing powder mixture of:

    70 to 80 percent Al_2O_3
     8 to 10 percent Al
    20 to 30 percent Cr
    about 0.25 percent halide activator (NH_4Cl or NH_4I)

6. Seal in double-wall retort (glass seal).
7. Heat to 1600° to 1800°F and hold for 15 to 18 hr.
8. Remove retort from furnace and air-cool.
9. Light dry hone to remove any adhering pack powder.

SLURRY METHOD

The slurry method[18,19] of applying protective coatings to superalloy components is advantageous for coating large sizes and reasonably complex shapes. The slurry, formed by mixing the coating metal(s) in powder form with some liquid carrier, is applied to the substrate by brushing, dipping, or spraying. After the coating suspension (bisque) has dried, it is given a diffusion anneal in an inert atmosphere (or vacuum) at some elevated temperature (the bisque may or may

---

*Such use would permit the deposition of different elements using the optimum halogen and deposition temperature for each metal.

not melt, depending on the process used). A liquid carrier that will decompose on heating without the formation of undesirable carbides is generally used. Organic liquid carriers that have been used successfully include acetone, collodion-butyl acetate, and nitrocellulose lacquers (a mixture of 2 parts collodion and 1 part butyl acetate will suspend aluminum compositions satisfactorily). To avoid excessive agglomeration and surface roughness problems, it is usually recommended that the particle size of the coating metal(s) be relatively small (e.g., –325-mesh).

For superalloy substrates, the slurry technique has been successfully used to apply a number of coating metals, including Al, Cr, Al–Cr, and Al–Cr–X, where X is a third element. A typical sequence of events used in aluminizing a nickel-base superalloy substrate is given below.

1. Vapor-degrease substrate.
2. Sandblast with 100-grit garnet.
3. Apply 0.006 to 0.008 in. of slurry.
4. Dry at 200°F.
5. Fire 10 min at 1400°F in N_2.
6. Remove flux by immersion in hot water (160° to 200°F) for 5 to 30 min.
7. Water–air blast.
8. Dip in $10^v/o$ HNO_3 (140°F) for 15 to 60 sec.
9. Water–air blast.
10. Diffuse 4 hr at 2100°F in argon.
11. Vapor-hone with 200-grit Alundum.

HOT-DIPPING

Hot-dip coatings[1,18,19] are applied by immersing the substrate in a molten bath of a coating metal or alloy that melts at a temperature significantly below that of the object being coated. The molten bath is covered with a flux (frequently a halide type) to prevent oxidation. Excessive dissolution of the substrate can occur through prolonged exposure to the molten metal, so the immersion time is generally short (e.g., up to 5 min). Since the short immersion time does not permit sufficient diffusion of the coating metal(s) into the substrate, the coated object is normally given a subsequent diffusion anneal at some elevated temperature. Usually, the diffusion anneal is accomplished in an inert atmosphere or vacuum.

Poor reliability of hot-dip coatings has been attributed to (a) difficulty in obtaining uniform coatings, (b) poor coverage on corners, (c) difficulty in coating deep recesses, and (d) inclusions in the coating caused by the dipping operation. The quality of these coatings is strongly related to the skill of the operator. For sheet material and wire, the hot-dipping process is quite successful and is considered one of the simplest and least costly coating methods.

The hot-dip coating technique has been used successfully in applying aluminum coatings to superalloy substrates. It reportedly can also be used to apply aluminum-alloy coatings. The process outlined below is typical of those used in applying a hot-dip aluminized coating to a nickel-base superalloy substrate.

1. Vapor-blast substrate.
2. HCl clean.
3. Water rinse.
4. 4 min in salt flux at 1300°F.
5. 8 to 10 sec in 2-S aluminum bath at 1325°F.
6. 15 sec in salt flux at 1300°F.
7. Spin in air to remove surplus aluminum.
8. Heat 3 min at 1325°F; air cool.
9. Scrub with fiber brush.
10. Immerse 30 to 40 sec in 10 to $15^w/o$ HNO_3 at 100° to 185°F.
11. Immerse 1½ hr in $5^w/o$ HCl.
12. Diffuse: 4 hr at 2100°F; air cool.
             4 hr at 1975°F; air cool.
             16 hr at 1400°F; air cool.
13. Light vapor blast.

CHEMICAL VAPOR DEPOSITION (CVD)

Chemical vapor deposition[18-20] (sometimes referred to as vapor plating) is accomplished by heating the object to be coated in a chamber containing an inert gas at reduced pressure. When the substrate reaches the desired temperature, a flowing stream of gaseous metal halide or organometallic compound mixed with an inert carrier gas (argon or nitrogen) is introduced into the chamber. The halide or organometallic compound thermally decomposes at the substrate surface, depositing the coating metal. Frequently, the deposition temperature is sufficiently high to permit diffusion of the coating metal into the surface of the substrate. Otherwise, the substrate must be given a subsequent diffusion anneal. A modification of this process is accomplished by chemical reduction instead of thermal decomposition. In this process, purified hydrogen is mixed with the coating-metal gas.

The basic requirement in producing coatings by CVD is that the coating metal form a compound that can be vaporized at some low temperature without deposition, and then be sufficiently unstable at elevated temperatures to thermally decompose or reduce at the heated substrate surface. A significant amount of information on this coating process has been compiled in a Defense Metals Information Report prepared by Krier.[20] The process is capable of producing reasonably smooth and uniform coatings on complex shapes; however, some investigators report that a coating buildup is frequently observed on sections of the substrate that face the flowing gaseous mixture. "Dead-spots" in the coating can also occur where there is no flow of gas. The main disadvantage of the CVD process is that it requires complex equipment and is presently limited to reasonably small objects. Although the process is not widely used for producing coatings on superalloy substrates, a typical CVD process for depositing aluminum on a nickel-base alloy is described below.

1. Vapor-blast substrate.
2. HCl clean.
3. Water rinse and air dry.
4. Place specimen in CVD chamber.
5. Preheat specimen inductively for 5 min at $500°F$ in flowing argon.
6. Maintain specimen at $500°F$ and introduce tri-isobutyl aluminum vapor into chamber.
7. Deposit aluminum for 5 to 10 min.
8. Discontinue deposition and begin diffusion anneal.
9. Diffuse 4 hr at $2100°F$; cool to $1975°F$ and hold 4 hr; cool to $1400°F$ and hold 16 hr; air cool to room temperature.
10. Lightly vapor-blast coated specimen.

FUSED-SALT PROCESSES

A fused-salt bath[19,21,22] can be used to deposit various coatings on superalloys without the use of an external current. In many cases, an external current is used to cathodically deposit the coating metal(s) on the substrate. Both of these fused-salt techniques will be described.

When an external current is not used, the fused-salt coating technique is somewhat similar to the basic pack-cementation process. Coatings are applied by preparing a fused-salt bath containing a metal halide salt of the coating metal (e.g., $CrCl_2$ or $TiCl_2$) and selected salt additives (e.g., NaCl and $BaCl_2$). Metal chips of the coating metal are generally added to the bath to replenish the metal deposited and to keep the salt in a divalent state. The bath is protected by an inert atmosphere. Sometimes the bath is agitated by bubbling argon through the melt. The deposition mechanism probably occurs by a disproportionation reaction of the type

$$MCl_2 \rightarrow M + MCl_x,$$

where M is the depositing metal and x is a higher valence state. Advantages claimed for this process, compared to the pack-cementation method, include good coating uniformity, versatility of coating composition, and possibly faster deposition rates. Disadvantages include scale-up problems, the need for special handling fixtures, and possibly slower diffusion rates. A typical sequence of events for chromizing pure nickel using a nonelectrolytic fused-salt bath follows.

1. Properly deburr and clean substrate.
2. Prepare fused-salt by melting in crucible under argon atmosphere:
    30 percent $CrCl_2$ (could vary from 5 to 30 percent)
    49 percent $BaCl_2$
    21 percent NaCl
    Sufficient Cr flake to cover crucible bottom
3. Introduce specimen under argon atmosphere when fused salt reaches operating temperature (about $1200°F$).
4. Hold specimen in fused-salt bath for 1 to 6 hr.
5. Remove specimen and cool to room temperature under argon.

A fused-salt bath can also be used to electrodeposit diffusion-type coatings on superalloy substrates. With a high-temperature fused-salt bath, aluminum can be electrodeposited on nickel by using a molten bath consisting of 440 g NaCl, 560 g KCl, and 150 g cryolite (sodium aluminum fluoride) at a temperature of about $1830°F$. Using a current density of 100 to 200 amp/ft^2 and a tungsten anode, aluminum is readily deposited from this bath. The high temperature of the bath permits the aluminum to diffuse into the nickel and form an Al–Ni alloy diffusion coating. For substrates other than pure nickel, the surface of the object to be coated may need nickel-enrichment treatment to obtain the desired coating composition.

Recent work at the General Electric Research and Development Center provides the ability to form diffusion coatings on metals through the use of molten alkali and alkaline-earth fluorides in an electrolytic process operated at $600°$ to $1200°C$ in an inert atmosphere.[23] Among the metals that can be diffusion-coated on a variety of substrates by the G.E. process are Be, B, Al, Si, Ti, V, Cr, Mn, Ni, Ge, Y, Zr, and the rare earths. Reportedly, combinations of these metals can be deposited simultaneously or sequentially. Advantages claimed for the G.E. technique, compared to pack-cementation and hot-dip techniques, are (a) accurate control of coating thicknesses, (b) excellent coating uniformity, (c) fast diffusion rates, (d) freedom from restriction to metals that form gaseous compounds or that have low melting points, and (e) continuous-process possibilities.

VITREOUS OR GLASSY-REFRACTORY TECHNIQUE

High-temperature vitreous or glassy-refractory coatings are applied to superalloy substrates by using modifications of existing enameling methods.[3,18,19,24] Vitreous coatings are smooth, nonporous, bonded coatings that are resistant to temperatures up to about $1800°F$ for long periods. Improvements in vitreous coatings are made by adding refractory oxides (e.g., ZrO_2, TiO_2, Cr_2O_3, or Al_2O_3) to the basic coating formulation. This results in "matte" finish coatings that can protect from temperatures approaching $2200°F$.

Frits are prepared by smelting and quenching in water. The frit, with or without refractory-oxide additions, is ball-milled in water to the proper particle size (200 to 325 mesh). (During milling, additives such as barium chloride, citric acid, chromic acid, and magnesium sulfate are generally added to the mix in order to stabilize the slip or increase the "set" of the coating formulation.) The coating is applied to the prepared substrate by spraying or dipping and allowed to air-dry (it is sometimes dried artificially at temperatures up to $450°F$). The coating is produced by firing (generally in air) at some elevated temperature for 1 to 30 min. Vitreous coatings are usually fired at $1050°$ to $1600°F$, while the glassy-refractory coatings are fired at temperatures above $1500°F$. Usually, several layers of coating are built up on the substrate surface by additional coating–firing operations. A typical coating process for applying a glassy-refractory coating to a nickel-base superalloy is described below.

1. Lightly sandblast substrate.
2. Clean in acetone or benzene.
3. Clean in boiling solution of commercial cleaner.
4. Rinse in hot water ($100°$ to $120°F$).
5. Pickle for 20 min at $120° ± 5°F$ (100 g $FeCl_3 \cdot 6H_2O + 50$ ml conc. $HCl + 850$ ml H_2O).
6. Rinse in hot running water ($100°$ to $120°F$).
7. Dry at $220°F$.
8. (Possibly give substrate a 10-min heat treatment in an oxidizing atmosphere at $1800°F$ to develop a thin oxide layer that sometimes improves adherence.)
9. Ball-mill coating mixture in water to proper particle size.

Typical Coating Composition, in parts by weight
70 NBS-332 frit
30 chromic oxide
5 Green Label Clay

NBS-332 Frit Composition, in parts by weight
1.5 aluminum hydrate
56.6 barium carbonate
11.5 boric acid
6.3 calcium carbonate
37.5 quartz
5.0 zinc oxide
2.5 zirconium dioxide

10. Add water to adjust specific gravity of coating mix to about 1.7.
11. Apply about 0.001-in. deposit of coating to substrate by spraying (or dipping).
12. Fire at $1900°$ to $1950°F$ for about 5 to 7 min.
13. Repeat steps 11 and 12 to apply second layer.
14. Repeat step 13, as desired.

PHYSICAL VAPOR DEPOSITION (PVD)

Sometimes referred to as vacuum metallizing, physical vapor deposition[19,25,26] is performed in a chamber evacuated to a pressure of 10^{-3} to 10^{-6} mm Hg. A sufficient vacuum would be one where the mean free path of residual gas molecules is greater than the distance from the evaporation source to the object being coated. The coating metal (or alloy) is wrapped around a helical refractory-metal filament (or placed in a refractory boat or crucible) and heated (using external heating coils, induction, electron-beam bombardment, or by passing current through coating-metal container) to a temperature where its vapor pressure is sufficiently high to permit its evaporation. Temperatures creating coating-metal vapor pressures greater than about 0.01 mm Hg are generally not used because they tend to produce spongy, nonadherent deposits. (For this reason, the evaporation temperature will be defined as the temperature at which the vapor pressure becomes 0.01 mm Hg.) Evaporation temperatures and deposition rates for elements that can be deposited on superalloy substrates in order to develop protective coatings are given below:

Element	Evaporation Temperature ($°C$)	Deposition Rate, $g/cm^2/sec$
Cr	1392	1.03×10^{-4}
Al(l)	1207	7.88×10^{-5}
Ni	1497	1.06×10^{-4}
Li	1547	7.24×10^{-5}

Since transfer of the coating material is accomplished by physical means, physical vapor deposition coatings must subsequently be heat-treated to provide bonding with the substrate. Coatings are generally quite thin, which is one significant limitation of this coating process, Another important limitation of PVD is that it is a "line-of-sight" coating process. Maneuvering of the part within the chamber can improve exposure, but complicated shapes are usually not coated by the physical vapor deposition technique. One advantage of this coating process is that certain oxidation-resistant alloys (including Nichrome Inconel and Fe–Cr–Ni) can be deposited by vacuum metallizing. A typical physical vapor deposition process for aluminizing a nickel-base superalloy substrate follows:

1. Degrease using alkaline cleaner, water rinse, and scrub with hot water.
2. Place in vacuum chamber and evacuate to a pressure of 10^{-4} to 10^{-6} mm Hg.
3. Heat coating metal to about $1200°C$, using helical coil of W or aluminum nitride boat to hold aluminum.

4. Coat to desired thickness (deposition rate is about 7.88×10^{-5} g/cm^2/sec).

5. Remove specimen and diffusion anneal using time–combination similar to that reported for hot-dipping, CVD, or electrophoresis techniques.

ELECTROPHORESIS

Electrophoretic coatings[19,27,28] are produced when colloidal particles of the coating material in a liquid medium migrate in an electric field and deposit. The liquid suspension is generally a solution of various organic dielectric solvents and usually contains selected proprietary additives for improving the green strength of the as-deposited coating. Organic liquids are used in order to prevent gas evolution at the electrodes. The suspension should have the following characteristics: low viscosity, low electrical conductivity, low evaporation rate, and high dielectric constant.[27] Particles of the coating material added to the suspension are generally prepared by ball-milling them to a diameter of 0.5 to 40 μ.

The substrate to be coated is properly cleaned and chemically etched or sandblasted to improve coating adherence. Proper polarity is selected, depending upon the electrical charge of the colloidal particles, and a potential of 50 to 150 Vdc (current density may be 0.05 to 0.30 amp/ft^2) is applied between the specimen and an auxiliary electrode. During deposition, the liquid bath is vigorously agitated to provide a uniform distribution of colloidal particles. After coating to the desired thickness, the specimen is removed from the bath and dried. Since the coating as deposited generally has poor green strength and is of lower than desired density, it is densified by isostatic pressing and sintering. Frequently, the coating is sintered without isostatic pressing.

Advantages of coating by electrophoresis include uniformity; thickness control; deposition of metals, alloys, cermets, oxides, and other nonmetallic substances with composition control; metallurgical bonding (after sintering); high-density coatings (after isostatic pressing); rapid deposition of thick coatings; low power consumption; and excellent throwing power.[19]

In one electrophoresis process, coating of Ni-Cr alloys has been accomplished by first depositing oxides of chromium and nickel on the substrate surface.[27] The alloy is produced by heating the specimen to 1100°C and reducing the oxides in purified hydrogen.

Mostly aluminum-rich coatings are being electrophoretically deposited in order to provide oxidation–sulfidation resistance to current nickel-base superalloys. Although the electrophoresis process is generally limited to relatively small components (e.g., bolts and nuts), it reportedly has been used successfully to coat combustion chambers of aircraft turbine engines. A typical electrophoresis process for producing an aluminum-rich coating on a nickel-base superalloy turbine engine component is described below.

1. Clean substrate as for conventional electroplating.
2. Chemically etch or sandblast substrate.
3. Place specimen in agitated, colloidal suspension of volatile organic solvents containing about 2 to 5 ^{w}/o solids of coating material (average particle size of about 10 μ).
4. Apply 50 to 150 volts between electrodes for about 40 sec.
5. Remove specimen from solution and air dry.
6. Diffusion anneal for about one hour at 1400° to 1600°F.

The green coating may be isostatically pressed at 10,000 to 100,000 psi prior to sintering.

EXISTING COATINGS FOR SUPERALLOYS

Some of the various metals and metal combinations that have been or are being developed for coating superalloys[18] include:

Aluminum by hot-dipping, chemical vapor-deposition, and pack-cementation processes

Aluminum with unidentified nonmetallics by a pack-cementation technique

Aluminum with unidentified alloying elements by a pack-cementation process

Aluminum–chromium by various pack-cementation processes

Aluminum–chromium–nickel by a pack-cementation process

Aluminum–chromium–silicon by pack-cementation and by electroplating plus hot-dipping techniques

Aluminum–cobalt by a spray plus diffusion process

Aluminum–iron by slurry and pack-cementation processes

Aluminum–nickel by slurry, fused-salt, and pack-cementation processes

Aluminum–silicon by a slurry process

Aluminum–tantalum by a slurry technique

Beryllium with unidentified alloying elements by a pack-cementation process

Zirconium and zirconium–chromium by a pack-cementation technique

Aluminum-rich by electrophoresis techniques

The following industrial organizations are developing high-temperature protective coatings for superalloys:

Alloy Surfaces Co., Inc. (Mr. W. R. Hudson),
100 S. Justison St., Wilmington, Del. 19801

Chromalloy Division (Mr. Richard Wachtell)
   Chromalloy American Corp.
   West Nyack, N.Y. 10994
Martin Metals Co. (Mr. L. I. Kane)
   Martin-Marietta Corp.
   Wheeling, Ill. 60090
United Aircraft Corp. (Mr. F. P. Talboom)
   Pratt & Whitney Div.
   East Hartford, Conn. 06108
High Temperature Composites Lab. (Mr. L. Sama)
   Sylvania Electric Products, Inc.
   Cantiague Road, P.O. Box 35
   Hicksville, N.Y. 11802
National Bureau of Standards (Mr. J. C. Richmond)
   Department of Commerce
   Washington, D.C. 20234
Wall Colmonoy Corp. (Mr. Forbes Miller)
   19345 John R. St.
   Detroit, Mich. 48203
General Electric Co. (Mr. M. A. Levenstein)
   Advanced Engine Technology Dept.
   Cincinnati, Ohio 45215
Allison Div., General Motors Corp. (Mr. George Sipple)
   Materials Laboratory
   Indianapolis, Ind. 46206
Misco Precision Casting Co. (Mr. R. A. Martini)
   116 Gibbs Street
   Whitehall, Mich. 49461
Solar Division (Mr. A. R. Stetson), International
   Harvester Co.
   2200 Pacific Highway
   San Diego, Calif. 92112
The Calorizing Corp. (Mr. B. J. Sayles)
   400 Hill Ave.
   Pittsburgh, Pa. 15221
TRW Equipment Laboratories (Dr. J. D. Gadd)
   23555 Euclid Ave.
   Cleveland, Ohio 44117
Union Carbide Corp., Whitfield Works (Mr. Robert Puyear)
   1 Paul Street
   Bethel, Conn. 06801
Lycoming Division AVCO Corp. (Dr. S. Baranow)
   550 Main Street
   Stratford, Conn. 06497
Vitro Laboratories (Mr. Martin Ortner)
   200 Pleasant Valley Way
   West Orange, N.J. 07052

The important coatings available from these firms at the present time are described in Table 11. Inspection of Table 11 reveals that (excluding the vitreous or glassy-refractory coatings) nearly all of the existing coatings are based on aluminum as the primary coating metal. The ex-

ceptions are the WL-6 beryllium coating produced by Union Carbide Corporation and the two Nicrocoats of Wall Colmonoy Corporation. Processes used to obtain these aluminum or aluminum-rich coatings are pack-cementation, hot-dipping, slurry, and electrophoresis. There probably is some redundancy among the 35 aluminum-rich coatings available from the 14 vendors producing these coatings. Such overlap cannot be determined, however, because of the problems associated with handling proprietary information.

Many of the coatings described in Table 11 are being used to coat a variety of high-temperature superalloy components in commercial and military gas-turbine engines. Although they have successfully increased the useful life for many of these components, further extensions are desired. Accomplishing this objective requires that existing coatings be employed, or that new coatings be developed.

EVALUATION OF EXISTING COATINGS

Extensive testing of existing (and experimental) coatings has been conducted in a large number of laboratories. The majority of this evaluation work has consisted of testing static or cyclic oxidation (weight-gain or weight-loss), mechanical properties, thermal fatigue, thermal shock, hot-gas erosion, sulfidation, and impact. Many of these data are summarized in the report by Jackson and Hall.[18] Comparison of these data is virtually impossible since the various investigators used different testing techniques and rarely reported on how their coatings compared with others—without coding the names of the competitive coatings.[2] Also, comparisons are frequently unrealistic because a coating developer will usually optimize his coating process based on a given test technique and then evaluate his competitors' coatings using the same test.

Some meaningful comparative test data are available, however, for environmental conditions where coated superalloys may be used in service. Using an oxidation-erosion test apparatus, Allison Division of General Motors has obtained results for several coating systems on a Hastelloy X substrate and has investigated the effect of applying the Alpak coating (aluminized coating applied by a pack-cementation process) to a number of superalloy substrates.[29] Figure 18 describes the oxidation-erosion of various bare and Alpak-coated turbine-blade superalloys when they were subjected to high-velocity air (2,000 ft/sec) at 2000°F. The oxidation-erosion characteristics of various commercial coatings on a Hastelloy X substrate exposed to high-velocity air (2,000 ft/sec) at 2100°F are given in Figure 19*. Commercial coating systems compared in this fig-

---

*Test conditions simulate service conditions for thermocouple probes used to monitor temperatures in gas turbine engines.

TABLE 11   High-Temperature Coatings for Superalloys

Designation	Vendor	Coating Type	Process
NC101A	Sylvania	Al-rich	Pack-cementation
ASC-HI-15	Alloy Surfaces	Al-Cr	Proprietary ("vapor deposition")
Nicrocoat 110	Wall Colmonoy	Ni-Cr-TiB$_2$	Slurry
Nicrocoat 130	Wall Colmonoy	Ni-Cr-Si-TiSi$_2$-TiN	Slurry
Type 701	Lycoming Div. AVCO	Al-rich	Vacuum pack-cementation
Type 606B	Lycoming Div. AVCO	Cr-Al-Si	Cr-plate + hot-dip Al-Si
WL-1	Union Carbide	Al-rich	Pack-cementation
C-9	Union Carbide	Al-rich (Al-Fe)	Pack-cementation
C-12	Union Carbide	Al-rich	Pack-cementation
WL-8	Union Carbide	Al-rich	Pack-cementation
WL-14	Union Carbide	Al-rich	Pack-cementation
WL-4	Union Carbide	Al-Si-Cr	Pack-cementation
C-20	Union Carbide	Al-Cr	Pack-cementation
C-3	Union Carbide	Al-Ni	Pack-cementation
WL-9	Union Carbide	Al-Si	Pack-cementation
1AD	Union Carbide	Al	Hot-dip
WL-6	Union Carbide	Be	Pack-cementation
Calorizing	Calorizing	Al	Pack-cementation
Codep A	General Electric	Al-Ti	Pack-cementation
Codep C	General Electric	Al-Ti	Pack-cementation
Aldip	Allison Div., GMC	Al	Hot-dip
Alpak	Allison Div., GMC	Al	Pack-cementation
A-11	Martin Metals	Al	Diffusion-annealed CVD
PWA-47	Pratt & Whitney	Al-Si	Slurry
MDC-1	Misco Precision Casting	Al-rich	Pack-cementation
MDC-1A	Misco Precision Casting	Al	Pack-cementation
MDC-6	Misco Precision Casting	Cr-Al	Pack-cementation
MDC-7	Misco Precision Casting	Al-rich	Pack-cementation
MDC-9	Misco Precision Casting	Al-rich	Pack-cementation
UC	Chromalloy	Al-Cr	Pack-cementation
SUD	Chromalloy	Al-rich (Al-Fe)	Pack-cementation
SAC	Chromalloy	Al-Cr-Si	Pack-cementation
UDM	Chromalloy	Al-rich	Pack-cementation
—	TRW	Al-rich	Vacuum pack-cementation
—	TRW	Al-Cr	Slurry
—	TRW	Al-Cr	Vacuum pack-cementation
S13-53C	Solar Div. Int. Harvester	Al-Fe	Slurry
S5210-2C	Solar Div. Int. Harvester	Barium-Silicate Glass + Additives	Glassy-refractory
S6100M	Solar Div. Int. Harvester	Modified S5210-2C	Glassy-refractory
NBS A-418	NBS	Frit 332 + Cr$_2$O$_3$	Glassy-refractory
MAL-2	Vitro	Al-rich	Electrophoresis
Sermetel J	Vitro	Al-rich	Electrophoresis

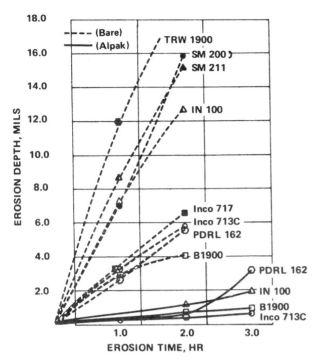

FIGURE 18   Oxidation-erosion of various bare and Alpak-coated turbine-blade superalloys exposed to 2,000 ft/sec air at 2000° F.[29]

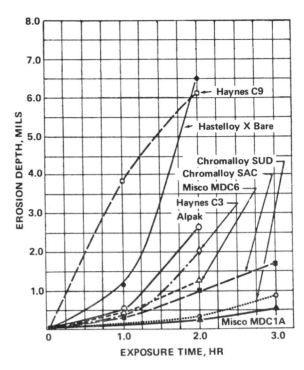

FIGURE 19   Oxidation-erosion of various commercial coatings on Hastelloy X substrate exposed to 2,000 ft/sec air at 2100° F.[29]

ure are Misco MDC1A (Al, applied by pack cementation), Alpak, Haynes C-9 (Al-Fe applied by pack cementation), Haynes C-3 (Al-Ni applied by pack cementation), Chromalloy SUD (Al with undisclosed alloying elements applied by pack cementation), Chromalloy SAC (Al-Cr-Si applied by pack cementation), and Misco MDC6 (Cr-Al applied by an undisclosed process).

RECENT RESEARCH ON COATINGS FOR SUPERALLOYS

The National Aeronautics and Space Administration (Lewis Research Center) has a number of active programs to develop or evaluate coatings for superalloy substrates. These include an in-house evaluation of available coatings; an evaluation program at the Solar Division of the International Harvester Company, where commercial coatings are being tested using an engine-type test rig; a program with Westinghouse Electric Corporation to develop controlled-viscosity glasses; a program with Battelle Memorial Institute to develop Nichrome-type claddings for superalloys; and programs with Illinois Institute of Technology and Sylvania Electric Products, Inc., to develop, respectively, cladding materials and techniques to apply them to dispersion-strengthened alloys.

Pratt & Whitney Aircraft Division of United Aircraft

Corporation has completed research on experimental coatings for various superalloys.[30] This research was sponsored by the Air Force Materials Laboratory. A Ta Co–Al coating (Ta was applied by a variety of techniques including vapor deposition and the Co-Al was applied by a slurry process) appeared to exhibit an increased temperature capability when compared with other coatings, but the coating was not sufficiently reproducible to warrant further investigation. The Air Force Materials Laboratory presently sponsors no research on high-temperature coatings for superalloy substrates. It does plan to support research in the immediate future to develop improved coatings that will have both increased useful lives at current operating temperatures and increased temperature capabilities.

Research sponsored by the United Kingdom's Ministry of Aviation at the National Gas Turbine Establishment (Pyestock, Farnborough, Hants) is directed at determining the upper useful operating temperatures for aluminized and chromized coatings on nickel-base alloys in simulated gas-turbine exhaust. Although this work started in 1965, no known reports were available as of November 1967.

Phillips Petroleum Company, under contract with the U.S. Naval Air Systems Command, has studied the effect of JP fuel composition on the hot corrosion of superalloys, including several coated alloys. Coatings involved were Misco MDC-1 on Inconel 713C[31] and Misco MDC-1, MDC-7, and MDC-9 on Inconel 713C.[32]

In addition, it is known that various coating vendors are developing improved coatings in their own laboratories using corporate funds. Engine manufacturers are evaluating many of these coatings for possible use in advanced propulsion systems. Unfortunately, because of the highly proprietary nature of the coatings industry, these data are not available for this report.

REFERENCES

1. Puyear, R. B. 1962. High-temperature metallic coatings. Mach. Des. 34.
2. Lane, P., and N. M. Geyer. 1966. Superalloy coatings for gas turbine components. J. Met. 18:186.
3. Long, J. V. 1961. Refractory coatings for high-temperature protection. Met. Progr. 79(3):114.
4. Kelley, F. C. 1921. Surface-alloyed metal. United States Patent No. 1,365,499, January 11.
5. Kelley, F. C. 1923. Chromizing. Trans. Amer. Electrochem. Soc. 43.
6. Naugle, C. A. May 1948. Investigation of the German BDS chromizing process. Air Documents Division Intelligence Department Report No. F-TR-1183-ND, Wright-Patterson AFB, Ohio.
7. Galmiche, P. 1950. A new process of thermal chromizing and the formation of mixed diffusion alloys. Rech. Aeron. 14: 55–63.
8. Samuel, R. L., and N. A. Lockington. 1951. The protection of metallic surfaces by chromium diffusion: Part I–Survey of the chromizing process. Met. Treat. Drop Forg. 18:354.
9. Samuel, R. L., and N. A. Lockington. 1951. The protection of metallic surfaces by chromium diffusion: Part II–Theoretical considerations. Met. Treat. Drop Forg. 18:407.
10. Samuel, R. L., and N. A. Lockington. 1951. The protection of metallic surfaces by chromium diffusion: Part III–Structure of chromized surfaces. Met. Treat. Drop Forg. 18:440.
11. Samuel, R. L., and N. A. Lockington. 1951. The protection of metallic surfaces by chromium diffusion: Part IV–Modern chromizing methods. Met. Treat. Drop Forg. 18:495.
12. Samuel, R. L., and N. A. Lockington. 1951. The protection of metallic surfaces by chromium diffusion: Part V–Properties of chromized steel. Met. Treat. Drop Forg. 18:610.
13. Samuel, R. L., and N. A. Lockington. 1952. The protection of metallic surfaces by chromium diffusion: Part VI–Applications of chromizing. Met. Treat. Drop Forg. 19.
14. Samuel, R. L., and N. A. Lockington. 1952. The protection of metallic surfaces by chromium diffusion: Part VII–Applications of chromizing. Met. Treat. Drop Forg. 19.
15. Samuel, R. L., and N. A. Lockington. 1954. The diffusion of chromium and other elements into non-ferrous metals. Trans. Inst. Met. Finish. 31.
16. Samuel, R. L. 1958. A survey of the factors controlling metallic diffusion from the gas phase. Murex Rev. 1.
17. Aves, W. A., and G. M. Ecord. 1964. Recent high-temperature coating developments at LTV-Vought Aeronautics. Presented at the 8th Refractory Composites Working Group Meeting, Fort Worth, Texas, January 14–16.
18. Jackson, C. M., and A. M. Hall. April 20, 1966. Surface treatment for nickel and nickel-base alloys. NASA Technical Memorandum No. NASA TM X-53448 (AD-634076). George C. Marshall Space Flight Center, Huntsville, Alabama.
19. Withers, J. C. 1963. Methods for applying coatings. Chapter 4 in J. Huminik, Jr., ed. High-temperature inorganic coatings. Reinhold, New York.
20. Krier, C. A. June 4, 1962. Chemical vapor deposition. DMIC Report No. 170. Battelle Memorial Institute, Columbus, Ohio.
21. Campbell, I. E., V. D. Barth, R. F. Hoeckelman, and B. W. Gonser. 1949. Salt-bath chromizing. Trans. Amer. Electrochem. Soc. 96.
22. Anonymous. 1960. Nickel-aluminum coatings protect metals up to 1830°F. Mat. Des. Eng. 52(5).
23. Geyer, N. M. Oct. 27, 1967. Air Force Materials Laboratory, Wright-Patterson AFB, Ohio, personal communication to J. R. Myers, Air Force Institute of Technology.
24. Cook, T. E. April 25, 1959. Glass-bonded refractory coatings for iron or nickel-base alloys. DMIC Memorandum No. 16. Battelle Memorial Institute, Columbus, Ohio.
25. Powell, C. F. 1966. Physical vapor deposition. Chapter 8 in C. F. Powell, J. H. Oxley, and J. M. Blocher, Jr., eds. Vapor deposition. John Wiley & Sons, Inc., New York.
26. Holland, L. 1956. Vacuum deposition of thin films. John Wiley & Sons, Inc., New York.
27. Shyne, J. J., H. N. Barr, W. D. Fletcher, and H. G. Scheible. 1955. Electrophoretic deposition of metallic and composite coatings. Plating 42.
28. Lamb, V. A., and W. E. Reid, Jr. 1960. Electrophoretic deposition of metals, metalloids, and refractory oxides. Plating 47.
29. Sippel, G. R., and J. R. Kildsig. 1966. Laboratory tests for evaluation of surface deterioration in gas turbine engine materials. Allison Research and Engineering, 2nd Quarter. Allison Energy Conversion Division of General Motors Corporation, Indianapolis, Indiana.
30. Talboom, F. P., and J. A. Petrusha. Feb. 1966. Superalloy coatings for components for gas turbine engine applications. Air Force Materials Laboratory Technical Report No. AFML-TR-66-15. Wright-Patterson AFB, Ohio.
31. Quigg, H. T., and R. M. Schirmer. Sept. 1966. Effect of JP fuel composition on hot corrosion. Phillips Petroleum Company Research and Development Report 4493-66R (Naval Air Systems Command Contract NOw 65-0310-d). Bartlesville, Okla.
32. Quigg, H. T., and R. M. Schirmer. June 1967. Effect of sulfur in JP-5 fuel on hot corrosion of coated superalloys in marine environment. Phillips Petroleum Company Research and Development Report 4792-67R (Naval Air Systems Command Contract NOw 66-0263-d). Bartlesville, Okla.

CHROMIUM*

INTRODUCTION

Chromium alloys are potentially attractive for high-temperature use in aircraft gas turbines as turbine buckets and vanes. The melting point of chromium is about 3400°F. This is about 700°F higher than that of either nickel (2650°F) or cobalt (2720°F), which are the base elements for the superalloys. Thus, while superalloy engine components begin to lose useful strength near 2000°F (approx. 0.7 mp), chromium alloy components could be expected to retain useful properties to at least 2400°F and provide a margin of safety in case of engine over-temperature. Furthermore, chromium alloy components offer a savings in engine weight because the density of chromium is 20 percent less than that of either nickel or cobalt.

Below a certain critical temperature, however, chromium alloys are brittle.[1] This ductile-to-brittle transition temperature (DBTT) is slightly above room temperature for commercially pure chromium. Unfortunately, nitrogen contamination increases the DBTT to considerably higher values. For example, chromium alloys exposed to air at high temperatures may have DBTT's in excess of 1100°F.[1] Thus, successful long-time cyclic service of chromium components in an aircraft gas turbine is primarily dependent on the solution of the nitrogen embrittlement problem.

Rare earth and yttrium additions help minimize nitrogen embrittlement in chromium and chromium alloys.[2] These metals preferentially react with nitrogen and remove it from solution in chromium. They may also improve oxide scale adherence and decrease the nitrogen permeability of the scale. Unfortunately, the solubility of these desirable additions in chromium is on the order of only 0.5 wt%.[3] Greater amounts cause chromium alloys to have poor high-temperature strength. Thus, in allowable levels, such additions do not provide adequate capacity to scavenge nitrogen from chromium during long-time high-temperature air exposures and thereby eliminate nitrogen embrittlement.

Coatings and claddings offer possible ways to prevent nitrogen from contaminating chromium. Such protection systems also offer a means of minimizing the oxidation of chromium alloys, which becomes significant above 2100°F. If by interdiffusion the coating or clad itself embrittles the chromium alloy, its primary purpose is, of course, defeated even though it protects against oxygen and nitrogen.

This section reviews the status of protection systems for chromium. The bulk of the work has been conducted at the Lewis Research Center of NASA or has been supported contractually by the Center. The protection system studies are part of a larger effort to determine the feasibility of fabri-

cating, protecting, and using chromium alloys in advanced jet engines. Since limited prior work existed on the protection of chromium, metal clads, metal coatings, aluminide coatings, and silicide coatings were all studied. The initial goals for protecting chromium alloys against air oxidation and nitrogen embrittlement were set at 600 hr for temperatures up to 2400°F, based on anticipated needs. Results to date have shown that none of the coating systems studied adequately met the goal of preventing nitrogen embrittlement without at least partially embrittling the substrate themselves. One purpose of this paper is to document the many approaches that have been tried in order to indicate their shortcomings and to stimulate further research to solve these problems.

THE STATUS OF CHROMIUM ALLOY DEVELOPMENT

Currently there are no commercially available chromium alloys on which to examine protective coating systems. However, experimental chromium alloys with properties of interest for turbine-engine applications have been developed by researchers in both Australia and the United States.[4,5] In Figure 20, a current experimental chromium alloy (Cr-4Mo-0.05Y-0.6Cb-0.4C) described by Clark[5] is compared at various temperatures to one of the strongest nickel-based commercial superalloys (IN-100) and to dispersion-strengthened nickel (TD-Ni) on the basis of 1000-hr stress rupture values normalized for density. In this figure the shaded areas represent the approximate strength ranges needed for turbine blade and vane requirements. Figure 20 shows that for the turbine bucket application, the wrought chromium alloy offers a potential temperature advantage of almost 200°F over the cast superalloy. For vane applica-

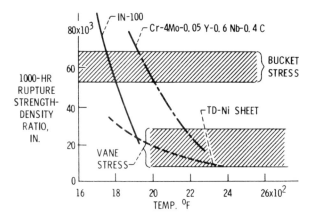

FIGURE 20 Comparison of high-temperature strengths of chromium- and nickel-base alloys.

*Prepared by S. J. Grisaffe.

tions, it is stronger than rolled dispersion-strengthened nickel up to at least 2300°F.

The tensile DBTT of the above chromium alloy was approximately 350°F in the stress-relieved condition and about 400°F in the as-recrystallized condition. This lack of ductility at low temperatures could lead to problems in engine operation. Above 400°F, however, this alloy has quite satisfactory ductility. At 600°F it has a tensile elongation of 10 percent, and at 2100°F the elongation is 26 percent. Thus, the current stage of chromium alloy development is capable of achieving competitive strength with current super-alloys, etc., in the 1900° to 2300°F range. Further development work is still needed to improve ductility in the low-temperature region.

PROTECTION SYSTEMS

The high-strength alloys were being developed concurrently with the coatings, and since no chromium alloys are available commercially, almost all NASA-supported work to date used a simple, easy-to-roll prototype experimental alloy of nominal composition Cr-5 ^{w}/o W - 0.1 ^{w}/o Y.[6] As-rolled to 60-mil sheet, its bend DBTT was between 400° and 650°F, depending on the heat tested.

In general, the coating development effort can be divided into two main categories: ductile metal clads or coatings, and intermetallic compound coatings. Each area is briefly reviewed in the following paragraphs.

DUCTILE METAL CLADS AND COATINGS

Since it was expected that chromium alloys would be notch-sensitive, ductile oxidation-resistant clads and coatings were studied to minimize the effects of surface notches.

Nickel-Chromium-Type Clads

Here, rather elaborate systems were studied[7] that involved an oxidation-resistant clad and several intermediate layers of metal foils, all isostatically hot-pressure-bonded to each other and to the substrate. Two nickel-chromium clads were selected for study. A Ni-30 Cr alloy was selected because of its good oxidation resistance and because it remains ductile after long-time oxidation exposure. A Ni-20 Cr-20W alloy was also selected, even though it has only modest oxidation resistance, because of the possibility that it might provide a better barrier to nitrogen than Ni-30 Cr.

Because of anticipated interdiffusion between the nickel-chromium alloys and the substrate and the possibility of forming hard solid solutions, or brittle phases, refractory metal diffusion barriers were employed. Their selection was made on the basis of simple diffusion calculations. Their function was to limit interdiffusion between the nickel-based clads and the chromium substrate for at least 600 hr.

In some cases, compatibility layers of either platinum or vanadium were inserted between the clad and the diffusion barrier. The purpose of the platinum compatibility layer was to retard the diffusion of tungsten into the clads because tungsten lowers the solubility of chromium in nickel-chromium alloys. It was thought that if too much tungsten were present, brittle chromium phases might precipitate. Vanadium was introduced to serve as a sink for nitrogen.

In some cases the clads were pack-aluminized after system assembly to form a surface aluminide intermetallic. A homogenization anneal served to dissolve this layer and cause the entire clad to contain 2 to 4 percent aluminum. These aluminum-enriched clads were also ductile but had improved oxidation resistance as compared to the original clads.

The arrangement of the various layers in these clad systems is presented in Table 12.

TABLE 12 Description of Coatings on a Chromium Alloy

Substrate	Diffusion Barrier	Compatibility Layer	Oxidation-Resistant Ductile Clad	Surface Treatment
Cr-5W-0.1Y	Usually W. Also tried were W-2 Percent ThO_2, W-25Re, or Mo	Usually Pt when used. V also studied	Either Ni-30Cr, or Ni-20Cr-20W	Aluminize, homogenize
	—	—	—	—
	1–2 mils	1/2 mil	5 mils	1 mil

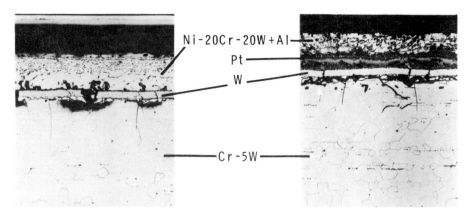

(A) SYSTEM I
(Cr-5W):W:Ni-20Cr-20W+Al
5 → 9 MILS BETWEEN PITS
0 → 4 MILS SUBSTRATE
CONTAMINATION

(B) SYSTEM II
(Cr-5W):W:Pt:Ni-20Cr-20W+Al
4 → 8 MILS BETWEEN PITS
2 → 6 MILS SUBSTRATE
CONTAMINATION

FIGURE 21   Microstructures of clad-substrate cross sections after 600 hr in air at 2100°F. 100×.

When the clad alloys were tested by themselves for 200 hr at 2100°F or 100 hr at 2300°F in air, the aluminized and homogenized Ni-30 Cr and Ni-20 Cr-20 W alloys and the Ni-30 Cr alloy without aluminum retained 3T bend ductility at room temperature after test. These materials gained less than 4 mg/cm² even at the higher temperature.

After 600 hr in air at 2100°F (20-hr cycles), some of the best clad-substrate systems showed similarly low weight gains. As shown in Figure 21, the substrates showed no nitriding. However, the thermal cycling caused cracks to develop in the tungsten diffusion barriers.

The extent of the pitting beneath these cracks and the substrate contamination due to interdiffusion are also shown in Figure 21. Both pitting and substrate contamination are more severe in System II, which contains platinum. At these discontinuities, electron microprobe analyses show that nickel and/or platinum had diffused several mils into the substrate. Where no cracks existed, only small amounts of nickel, chromium, and/or platinum had diffused into the tungsten, and little nickel or platinum had reached the substrate. Figure 21 also shows separation at the clad-barrier interface in System I and at the barrier-substrate interface in System II.

Figure 22 shows that both the unprotected substrate and the clad systems are embrittled after exposure in air at 2100°F. The as-received material had a DBTT of around 600°F, but air exposure raised it to 1200°F after only 25 hr. The clad substrates were more resistant to embrittlement than the substrate alone. Still, both System I and II showed significant embrittlement after 300 hr in air at 2100°F. System II, which showed more embrittlement, was the one that

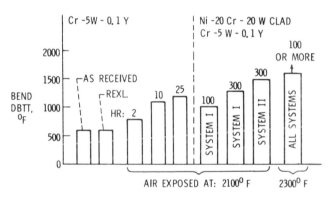

FIGURE 22   Bend DBTT of Cr-5W-0.1Y, bare and clad after various high-temperature air exposures.

contained platinum. In both cases, system embrittlement appears related to cracking of the diffusion barrier layer and the more cracking observed (System II), the greater the embrittlement. Similar DBTT levels were obtained after argon exposures, indicating that interdiffusion and not nitrogen must be the prime cause of embrittlement.

After 2300°F exposure for times of from 40 to 240 hr, gross substrate nitriding was observed in all clad systems. This is reflected in Figure 22 by the fact that all systems were brittle at 1600°F.

These nickel–chromium-type clads with tungsten diffusion barriers have good air oxidation and nitrogen resistance but do not confer protection from substrate embrittlement at 2100°F. They apparently are not satisfactory for potential turbine service. The localized failure of the diffusion barriers and the resultant diffusion-induced embrittlement

TABLE 13   Protection and Interaction Between Fe–25Cr–4Al–1Y Clads on Cr–5W–0.1Y[a]

Condition	Exposure Temp, °F	Exposure Time, Hr	ΔW, mg/cm²	Bend DBTT, °F	Diffusion Zone Thickness, mils	Hardness, KHN$_{200\,g}$ Clad	Hardness, KHN$_{200\,g}$ Diffusion Zone avg	Hardness, KHN$_{200\,g}$ Core
As-bonded	2150	2	–	625	0.8	265	405	207
Cyclic furnace air, 5 each 20-hr cycles	2100	100	+0.84	~1000	3.2	260	427	205
Cyclic furnace air, 8 each 20-hr cycles and 2 each 70-hr cycles	2100	300	+1.43	~1050	5.0	282	440 (500 near substrate)	210

[a]As-received DBTT = 425°F.

eventually lead to DBTT's that are not much lower than those of the unprotected substrate after high-temperature exposure.

Iron–Chromium–Aluminum-Type Clads

Preliminary studies at Lewis Research Center[8] showed that 10-mil-thick Fe–25 Cr–4Al–1Y alloy clads are protective against oxidation and, as judged from visual evidence, against nitrogen contamination for times up to 300 hr at 2100°F. These clads are less embrittling than the Ni–Cr clads.

Table 13 indicates that the isostatic hot pressure bonded specimens have a DBTT about 200°F higher than that of the as-received material used in the study. Some of this increase is, of course, due to recrystallization during the cladding process. This as-clad condition exhibited only a small interdiffusion zone with some hardening in that zone.

After 100 hr of cyclic exposure to air at 2100°F, the weight gain was only 0.84 mg/cm². However, the system bend DBTT increased to approximately 1000°F, and the diffusion zone became thicker and harder. Similarly, after 300 hr at 2100°F, nitridation was still minor: a weight gain of only 1.4 mg/cm² was measured. The diffusion zone became still thicker and harder. From the zone thickness data the growth rate appears parabolic. Similar indications of hardening in iron–chromium binary alloys have been reported previously.[9]

Figure 23 shows a photomicrograph of a cross section of this system after 300 hr in air at 2100°F. The clad still con-

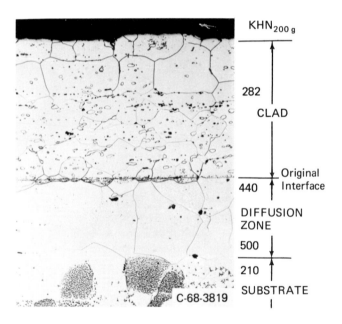

FIGURE 23   Iron-25 Cr-4 Al-1Y clad Cr-5W-0.1Y after 300 hr at 2100°F. 250×.

tains numerous yttrium-rich particles that were also observed in the as-received clad material. The diffusion zone appears single-phased, and this point was confirmed by electron microprobe analysis. Etch pits can be seen in the substrate.

Examination of the bend-tested specimens showed crack

development in the hardened diffusion zone on the tensile side of the bend. These cracks apparently were the source of substrate cracking at failure.

Here, too, interdiffusion of the protective system, rather than nitrogen, produced system embrittlement, even though the protection system minimized oxidation, and no visual evidence of nitrogen contamination was seen.

Metal Coatings

Platinum and palladium were considered as protection systems for chromium in a study at the Lewis Research Center.[10] These noble metals appeared attractive due to their inherent oxidation resistance and their extremely low solubility for nitrogen.

Foils of these metals withstood a 100-hr exposure at 2000°F in nitrogen with little or no attack, confirming their inertness to nitrogen. Thin platinum and palladium coatings were satisfactorily deposited by electroplating. Screening tests showed that platinum readily dissolves in the Cr-5 W-0.1 Y substrate at temperatures as low as 2000°F. Palladium appeared much less reactive; thus, only palladium coatings were studied completely.

On a weight gain basis, the air exposure resistance of 1 mil palladium electrodeposits, diffusion annealed to the substrate prior to test, was satisfactory at 2000°F. At 2200°F, after 100 hr, spalling of the heavy surface oxide caused a 3 mg/cm^2 weight loss. At 2400°F, serious nitriding produced a 9 mg/cm^2 weight gain even after the oxide scale spalled.

Based on microstructures and ductility retention, however, this coating was unsatisfactory even at 2000°F. A weight gain of 3.5 mg/cm^2 in 100 hr was measured. At 2000°F, chromium penetrated the palladium layer and Cr$_2$O$_3$ formed both at the surface and within the palladium. Also at 2000°F, the ductility retention was not satisfactory, since some nitrogen contamination occurred, as shown in Table 14. This table shows that diffusion annealing in argon increased the bend DBTT about 150°F. During this anneal, there was a slight increase in the nitrogen content to 47 ppm as a result of reaction with nitrogen impurity in the argon used. Air exposure at 2000°F for only 100 hr raised the DBTT to 1250°F and the nitrogen level to 86 ppm. Thus, the 1-mil palladium coating was not protective even at 2000°F. Since even a 1-mil palladium is expensive, it does not appear that the relative merits of a 7- to 10-mil-thick palladium coating (similar in thickness to the clads previously described) warrant study.

Potential Clads or Metal Coatings

Generally, the ductile clad or coating systems studied offered some oxygen and nitrogen protection to the chromium alloy but embrittled the substrate through interdiffusion. The least embrittling of these systems was the one that con-

TABLE 14 DBTT and Nitrogen Content of 1-Mil Palladium-Coated Cr–5W–0.1Y

Condition	Bend DBTT, °F	Nitrogen Content,[a] ppm
As received	450	27
As plated	450	27
Plated and argon-annealed, 20 hr at 2000°F	600	47
Plated, argon-annealed, and exposed for 100 hr in air at 2000°F (20-hr cycles)	1250	86
Uncoated and exposed for 100 hr in air at 2000°F (20-hr cycles)	>1300[b]	205

[a]After coating or scale mechanically removed.
[b]Limit of DBTT apparatus.

tained the most chromium and had a body-centered cubic crystal structure similar to the substrate, i.e., the Fe–Cr–Al–Y clad. Extended further, cladding with very high chromium alloys may further minimize interdiffusion. There are some relatively weak chromium alloys that retain ductility after exposure to high-temperature air and also do not oxidize too severely. These were also studied in the contracted chromium alloy development program.[5]

Air oxidation resistance and nitrogen contamination resistance data for these alloys are presented in Table 15. The table shows that at least for 100 hr at 2100°F the Cr–Y alloys appear to offer some potential, with higher yttrium contents (still below the 0.5 percent solubility limit) appearing more resistant to both oxidation and nitrogen contamination. The Cr–Y–Hf–Th alloy appears to be very resistant to both oxidation and nitrogen contamination. This type of material warrants further study as a cladding for some of the advanced chromium structural alloys.

INTERMETALLIC COMPOUND COATINGS

Aluminide Coatings

The protective ability of simple and modified aluminide surface conversion coatings on chromium has also been studied.[11] Modifiers included iron and titanium. Iron was studied in an effort to improve coating toughness and minimize the inward diffusion of aluminum. Titanium was incorporated in an attempt to improve the ductility of chromium alloys by scavenging nitrogen either from the substrate or as it diffused inward to the substrate from the air environment.

TABLE 15   Potential Clads for Chromium Alloys

Alloy	Air Exposure		Total ΔW mg/cm²	Nitride Layer, mils	Depth-Hardened, mils	Hardiness, DPH₁₀₀g	
	Temp, °F	Time, hr				Surf.	Core
Cr–0.13 Y	1500	100	0.7	0	0	164	164
	2100	100	2.3	1–4	1–4	>1,000	181
	2400	25	10.6	2.5	4	1,500	156
Cr–0.25 Y	1500	100	0.3	0	0	172	159
	2100	100	1.8	0	0	201	186
	2400	25	25.1	1–2	2	1,500	150
Cr–0.13 Y–0.05	1500	100	0.04	0	0	168	156
Hf–0.3 Th	2100	100	1.9	0	0	201	190
	2400	25	1.9	0	0–1	259	170

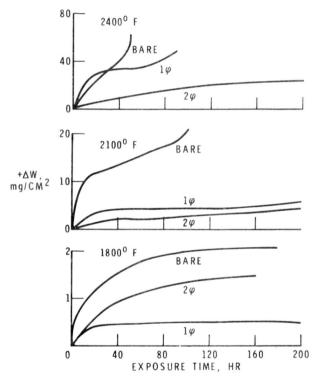

FIGURE 24   Weight change of uncoated and aluminide coated Cr-5W-0.1Y at 1800°, 2100°, and 2400°F for times to 200 hours. 1φ = single-phase solid solution, 2φ = surface aluminide plus solid solution.

TABLE 16   Bend DBTT and Nitrogen Content of Uncoated and Aluminide-Coated Cr-5W-0.1 Y

Material and Condition	Average Bend DBTT (105° <) °F	N₂, ppm
Cr–5W–0.1 Y		
as received	750	60
2100°F/100 hr/air	900	–
2400°F/100 hr/air	Specimen disintegration	–
1-Phase coating		
as coated	900	–
2100°/100 hr/air	>1600	–
2400°/100 hr/air	>1600	–
2-Phase coating		
as coated	825	–
coated + stripped	800–1000	63
coating thermal ~	800–1000	80
2100°/100 hr/air	>1600	–
2100°/100 hr/argon	>1600	83
2400°/100 hr/air	>1600	–
2400°/100 hr/argon	>1600	130
Fe–Al coating		
as coated	800–1000	–
coated + stripped	800–1000	93
coated thermal ~	1000–1200	120
2100°/100 hr/air	–	–
2100°/100 hr/argon	1600	74
2400°/100 hr/air	–	–
2400°/100 hr/argon	>1600	73

All systems were applied by pack-cementation techniques. In some cases a two-step process was employed to first deposit a modifier element and then aluminize the pre-alloyed surface.

Cyclic furnace air exposure testing at 1800°, 2100°, and 2400°F produced the weight change information plotted in Figure 24. These plots include uncoated Cr–5W–0.1Y data as well as data for a single-phase aluminum solid solution coating, designated by 1φ, and a Cr₅Al₈ coating, designated by 2φ. (The latter data are quite similar to those for an Fe-Al system deposited in two steps.) At 1800° and 2100°F, both the 1φ and 2φ coatings were protective. Even at 2400°F, the air exposure weight gains of at least the 2φ

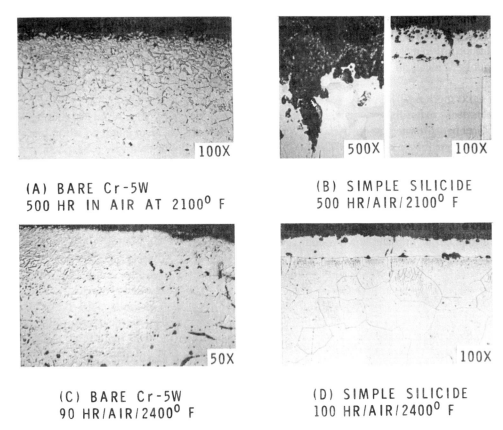

(A) BARE Cr-5W
500 HR IN AIR AT 2100° F

(B) SIMPLE SILICIDE
500 HR/AIR/2100° F

(C) BARE Cr-5W
90 HR/AIR/2400° F

(D) SIMPLE SILICIDE
100 HR/AIR/2400° F

FIGURE 25   The effects of air oxidation on uncoated and simple silicide-coated Cr-5W-0.1Y.

coating were not prohibitive in 200 hr of furnace test. High-velocity gas stream oxidation tests at 2100°F, however, indicated that these coatings degraded much more rapidly than in furnace testing.

As with so many other systems, the bend ductility was not satisfactory (even though the Al_2O_3 oxide scales formed on the coatings prevented serious nitrogen contamination), as shown in Table 16. The uncoated material is compared with 1φ, 2φ, and Fe-Al systems. Generally, exposure in either argon or air caused serious embrittlement, as evidenced by the DBTT values at or above 1600°F. This embrittlement was primarily related to the rapid inward diffusion of aluminum (16 mils in 10 hr at 2400°F), as established by electron microprobe analyses.

Some earlier work[12] involved the deposition of nickel prior to aluminizing. The system was somewhat protective in the 1600° to 1900°F range, but problems were encountered with the interdiffusion of nickel and chromium and the resultant subsurface hardening.

Silicide Coatings

One simple silicide coating and a variety of complex ones have been evaluated on chromium alloys. A contractural program[13] explored the effects of simple siliciding and siliciding of sequentially deposited layers of Mo, V, and Ti.

These elements were selected for a number of reasons. Molybdenum improves the silicide expansion match with chromium alloys and forms an oxidation-resistant silicide. Vanadium and titanium preferentially react with nitrogen to prevent it from contaminating the substrate. Also, these elements improve the 1400°–1800°F oxidation resistance of silicide coatings. This is a range where simple silicide coatings on Mo, Ta, W, and Nb alloys tend to fail catastrophically because of silicide pest.

Molybdenum was deposited by vapor deposition from the chloride. The titanium and vanadium were either pack-cementation-deposited or applied by fused-salt electrolysis. Siliciding was generally accomplished by pack cementation.

The simple silicide coatings on chromium showed no nitrogen contamination and showed no pest problem at 1500°F. They offered protection from visible nitrogen contamination for 500 hr at 2100°F, but the nitrogen content rose from 22 ppm as coated to 67 ppm during test. At 2400°F, protection was unsatisfactory after only 100 hr, as shown in Figure 25. The bare material, however, is severely attacked by nitrogen at both time–temperature conditions. In Figure 26, weight change and selected nitrogen content data are presented as a function of time for tests of both simple and complex silicides after 1500°, 2100°, and 2400°F exposure. This figure shows that the simple silicide coatings

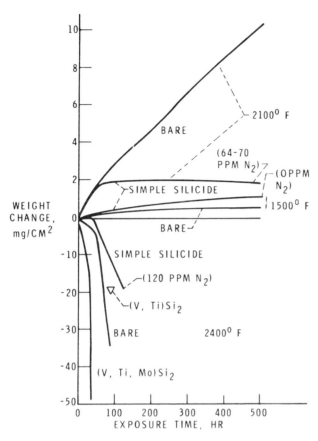

FIGURE 26 Weight change versus time for uncoated and silicide-coated Cr-5W-0.1Y.

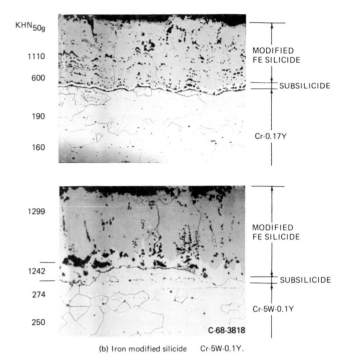

(b) Iron modified silicide   Cr-5W-0.1Y.

FIGURE 27 Microstructures of iron-modified silicides on two chromium alloys after 100 hr at 2100°F. 250×.

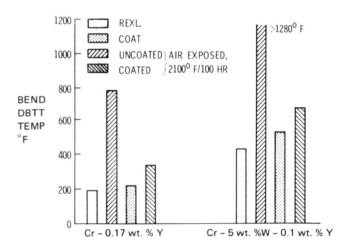

FIGURE 28 Effect of iron-modified silicide on DBTT of two chromium alloys.

were protective at both 1500° and 2100°F. Even at 2400°F the simple silicide was less severely attacked than any other system.

Although the concept of modifying silicides to impart improved properties is a good one, this layered-deposition approach of incorporating the desired modifiers was unsatisfactory. In those systems containing molybdenum or vanadium, any crack in the outer coating produced catastrophic oxidation of the coating and substrate.

The above program did not include DBTT tests, but subsequent tests on simply silicided Cr-5W-0.1Y have shown the as-silicided condition to have a DBTT somewhat above 1300°F—an unacceptably high value.

Subsequent in-house work on silicide coatings for chromium alloys has shown more promise.[14] This work has involved the initial slurry deposition of an element or mixture of elements, diffusion annealing to partially sinter and bond the material to the substrate, and subsequent siliciding—a technique originally developed for the protection of tantalum alloys.[15] A number of different elements and element mixtures have been applied and silicided. One of the better systems has involved an iron slurry plus siliciding. Cyclic fur-

nace oxidation tests at 2100°F for 100 hr have shown this system to gain only 2 mg/cm². Microstructures of this coating after the above test can be seen in Figure 27. These photomicrographs show a hard, somewhat porous outer coating, a softer subsilicide layer, and the substrates: Cr-0.17Y and Cr-5W-0.1Y. These coatings are much less detrimental to the ductility of chromium alloys than the simple silicide coating described above. Figure 28 shows the bend DBTT of both alloys in the recrystallized, air-exposed, coated, and coated and exposed conditions. In both cases the DBTT of the coated material is no more than 100°F above the as-recrystallized DBTT. Air exposure raises this no more than another 180°F. Furthermore, the figure shows that chromium with only minor yttrium additions is less affected by the coating than the material with tungsten and yttrium. For this reason, it may be possible to combine the most promising clad with this type of silicide coating to produce a fail–safe protection system for chromium.

SUMMARY AND CONCLUSIONS

Many exploratory approaches have been taken in an effort to protect chromium alloys from both nitrogen embrittlement and high-temperature oxidation. Metal clads, metal coatings, and silicide and aluminide coatings have been deposited and evaluated. In most cases the systems studied offered partial protection but at least partially embrittled the chromium alloy through interdiffusion of one or more constituents. Thus, there are currently no systems available that appear ready for the jet engine environment.

Modified silicide coatings have provided the best combination of environmental protection (during 100-hr cyclic furnace oxidation tests at 2100°F) and subsequent ductility retention. Further work in this area appears worthwhile. Also of interest are chromium alloys containing very minor percentages of reactive elements such as Y, Th, and Hf. These, by themselves, appear to retain ductility after limited cyclic furnace oxidation exposure at temperatures up to 2400°F. Perhaps a combination of modified silicide coatings and clads of such materials would prove worthwhile.

To date, almost all evaluation of coatings for Cr alloys has involved only furnace tests in static or low velocity air. As promising protection systems are developed, they will be evaluated in facilities more closely simulating the high-velocity gas flow to which aircraft engine components will be exposed.

REFERENCES

1. Maykuth, D. J., and A. Gilbert. 1966. Chromium and chromium alloys. Battelle Memorial Institute. DMIC Report 234.
2. Ryan, N. E. 1964. An appraisal of possible scavenger elements for chromium and chromium alloys, J. Less-Common Met. 6:21.
3. Elliot, R. P. 1965. Constitution of binary alloys. first supplement. McGraw-Hill Book Co., Inc., New York.
4. Wain, H. L., and S. T. M. Johnstone. 1964. Chromium alloy development in Australia. Presented at the 93rd AIME Annual Meeting, New York.
5. Clark, J. W. 1967. Development of high-temperature chromium alloys. General Electric Co. NASA CR-92691.
6. Goetz, L. J., J. R. Hughes, and W. F. Moore. 1967. The pilot production and evaluation of chromium alloy sheet and plate. General Electric Co. NASA CR-72184.
7. Williams, D. N., et al. 1968. Development of protective coatings for chromium-base alloys. Battelle Memorial Institute. NASA CR-54619.
8. Gedwill, M. 1968. Unpublished research on Fe-25Cr-4Al-1Y alloy clads on Cr-5W-0.1Y. NASA Lewis Research Center.
9. Hansen, M., and K. Anderko. 1958. Constitution of binary alloys. McGraw-Hill Book Co., Inc., New York.
10. Grisaffe, S. J. 1967. Exploratory investigation of noble metal coatings for chromium. Presented at the 13th Meeting of the Refractory Composites Working Group.
11. Negrin, M., and A. Block. 1967. Protective coatings for chromium alloys. Chromalloy American Corp. NASA CR-54538.
12. Clark, J. W., and W. H. Chang. 1966. New chromium alloys. Presented at AIME High Temperature Materials Symposium, New York.
13. Brentnall, W. D., H. E. Shoemaker, and A. R. Stetson. 1966. Protective coatings for chromium alloys. Solar Division of International Harvester. Report RDR-1398-2. NASA CR-54535.
14. Stephens, J. R., and W. D. Klopp. April 1969. Exploratory study of silicide, aluminide, and boride coatings for nitridation-oxidation protection of chromium alloys. NASA-TND-5157.
15. Wimber, R. T., and A. R. Stetson. 1967. Development of coatings for tantalum alloy nozzle vanes. Solar Division of International Harvester. Report RDR-1396-3. NASA CR-54529.

COLUMBIUM*

INTRODUCTION

Columbium-base alloys are prime candidates as structural materials for advanced aerospace vehicles and flight-propulsion systems. With a melting point of 4379°F, a density nearly equivalent to that of steel, and strength potential to 3000°F, columbium provides the basis for a fabricable class of alloys suitable for a multitude of high-temperature applications. However, columbium and its alloys suffer severe degradation in elevated-temperature, oxidizing environments; and columbium oxide (Cb_2O_5) melts as low as 2690°F. Protective coatings are therefore required to permit the exploitation of the high-temperature structural properties of columbium alloys. Depending upon the application, these coatings may be required to provide resistance to: oxidation, thermal fatigue, hot-gas erosion, particle abrasion, impact damage, strain-induced cracking, and other phenomena in both atmospheric and reduced pressure environments.

A review is presented of the coating systems and coating processes developed for the high-temperature protection of

*Prepared by J. D. Gadd.

columbium-base materials. The bulk of the effort on coatings for columbium has been conducted over the past 10 years, and the work has been predominantly government-sponsored. Materials requirements for re-entry vehicle applications have motivated the majority of these programs, although a number of the more recent activities have been directed to the aircraft gas turbine.

COATING FORMATION METHODS

Many techniques have been utilized to form protective coatings on columbium alloys. Each method is described below, and reference is made to these process definitions in the subsequent discussion of specific coating systems. The processes are:

Pack cementation—immersion of the material to be coated in a pack consisting of inert filler, metal coating elements, and halide activator. Heat treatment is conducted in hydrogen or an inert atmosphere. Coating is formed by vapor transport and diffusion.

Vacuum pack—immersion of material to be coated in an all-metal pack of coating elements plus halide activator. Heat treatment is performed in dynamic vacuum or under a partial atmosphere of inert gas. Coating is formed by vapor transport and diffusion.

Slip pack—similar to the two processes above—substitution of thin disposable bisque for massive pack. Bisque is sprayed, dipped, or otherwise applied to the substrate surface.

High-pressure pack cementation—similar to the first two methods, with utilization of inert gas pressure in excess of 1 atm.

Fused slurry—uniform application of a metal (alloy) slurry to the substrate surface, followed by fusion of the slurry in an inert gas or vacuum environment. Coating forms by liquid–solid diffusion.

Slurry–sinter—spray or dip application of metal particles (plus binder) to substrate surface, followed by solid- or liquid-phase sintering in vacuum or inert environment.

Fused salt—electrolytic or nonelectrolytic deposition of metal ions from a fused-metal-salt solution (fluorides, chlorides, bromides, etc.) on the substrate material. Pure or alloy coatings are formed.

Electrophoresis—deposition of charged metal particles from a liquid suspension onto a substrate surface of opposite electrical potential. Synthesis of particulate deposit generally involves isostatic compaction followed by vacuum or inert-atmosphere sintering.

Electroplating–aqueous—electroplating of metal ions from aqueous solution. Process can involve suspension of particulate materials in electrolyte and occlusion of these particles in metal deposit.

Fugitive vehicle—involves spray or dip application of metal slurry onto the substrate surface, followed by vacuum heat treatment. Slurry contains coating elements and a fugitive vehicle that is molten at the firing temperature. Fugitive vehicle has high solubility for coating elements and very low solubility for substrate. Coating forms by transfer of elements from the liquid solution to the substrate surface, with subsequent diffusion growth of the coating phase. Fugitive vehicle is eventually removed by evaporation.

Fluidized bed—immersion of the material to be coated in a heated fluidized bed of coating elements, using as the fluidizing gases mixtures of reactive halogen gases and hydrogen or argon. Coating forms by disproportionation of gaseous metal halides on substrate surface, followed by diffusion alloying.

Chemical vapor deposition (CVD)—entirely gaseous process. Metal halide gases (plus argon or hydrogen) pass over surface of metal to be coated. Coating forms by hydrogen reduction or thermal decomposition of metal halide at substrate surface, followed by diffusion-controlled coating growth.

Vacuum vapor deposition—evaporation of a metal (or alloy) from a filament, liquid bath, or other source, followed by condensation of the metal vapor onto a cold or hot substrate surface. Postdiffusion treatment is optional.

Hot-dipping—immersion of material to be coated in a hot liquid bath of the molten coating elements. Coating forms by liquid–solid diffusion.

Hydride and oxide reduction—spray or dip application of metal hydrides or oxides on substrate surface, followed by vacuum or hydrogen reduction, respectively.

Plasma arc—spray deposition of particulate metal or oxide particles using conventional plasma-arc facility.

Detonation gun—Linde patented gun-detonation process.

Gun metallizing—utilization of conventional wire or powder metallizing equipment.

Cladding—bonding of metal cladding (thin sheet) to substrate surface by diffusion bonding, forging, rolling extrusion, etc.

HISTORY OF DEVELOPMENTS

Development of protective coatings for columbium alloys has been pursued energetically for about 10 years. Information emanating from government-funded activities, with the exception of the AEC, is readily available in the literature, and several summary documents have been prepared that provide details of coating processes and related coating properties.[1-4] This review is a more up-to-date survey of the technology than is available in the referenced publications; however, the treatment is very general owing to the large quantity of technical data published on coatings for columbium.

Six categories of coating systems are defined: (a) heat-resistant metals and alloys of the iron, nickel, and cobalt class; (b) oxides and ceramic composites; (c) intermetallics other than aluminides and silicides, e.g., zinc base, beryllides, borides; (d) noble metals; (e) aluminides; and (f) silicides. Silicides have heavily dominated the development activity, and nearly all systems that have survived application-oriented evaluations have been of the silicide type. Table 17 is a concise summary of the coating development efforts; the various coating systems are discussed in more detail below.

HEAT-RESISTANT METALS AND ALLOYS

Hirakis's work at Horizons in 1959 represents the first significant program aimed at developing coatings for columbium.[5] A number of coating formation techniques were explored by Hirakis: electroplating, fused-salt deposition, hot-dipping, ceramic and metal spraying, pack deposition, electrophoresis, and gun metallizing. Electroplating was used to deposit Ni, Cr, and Fe coatings containing Si, Al, Cr, TiC, FeB, NiB, SiO_2, Al_2O_3, and ThO_2 occlusions. Gun metallizing (spraying) was employed to develop a number of systems including the components Ni, Fe, Cr, Al, B, Si, $CrSi_2$, $TiSi_2$, $TiCr_2$, and Mo. Electrophoresis was also explored as a means of depositing particulate materials, but with little success. The majority of these systems containing Fe, Ni, and Cr showed little potential above $1800°-2000°F$, owing to rapid oxidation, porosity, poor adherence, and a tendency to form low-melting phases and intermetallics with the columbium base. These coatings protected columbium from oxidation for only 4-6 hr at $2500°F$.

Wlodek employed gun metallizing, plasma-arc spraying, and the gun-detonation process to apply a variety of Fe, Ni, Cr, Si, and Al base materials as protective coatings.[6] Chromium, 302SS, and Nichrome V were most effective at temperatures below $1750°F$, while plasma-sprayed LM-5 (40Mo-40Si-8Cr-10Al-2B) was the most promising system at temperatures above $1750°F$. Chromium coatings tended to leak oxygen at temperatures above $1750°F$. The LM-5 coating, utilized in combination with a flame-sprayed Cb-Ti-Cr-Al-Ni intermediate layer, reportedly afforded up to 1,000 hr protection at $2100°F$ and 100 hr at $2700°F$. Major difficulties associated with these systems included difficulty of application to practical shapes, nonuniformity, relatively thick layers required, porosity, and the formation of deleterious phases with the columbium substrate. In spite of these deficiencies, the properties of LM-5 make it the most promising coating system emerging from the very early coating development efforts.

Beach and Faust explored the use of Cr, Cu, Fe, Au, Ni, and Pt plates from aqueous solutions as protective coatings.[7] Exfoliation of the metallic plates produced early failure of all coatings, while difficulties with chromium plating and a low-melting Au-Cb eutectic were additional problems.

No other significant research was found in the literature relative to the use of heat-resistant metals and alloys as protective coatings for columbium.

OXIDES AND CERAMICS

A relatively minor effort has been devoted to the utilization of oxide- and/or ceramic-type protective coatings for columbium. Spretnak and Speiser initiated an investigation into the use of niobates as protective oxide films on columbium.[8,9] This approach was shortly abandoned owing to the very low melting points (below $2000°F$) of all niobates for which data were available.

General Electric developed a glass-impregnated Al_2O_3 coating (System 400) produced by flame-spraying alumina powder, followed by sealing with a baria-alumina-silicate glass slurry.[1,10] System 400 demonstrated average protective lives of 500 hr at $2300°F$ and 100 hr at $2500°F$ on columbium alloys; however, the thickness requirement of 12-16 mils and the application limitations imposed by the flame spray operation rendered the coating relatively impractical. Poor resistance to impact and strain and a relatively high application temperature ($2700°F$) were additional disadvantages of System 400.

North American Aviation developed an Al_2O_3-aluminide composite coating (NAA-85) for the protection of columbium alloy aerospace components.[2,11] A slurry mixture of alumina and aluminum powders was cold-spray-applied and subsequently fired at $1900°F$ to produce a $CbAl_3$-bonded Al_2O_3 coating. Only short-term protection at temperatures above $2000°F$ was afforded by the NAA-85 coating system—less than 5 hr at $2600°F$.

Oxide- or ceramic-type coatings have been categorically unsuccessful for the protection of columbium alloys. Porosity, poor resistance to strain, permeability to oxygen, and application difficulties have rendered the systems poor candidates for the protection of columbium alloy aerospace and aircraft gas-turbine components.

INTERMETALLICS

Aluminides, silicides, and zincides comprise the intermetallic coatings investigated for the protection of columbium alloys. The aluminum- and silicon-base compounds are discussed later.

Sandoz, with Brown *et al*, has performed the only significant work on the utilization of zinc-base coatings for the protection of columbium alloys.[12-14] Vapor-plating, hot-dipping, electroplating, and cladding were employed to form zinc-base intermetallic coatings. An exceptional self-healing capability is exhibited by the zinc intermetallic, owing to zinc vapors migrating to a fissure in the protective zinc oxide and effecting repair. Zinc-base coatings provide reliable protection to columbium for hundreds of hours at temperature

TABLE 17 Coating Systems for Columbium

Basic Type	Designation	System Concept	Composition	Process	Developer	Literature Reference
Silicide	Cr–Ti–Si	Complex silicide multilayered	Cr, Ti, Si	a. Vacuum pack and vacuum slip-pack	TRW	28, 29, 31, 32
				b. Fused slurry and pack	Sylcor	46, 47
				c. Fluidized bed	Boeing	37, 39
					Pfaudler	33, 34
				d. Electrophoresis	Vitro	35
				e. Chemical vapor deposition	Texas Instruments	36
				f. Electrolytic fused salt	Solar	40, 41, 42
					Pfaudler	33, 34
Silicide	Disil[a] Disil 3	Modified silicide	Si+V, Cr, Ti Si	Fluidized bed Pack cementation (iodine)	Boeing	37, 39
Silicide	Vought I	Modified silicide	Si–B–Cr	Pack cementation–multicycle	Vought	25, 26
Silicide	Vought II	Modified silicide	Si–Cr–Al	Pack cementation–multicycle		
Silicide	Durak[a] KA	Modified silicide	Si+additives	Pack cementation–single cycle	Chromizing	1, 28
Silicide	N-2	Modified silicide	Si+Cr, Al, B	Pack cementation–single cycle	Chromalloy	1, 28
Silicide	S-2	Silicide	Si	Chemical vapor deposition	Fansteel	1, 27
Silicide	M-2	Molybdenum disilicide	Si+Mo	MoO_3 reduction and chemical vapor deposition		
Silicide	PFR[a]	Modified silicide	Si+additives	Pack cementation, fluidized bed, fused salt, slurry dip	Pfaudler	33, 34
Silicide	AMFKOTE[a]	Modified silicide	Si+additives	Pack cementation–single cycle	American Machine & Foundry	1, 24
Silicide	—	Liquid phase–solid matrix	Si, Sn, Al	Porous silicide applied by pack or CVD-impregnated with Sn–Al	Battelle Memorial Institute, Geneva	49
Silicide	LM-5	Multilayered complex silicide	40Mo–40Si, 10CrB–10Al	Plasma spray-diffuse	Linde–Union Carbide	1, 6
Silicide	—	Modified silicide	Si+additives	Pack cementation	Pratt & Whitney Aircraft–Canel	24
Silicide	R[a]	Complex silicide	Si–20Fe–20Cr Si–20Cr–5Ti Si+(Cr, Ti, V, Al, Mo, W, B, Fe, Mn)	Fused silicides Fusion of eutectic mixtures	Sylvania	46, 47
Silicide	V–Cr–Ti–Si	Complex silicide	V–Cr–Ti–Si	Vacuum and high-pressure pack	Solar	40, 42

Coating	Trade name	Type	Composition	Process	Developer	References
Silicide	—	Complex silicide	Mo–Cr–Ti–Si, V–Cr–Ti–Si, V–Al–Cr–Ti–Si, Mo–Cr–Si	Multicycle vacuum pack	TRW	43
Silicide	—	Glass-sealed silicide	Si+glass	Silicide by pack cementation or CVD+glass slip over-coat	NGTE	48
Silicide	TNV 12	Multilayered	Mo, Ti+Si and glass	1. Slurry sinter application of Mo+Ti powder 2. pack silicide 3. glass slurry seal	Solar	44, 45
Aluminide	AMFKOTE[a]	Modified aluminide	Al+B	Pack cementation	American Machine & Foundry	24
Aluminide	LB-2	Modified aluminide	$88Al-10Cr-2Si$	Fused slurry	General Electric, McDonnell Aircraft	10, 19, 20, 21, 22
Aluminide	—	Modified aluminide	Al–Si–Cr, Al–Si–Cr, Ag–Si–Al	Fused slurry, Hot dip	Sylvania	2, 23, 40
Aluminide	—	Multilayered systems	Fe, Cr, Al, Ni, Mo Si, VSi_2, $TiCr_2$, $CrSi_2$, B + Al	Powder metallize + Al hot dip	Horizons	5
Aluminide	—	Simple aluminide	Al	Pack cementation	Chromalloy	1
Aluminide	—	Multilayered systems	Cr, FeB, NiB, Si, Al_2O_3, SiO_2, ThO_2+Al	Electroplate dispersions in Ni+Al hot dip	Horizon	5
Aluminide	—	Modified aluminide	Al+(Si, Ag, Cr)	Silver plate + Al, Cr, Si hot dip	General Electric	18, 19
Aluminide	—	Multilayered systems	Al_2O_3+Ti+Al	(Al_2O_3 + TiH)–Spray–sinter + Al hot dip	General Electric	18, 19
Aluminide	Slurry ceramic	Oxide–Metal composite	Al_2O_3+Al	Slurry fusion of Al_2O_3–Al mixture	North American Rockwell	2, 11
Aluminide	Lunite 2	Aluminide	Al+additives	Fused slurry	Vac Hyd	23
Aluminide	—	Modified aluminide	Al+Sn	Hot dip	a. Pfaudler b. Sylcor	33, 34 45, 46
Zinc	—	Self-healing intermetallic	Zn and Zn+Al Ti, Co, Cu Cr, FeMg, Zr, Cu, Si	Vacuum distillation and hot dip	Naval Research Lab.	12, 13, 14
Oxide	System 400	Glass-sealed oxide	Al_2O_3+glass (baria, alumina, silica)	Flame spray Al_2O_3 + glass slurry	General Electric	10
Nichrome	—	Oxidation-resistant alloy	Ni–Cr	Flame spray, detonation gun, plasma arc	Linde–Union Carbide	6

TABLE 17 (Continued)

Basic Type	Designation	System Concept	Composition	Process	Developer	Literature Reference
Chromium carbide	—	Carbide	Cr-C	Plasma-sprayed	Linde—Union Carbide	6
Noble metal	—	Clad	Pt, Rh	Roll bonding and hermetic sealing	Metals and Controls	15, 16
Noble metal	—	Barrier-layer–clad	Pt, Rh+Re, Be, Al_2O_3, W, ZrO_2 MgO, SiC, Hf	Noble metal clad over barrier layer-diffusion couple study	Metals and Controls	15, 16
Noble metal	—	Pure metal	Ir	Fused-salt deposition	Union Carbide	17

[a]Denotes several trade designations.

to 1800°F; however, decomposition of the Cb–Zn inter-metallics establishes a service temperature limit of this system of 2000°F. No practical use of the zinc-base coating for columbium has been reported.

NOBLE METALS

The remarkable corrosion resistance of the noble metals make them excellent candidates for protecting columbium at elevated temperatures. Work is reported for the investigation of gold, silver, platinum, iridium, and rhodium as protective surface layers on columbium. The melting points of gold and silver are too low for their use as primary coatings above 1800°–2000°F; however, silver has been used in conjunction with aluminum and silicon.

Under Navy sponsorship, Texas Instruments conducted two programs studying the use of ductile noble metal coatings for the protection of columbium.[15,16] The columbium alloy material was encased in platinum and platinum–rhodium alloy sheet by a diffusion-bonding technique. Platinum or platinum alloy claddings 2–3 mils thick failed in less than twenty hours at 2550°F as the result of interdiffusion between the platinum cladding and the columbium substrate. Formation of platinum–columbium intermetallics produced paths for accelerated oxidation of the refractory metal substrate. Major emphasis was placed on exploring W, Re, Hf, Ir/W, SiC, ZrO_2, Al_2O_3, MgO, Au, Al, and BN as diffusion barrier layers; however, none was effective. Interdiffusion and difficulties with the application of the platinum claddings to useful shapes render these coatings relatively impractical for protecting columbium alloy hardware.

Iridium coatings for the protection of columbium have been studied by Union Carbide.[17] Fused-salt electrolytic deposition was employed to form iridium coatings on columbium. A nickel strike was required prior to iridium deposition because of reactivity of columbium with the fused-salt bath. Although life expectancies of several hundred hours at temperatures up to 3200°F were calculated by Union Carbide for 5-mil iridium coatings, factors such as porosity, defects, and oxygen permeability of the iridium layers rendered the coatings substantially less protective. Application of reliable iridium coatings to complex shapes is not currently feasible with this processing method.

ALUMINIDES

Aluminides rank second to silicides in their utility for the protection of columbium-base alloys. Although the aluminides exhibit shorter protective lives and a lower temperature ceiling than the silicides, their ease of application has made them useful for the protection of heat shield components on re-entry vehicles designed for relatively mild flight regimes.

Foldes[18] and Carlson[19] of General Electric conducted some of the earliest work on the application of aluminide coatings by hot-dipping.[18,19] Coatings were formed on columbium by dipping it in molten baths of Al–Si–Cr alloys, followed by diffusion treating to convert all free aluminum to Cb–Al intermetallics. An Al–11Si–2Cr bath proved optimum. Precoats of electroplated silver and sprayed and vacuum-heat-treated Al_2O_3 and TiH provided marginal improvement in the performance of the aluminides. Self-healing was affected by the silver addition. All of the aluminide coatings exhibited reliable short-term protection (2 hr) for columbium at 2500°F.

Luft continued the General Electric aluminide coating development effort in response to the coating needs of the McDonnell–General Electric Refractory Metals Structural Development Program.[10] Hot-dipping was pursued initially; however, the problems associated with dipping large complex sheet metal configurations into molten aluminum baths prompted General Electric to develop a slurry application technique. This latter development was designated the LB-2 process. The optimized LB-2 slurry composition 10Cr–2Si–Al was spray-applied and fired in vacuum at 1900°F to produce the aluminide surface compounds. Protective lives up to 24 hr at temperatures to 2200°F, 2 hr at 2500°F and ½ hr at 2750°F were exhibited by 2–3-mil-thick LB-2 coatings. This coating process proved adequate to meet the needs of the columbium structure development program. Air Force sponsorship of other columbium alloy structures programs continued at McDonnell Aircraft with the ASSET and BGRV vehicles. In support of these programs, McDonnell continued to utilize and improve the LB-2 aluminizing process for coating the very large structural segments of these vehicles.[20-22] Improved slurries, spraying techniques, and firing sequences and the utilization of vacuum-dipping for applying the slurry to large corrugated body sections produced a reliable manufacturing process for aluminizing columbium alloy components of all required configurations. The LB-2 process represents a practical manufacturing-scale coating method for protecting columbium alloys in an oxidizing environment for several hours at temperatures to 2500°F, and for shorter times at temperatures to 2800°F.

Vac Hyd Corporation now offers a commercial aluminide coating applied by a process similar to LB-2.[23] The coating is designated Lunite 2.

Sylvania Electric Products, Inc. conducted research on silver-modified aluminide coatings (Ag–Si–Al) for the protection of columbium alloys.[2,23,24] Fused-slurry and hot-dipping techniques were employed to form 6–8-mil-thick coatings that provided in excess of 100 hr protection for columbium in static air at 2500°F. The presence of a liquid phase in the aluminide and the excessive coating thickness make these coatings unattractive for aerospace and gas-turbine airfoil applications. Sylvania also investigated the use of the Sn–Al coating system developed for tantalum as a protective system for columbium.[2] Protective lives of the Sn–Al coating at temperatures to 2500°F were typical of slurry-applied aluminide coatings on columbium.

Boeing utilized the hot-dipping method to form aluminide coatings on columbium alloys.[1] Coatings formed in an Al–Si bath and diffused at 2200°F protected columbium for about 1 hr at 2600°F.

Chromalloy Corporation employed pack cementation to form aluminide coatings on columbium.[1] Protection was afforded for up to 25 hr at 2500°F; however, the aluminide coatings were reportedly permeable to oxygen.

Pratt and Whitney Aircraft Company CANEL also developed hot-dipping techniques for aluminizing columbium alloys.[1] Performance data for these coatings were not reported.

SILICIDES

Silicides have proved to be the most thermally stable diffusion alloy coatings available for the protection of columbium alloys. Virtually all of the coating application methods defined earlier have been employed to form silicide coatings on columbium. A chronological discussion of the various coating development activities directed to silicides is presented below.

Aves of Vought Astronautics (Ling-Temco-Vought) conducted two of the earliest programs concerning the pack cementation application of silicide coatings to columbium.[25,26] In the first program, two-step inert gas pack cementation processes were employed to deposit Si–Cr–Al and Si–Cr–B coatings on columbium. Silicon was deposited in a first cycle, followed by the codeposition of Cr–Al or Cr–B, to produce 2-4-mil-thick disilicide coatings. Average oxidation protective lives of 9 and 11 hr were realized at 2600°F with the Si–Cr–Al and Si–Cr–B coatings, respectively. The Si–Cr–Al system was found sensitive to the low-temperature silicide pest phenomena, while the Si–Cr–B coating was reportedly resistant to pest. In a second program, efforts were conducted by Vought to extend the utility of these pack-cementation silicides for coating columbium hardware. Alternative processing techniques such as slip pack, halogen gas streaming through the pack, segregation of the pack activator from the hardware, and the use of exothermic reaction heat sources for local coating application were investigated. Slip-pack processing was reportedly the most versatile technique developed for siliciding columbium hardware, owing to improved heat transfer, elimination of the bulky coating pack, and adaptability of the technique to local coating repair. Considerable work was also done on the codeposition of elements such as Ti, Hf, Ir, V, Y, Re, and Mn with silicon by the slip-pack method. Only titanium effected any improvement in silicide perfor-

mance; however, no data were available to confirm that deposition of the modifier elements with silicon was actually achieved.

Fansteel Metallurgical Corp. developed two silicide coatings for columbium, the S-2 conversion coating and the M-2 duplex system.[27] The S-2 silicide was formed by reacting the columbium surface with a $SiCl_4$–H_2 gas mixture at 2550°F, using induction heating of the substrate. The M-2 system was formed in three steps: (a) deposition of MoO_3 from a liquid bath at 1470°F, (b) hydrogen reduction of the MoO_3 at 1470°F, and (c) gas-phase siliciding ($SiCl_4$–H_2) to produce a $MoSi_2$ surface coating over the substrate silicides. The systems were typical of simple refractory metal silicides, providing 20–40 hr of life in air at 2300°F but poor resistance to pest oxidation.

Early work at Chromalloy Corporation produced the modified W-2 and Chromized N-2 pack-cementation silicide coatings for columbium.[28] Both systems were formed in a hydrogen atmosphere that embrittled the columbium substrate. Very little data were available for either system.

Chromizing Corporation produced a similar simple silicide system termed Durak KA.[28] Short-term life at temperatures in the range 2500°–2950°F and susceptibility to pest oxidation failure were characteristic of Durak KA.

The most successful and widely evaluated protective coating system for columbium alloys is the Cr–Ti–Si coating developed by Jefferys and Gadd of TRW.[28,29,32] The concept for the duplex Cr–Ti–Si coating system was motivated by the need to circumvent two problems associated with simple silicide coatings on columbium: (a) susceptibility to pest oxidation and (b) rapid substrate oxidation at the base of cracks in the brittle silicide. It was learned from early attempts to increase the oxidation resistance of columbium that columbium alloys containing chromium and titanium were among the most resistant compositions developed, although their utility as structural materials was nil. Alloying the columbium substrate beneath the surface silicide with chromium and titanium was therefore a logical step to achieving a subsurface defense against oxidation at the base of silicide cracks. Modification of the columbium disilicide with chromium and titanium was also a route to introducing pest resistance into the primary silicide layer. Hence, the two-cycle Cr–Ti–Si coating.

The TRW Cr–Ti–Si coating system is applied to columbium by a two-cycle vacuum-pack process, utilizing all-metal granular packs. The optimized parameters for producing the system involve codeposition of chromium and titanium from a halide-activated prealloyed 60Cr–40Ti alloy pack at 2325°F, followed by siliconizing in a halide-activated pure silicon pack at 2100°F. A multilayered 2-4-mil coating is formed, involving a primary (Cb, Cr, Ti . . .) Si_2 outer layer, two lower silicide intermediate layers, and a substrate region alloyed with substantial quantities of

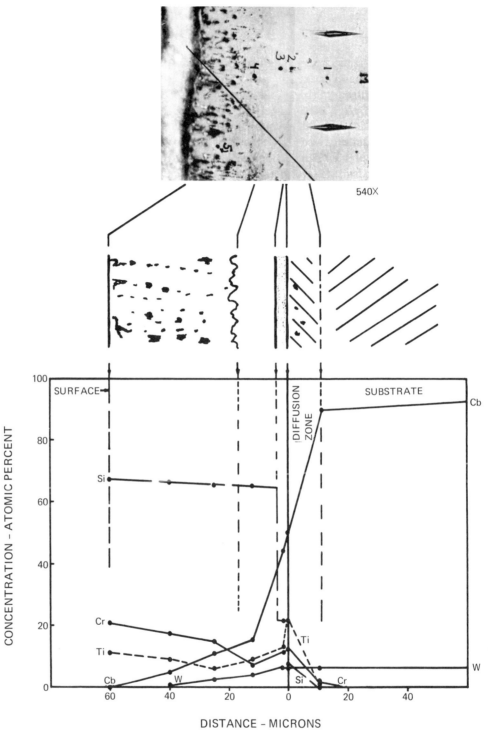

FIGURE 29   Concentration profile for Cr–Ti–Si-coated D-43 alloy.

chromium and titanium. In some cases, this diffusion zone contains residual Laves phase (Cb, Ti) Cr_2 remaining from the Cr–Ti precoat. Figure 29 is a photomicrograph and electron microprobe data for a typical Cr–Ti–Si coating.

In the first two programs,[28,29] TRW developed and optimized the vacuum-pack Cr–Ti–Si coating process. Emphasis was placed on simplifying the process to permit the utilization of conventional vacuum-heat-treating furnaces.

In a third program,[30] the process was advanced to a pilot production-scale facility. The protective reliability of the system was demonstrated in a series of statistically designed oxidation tests; and the performance of the Cr–Ti–Si system was assessed in a range of elevated-temperature-reduced-pressure environments. The fourth program[31] provided a detailed study of the basic factors associated with the diffusion formation and oxidation protection of the Cr–Ti–Si system: A thorough oxidation and mechanical property study was performed on Cr–Ti–Si-coated columbium alloys, and a slip-pack technique for application of the Cr–Ti–Si coating was developed. In the final program,[32] manufacturing processes were developed for forming the Cr–Ti–Si coating on large columbium alloy components employing both the vacuum-pack and vacuum-slip-pack coating techniques. A vacuum facility with hearth dimensions of 48 in. diameter × 54 in. high was employed in this latter program. The five development programs evolved a reliable and reproducible manufacturing coating process that will produce Cr–Ti–Si coatings capable of protecting columbium alloys for well in excess of 300 hr at temperatures to 2300°F, 100–200 hr at 2500°F, 50–100 hr at 2600°F, and up to 20–30 hr at 2700–2800°F. The very attractive properties of the Cr–Ti–Si coating composition prompted many investigators to seek alternative and/or simplified ways of producing the coating system on columbium alloys.

Pfaudler Company conducted two programs aimed at developing "practical" techniques for forming simple and Cr–Ti-modified silicide coatings on columbium alloys.[33,34] The fluidized bed was employed to deposit a simple silicide coating on columbium, and a nonelectrolytic fused-salt technique was used in an attempt to form the Cr–Ti–Si coating system. A laboratory fluidized-bed apparatus was developed for forming typical disilicide coatings on columbium. Efforts to form the Cr–Ti–Si coating in three steps by fused-salt deposition were unsuccessful because of corrosion of the titanium precoat during chromium and silicon deposition.

Electrophoresis was investigated by Vitro Corporation as a means of applying the Cr–Ti–Si coating to columbium.[35] The Cr–Ti precoat was electrophoretically deposited, isostatically pressed, and sintered in vacuum at approximately 2550°F. Silicon was electrophoretically deposited in a second cycle and treated in argon at 2300°–2400°F to produce the Cr–Ti-modified silicide. Cr–Ti–Si coatings resistant to pest in the 1200–2000°F region and capable of protecting columbium for average lives of 86 hr at 2600°F and 16 hr at 2700°F were produced by the Vitro techniques.[35] Thus, Cr–Ti–Si coatings equivalent in oxidation protection to the vacuum-pack version were produced by the electrophoretic method; however, the higher application temperatures and the isostatic pressing requirement render this process less practical than the vacuum-pack or slip-pack application methods.

Chemical vapor deposition was employed by Texas Instruments for the deposition of Cr–Ti–Si coatings on columbium.[36] Three-, two-, and single-cycle deposition procedures were developed, all of which produced Cr–Ti–Si coatings comparable in air oxidation protection to the vacuum-pack Cr–Ti–Si version. A single-cycle method capable of forming the entire coating system in 10 minutes was developed on a laboratory scale. The speed of the CVD process and its versatility for varying coating chemistry and morphology are unique assets of this coating method. However, the difficulties associated with uniformly coating large complex shapes exhibiting sharp edges and re-entrant surfaces are practical limitations to the CVD coating of aerospace hardware.

Boeing developed both vacuum-pack and fluidized-bed methods for applying silicide coatings to columbium alloys.[37,39] Typical simple silicide coatings were formed by vacuum-pack siliconizing at temperatures as low as 1900°F. An 18-in. fluidized bed was successfully operated to form the Disil-simple silicide coating on columbium hardware. Fluidized beds 4 in. in diameter were operated to form vanadium and Cr-Ti-modified silicide systems. The V–Si system produced by presiliconizing, vanadizing, and postsiliconizing was three times more protective than the simple silicide. Codeposition of Cr–Ti in an initial cycle followed by siliconizing produced Cr–Ti–Si coatings comparable in 2600°F oxidation protection to vacuum-pack Cr–Ti–Si coatings. Excellent heat transfer, rapid coating deposition, and coating uniformity are advantages of the fluidized-bed process, while high construction and operating costs are major disadvantages. Boeing's primary effort on fluidized-bed coating was aimed at coating components for the Dynasoar (X-20) structure.

Under Air Force sponsorship, Solar (Division of International Harvester Co.) investigated four techniques for the formation of Cr–Ti–Si coatings on columbium: (a) high-pressure pack, (b) electroless fused salt, (c) slurry deposition, and (d) vacuum-pack cementation.[40] High-pressure pack processing involved the use of halide-activated Cr–Ti–Si metal powders similar to the TRW vacuum-pack process; however, the pack was heated in a sealed retort that had been purged with argon, and a static argon–halide pressure of 800 Torr was maintained throughout the coating cycle. Cr–Ti–Si coatings similar in microstructure to the vacuum-pack coatings, but inferior in performance, were produced by this method.

Electroless fused-salt deposition was accomplished by the sequential deposition of Ti–Cr–Si and by codepositing Cr–Ti and Ti–Si. Facilities limitations reportedly prohibited the use of sufficiently high bath temperatures; consequently, very inferior Cr-Ti-Si coatings were produced. Solar continued its development efforts on fused-salt deposition of the Cr-Ti-Si system under separate government sponsorship; however, a reliable and reproducible technique for forming the Cr-Ti-Si coating was not developed.[41] Corrosion during the deposition process, and nonuniformity of the deposit were specific problems.

Solar's slip-pack or slurry deposition of the Cr-Ti-Si coating was generally unsuccessful owing to failure to achieve the proper coating chemistry and morphology for the Cr-Ti-Si coating. Solar's final approach to applying the Cr-Ti-Si system, the vacuum pack method, produced coatings comparable in microstructural characteristics and performance to the TRW vacuum-pack Cr-Ti-Si coatings.

A recently completed Solar program was directed to the problem of the extremely low ductility exhibited by the available silicide coatings for columbium.[42] The program approach was to produce protective coatings that included at least one ductile layer that would absorb strain induced by impact, deformation, or thermal stress. Ductile layers based on Cb-Al and Ti-Al binaries and bcc alloys of the Fe-Cr-Al type were considered. For silicides, potentially ductile intermediate layers such as Cr-V-Cr, Cb-Ti-Mo, Cb-Ti-W, and Cb-Ti-Cr were investigated. Deposition difficulties prevented the formation of the majority of these silicide coating systems, particularly those involving molybdenum or vanadium without titanium. Diffusional instability of the Fe-Cr-Al/Cb couple at $2400°$-$2500°$F rendered the bcc system incapable of protecting columbium. Vanadium modification of the Cr-Ti-Si system proved to be the only effective coating system evolving from the program. Vanadium served as a substitute for titanium, thereby reducing formation of the brittle $TiCr_2$ intermetallic phase; vanadium is also a less active interstitial sink than is titanium, exhibiting a lesser tendency to getter interstitials from the substrate by reducing the oxide and carbide hardening precipitates in the columbium alloy matrix.

A follow-on effort to the Solar program was conducted by TRW, with emphasis placed on developing coatings for improved performance at intermediate temperatures.[43] Molybdenum and vanadium deposition problems encountered by Solar were pursued initially. A reproducible vacuum-pack method for vanadizing columbium was developed; however, no practical technique other than CVD was developed for depositing molybdenum. Several modified silicides including V-Cr-Ti-Si were produced and evaluated. Only the V-Cr-Ti-Si system proved comparable in protective capabilities to the Cr-Ti-Si system, and in view of the added vanadizing cycle, this modification of the basic Cr-Ti-Si chemistry was not attractive.

Work by Solar on coatings for tantalum nozzle vane applications has brought forth a unique coating system that shows excellent potential for protecting columbium.[42,44] Designated TNV-12, the process involves applying a particulate mixture of Mo-5Ti to the substrate surface, followed by vacuum sintering to produce a porous product and subsequent pack siliciding to eliminate interconnected porosity and produce $MoSi_2$. The surface is subsequently sealed with a barium borosilicate glass, producing a multilayered coating of approximately 6-8 mils thickness. The system has looked promising for gas-turbine applications, based on good performance in erosion-oxidation and thermal fatigue tests.[45]

Sylvania has recently developed one of the more successful coating systems for columbium, the fused silicides.[46,47] A number of elements and intermetallics have been studied as modifiers to the basic Si-20Ti and Si-20Cr eutectic silicide compositions. Fusion of the sprayed or dipped silicide slurry requires vacuum firing at $2500°$-$2600°$F. The compositions Si-20Cr-5Ti and Si-20Cr-20Fe have looked most promising. The fused silicide process is particularly adaptable to coating hardware and for coating repair, and the process is scaled up for coating large hardware components.

The National Gas Turbine Establishment of Great Britain (NGTE) has reported good success with a glass-sealed silicide coating as a potential system for gas-turbine applications.[48] The silicide is applied by a pack or CVD method and is subsequently sealed with a fused glass slip. The final coating consists of approximately 4-5 mils of silicide plus 2-3 mils of glaze.

Battelle-Geneva has also explored the use of silicide coatings for protecting columbium in gas-turbine applications.[49] The most unusual system consists of applying a porous columbium or Cr-Ti-modified disilicide layer by a pack or CVD process, followed by impregnating the porous matrix with a Sn-Al slurry. The Sn-Al alloy remains fluid at the use temperature (above $2000°$F), thereby providing a self-healing capability to the coating. The porous matrix is effective in preventing washing of the liquid phase during shear loading of the surface.

Work on silicides for protecting columbium is also reported by American Machine and Foundry and Pratt and Whitney CANEL.[24] Both laboratories employed pack-cementation techniques to form modified silicides coatings; however, no data are available at this time to describe the capabilities of these coatings.

COMPARATIVE PERFORMANCE CHARACTERISTICS

Owing to the several anticipated applications for coated columbium alloys, many performance characteristics have been evaluated by coating developers and potential users.

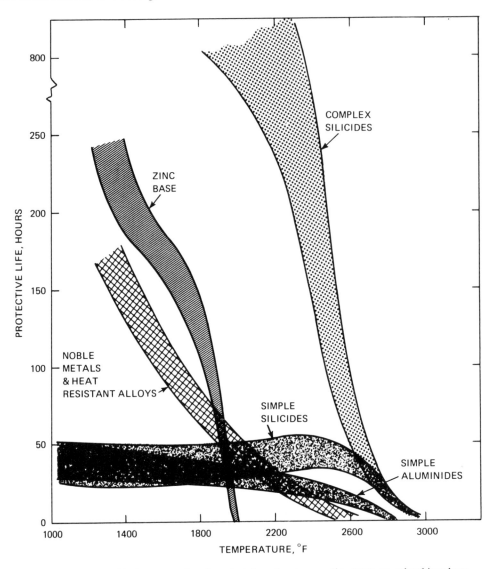

FIGURE 30   Cyclic oxidation-protective characteristics of various coating types on columbium-base materials.

These include oxidation resistance, mechanical properties, impact resistance, low-pressure behavior, erosion resistance, thermal fatigue tolerance, and others. It is beyond the scope of this report to present comparative data for all of these performance criteria; however, an effort is made to delineate those trends in performance that generally describe the state of the art of coatings for columbium alloys.

CYCLIC OXIDATION

Cyclic or static oxidation resistance in 1-atm air has been the yardstick for screening protective coatings for columbium. Since one segment of a coated component may experience low-temperature service while other regions are exposed to a very high-temperature environment, coated

columbium has been evaluated at temperatures over the range 1200°–3000°F. The general protective behaviors of the various categories of coating systems are presented schematically in Figure 30. The simple aluminides exhibit reliable short-term protection for columbium at temperatures from 1200°–2800°F, with a gradual decrease in life with increasing temperature.[1-4,10,11,18-22,50-53] Permeability of aluminides to oxygen and a pest-type oxidation behavior at lower temperature produce the slight perturbation in protective capability with decreasing temperature.

The simple silicides are particularly susceptible to pest oxidation at temperatures below about 2000°F. Consequently, these systems will generally afford less protection at lower temperatures than in the range 2000°–2600°F.[1-4,23-24,50-56] The formation of a continuous silica film at temperatures above 2000°F and the increasing ductility of the silicide–silica composite contribute to the improved higher temperature life.

The zinc-base coatings exhibit a remarkable self-healing capability, and therefore a very reliable protective performance, for hundreds of hours at temperatures below 2000°F.[12-14] However, decomposition of the Cb–Zn intermetallics at temperatures above 2000°F establishes a temperature limit for this system.

Work on the heat-resistant alloys of the iron, nickel, and cobalt class, and also the platinum-group metals, has been relatively limited in comparison with that on the intermetallic systems. Many of these materials show potential for protecting columbium at temperatures below 2000°F, while at higher temperatures intermetallic formation with the columbium substrate, porosity, application problems, and other factors have generally produced coatings inferior to the silicides.[5-7,15-17] The potential use of ductile, noble metal, or conventional alloy coatings at lower temperatures, particularly in the gas-turbine industry, is an area that shows promise and requires exploration.

The complex silicides such as Cr–Ti–Si, V–Cr–Ti–Si, and the fused silicides have been by far the most effective protective coatings for columbium. Absence of the silicide pest problem with these systems has provided coatings that exhibit increasing protective life with decreasing exposure temperature, producing lives increasing from 1–4 hr at 3000°F to 100–200 hr at 2500°F to many hundreds of hours at temperatures below 2200°F.[1-4,23,24,28-32,35-58] Complex silicides containing Cr, Ti, V, Fe, and Cb have proved to be the most oxidation-resistant coatings available for protecting columbium. Studies directed at identifying the oxidation products and modes of oxidation degradation of these complex silicide systems are reported by Bracco et al.[68], Gadd[31], Perkins and Packer[59] and others.[40,42,43,47] The four protective coating systems now most widely used, or considered for use, in the aerospace and gas-turbine areas are:

System	Process	Developer
Cr-Ti-Si	Vacuum pack	TRW
V-Cr-Ti-Si	Pack	Solar
20Cr-5Ti-Si	Fused silicide	Sylvania
10Cr-2Si-Al	LB-2 slurry	McDonnell-Douglas

Reliability of silicide coatings has been a point of major concern to potential columbium alloy users, and the low ductility of silicides, coupled with the relatively high incidence of structural defects, has discouraged their use. Many statistical studies have been performed on the oxidation resistance of silicide-coated columbium; examples of the

Weibull method of plotting failure data are shown for Cr-Ti-Si-coated D-43 alloys in Figures 31 and 32.

Efforts to compare large populations of oxidation test data generated on various coating/base metals by different laboratories has been found virtually impossible. Batch-to-batch variations in coating quality and significant deviations in oxidation test procedures have produced very wide variations in performance for individual coating systems. The relative performance trends shown in Figure 32 are real; however, it has been difficult to delineate precise, reliable values for the protective capabilities of the various coating systems on columbium alloy hardware.

REDUCED-PRESSURE OXIDATION

The use of coated columbium alloys in aerospace applications will involve high temperature exposure in reduced-pressure oxidizing environments. Upon exiting from the earth's atmosphere, aerodynamic heating can increase the skin temperature of a coated columbium alloy heat shield to well in excess of 2000°F. Based upon the findings of Perkins,[59] loss of silicide coatings by silicon monoxide vaporization will occur at substantial rates under the temperature–pressure conditions subsequently experienced by the vehicle in outer space.

Reduced-pressure oxidation studies have been performed by several laboratories on the major silicide coatings applied to columbium alloys, including both isothermal exposure and temperature–pressure profile tests.[31,40,42,43,46,47,59,60] Degradation of the protective silica film via SiO vaporization leads both to removal of the silicide reservoir and to internal contamination by the ingress of oxygen through the intermetallic matrix. In the case of the pack-applied Cr–Ti–Si and V–CrTi–Si coatings, substantial chromium loss is also experienced by vaporization, a factor which both disrupts the surface oxide film and opens paths for the accelerated transport of oxygen through the silicide coating. Figure 33 is a photomicrograph showing this phenomena for a Cr–Ti–Si-coated Cb-752 alloy material exposed at 2500°F and 10^{-2} mm pressure.[30] Even though this mode of degradation proceeds at a significant rate, coatings of the Cr–Ti–Si type provide protection for times well in excess of those required for currently anticipated re-entry vehicle missions.

Reduced-pressure isothermal and simulated re-entry profile tests performed on fused silicide coatings have shown an excellent resistance of these coatings to degradation in reduced pressure environments. Figure 34 is an example of the behavior of the Sylvania R512E (20Cr–20Fe–Si) fused silicide on Cb-752 alloy.[61] At temperatures below 2600°F, reliable life for periods acceptable for re-entry applications are demonstrated at all pressures, while above 2600°F, protective life is significantly decreased, even though coating life generally increases with increasing environmental pres-

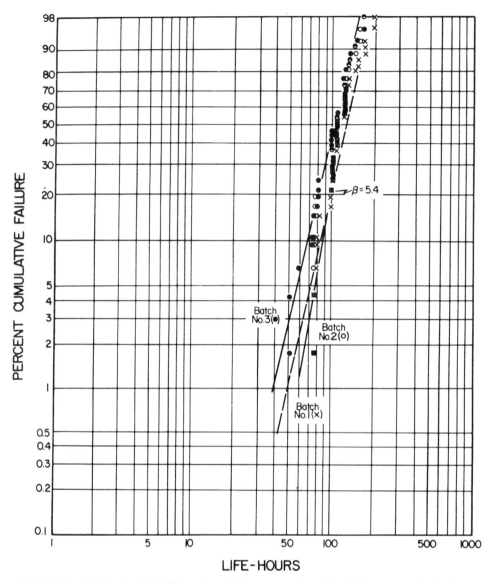

FIGURE 31   Weibull plot of 2500°F cyclic oxidation test results of Cr–Ti–Si coating on D-43 alloy.

sure (above 1 Torr). Liquation of columbium oxide above 2600°F is the major reason for the abrupt decrease in protective life.

The requirement for utilization of silicide coatings in high-temperature–reduced-pressure environments is unique to the aerospace industry. Based on current data, the complex silicide coatings and the LB-2-type aluminides for specific applications are capable coating systems for the protection of columbium alloy components in reduced-pressure re-entry environments.

MECHANICAL PROPERTIES

The mechanical properties of protective coated columbium alloys are a major concern to the designers of hardware for high-temperature applications. Silicide and aluminide coat-

ings exhibit negligible or no ductility at low temperatures. It has been shown that a brittle surface layer can influence the properties of substrate materials to a considerably greater degree than would be anticipated, based upon the percentage of the composite cross section that the coating assumes.[62] Owing to this recognized mechanical behavior of coated materials, a number of programs have been conducted to assess the mechanical properties of protective coated columbium alloys.[50–53,62,63,65,66] A typical representation of the mechanical property data for Cr–Ti–Si-coated B-66 alloy is presented in Figures 35, 36, and 37.

The influence of brittle protective coatings on the mechanical behavior of columbium alloys is quite consistent with theoretical considerations. At low temperature, where intermetallics are brittle, very sharp notches are generated in strained intermetallic coatings. These notches tend to

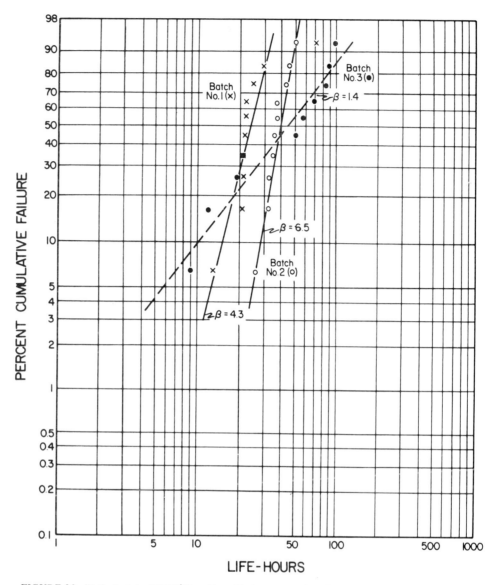

FIGURE 32 Weibull plot of 2700°F cyclic oxidation test results of Cr–Ti–Si coating on D-43 alloy.

propagate fracture paths through the columbium alloy matrix, particularly the notch-sensitive alloys such as Cb-132M, Cb-752, and XB-88, thereby reducing yield strength, tensile strength, and tensile ductility. This notch introduction from the brittle surface layer also significantly reduces the bend ductility of coated columbium alloys.

In stress rupture and creep, the influence of protective coatings is far less significant. At elevated temperatures, ductility is restored to the silicide and aluminide coatings, and their influence on fracture behavior is nil. However, the creep strength of columbium alloys may be significantly influenced by the presence of coating systems such as the Cr-Ti-Si composition. Titanium, owing to its strong affinity for interstitials, will operate as a "sink" for interstitials contained in the columbium matrix.[67] Oxides and carbides that

contribute to the high-temperature creep resistance of columbium alloys will be reduced by the titanium "sink," thereby seriously weakening the columbium alloy. This high-temperature phenomenon must be given careful consideration when designing with precipitation-strengthened alloys such as Cb-752, Cb-132M, and B-66.

APPLICATIONS OF COATED COLUMBIUM ALLOYS

Aerospace vehicles and aircraft gas turbines are the primary areas for the application of protective coated columbium alloys. About the earliest serious effort to develop columbium as a structural material was a joint program of McDonnell

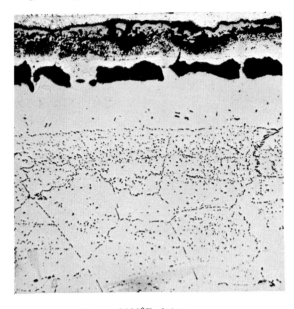

1 hour–2500°F–1 Atm
4 hours–2600°F–10⁻² mm

FIGURE 33   Cr–Ti–Si-coated D-14 alloy after indicated 1-atm and reduced pressure exposures. 250×.

Aircraft and General Electric, completed in 1961.[10] The program was directed at fabricating and testing a simulated structural component of a manned re-entry glide vehicle. The LB-2 slurry aluminide coating was a direct outgrowth

of this program, while this general philosophy of utilizing refractory metals for constructing space vehicles provided the impetus for the majority of the coating development programs conducted over the ensuing 6–8 years.

The ASSET (Aerothermodynamic/elastic Structural Systems Environmental Test) program followed at McDonnell Aircraft, and glide re-entry vehicles fabricated from columbium alloys and other refractory and nonrefractory materials were constructed and flight-tested.[69] LB-2-coated columbium alloy heat shields, Cr–Ti–Si-coated leading edge segments, and approximately 2,000 Cr–Ti–Si-coated columbium alloy fasteners were employed in each vehicle. Recovery of a flight-tested structure demonstrated the reuse capability of coated columbium alloys for aerospace vehicles.

A third structures program nearing completion at McDonnell involves the fabrication and flight-testing of columbium alloy boost glide re-entry vehicles (BGRV).[22] Truncated cone-shaped body sections up to 30 in. in diameter and 5 ft high, fabricated from Cb-752 alloy, are being protected by the LB-2 aluminide. Cr–Ti–Si-coated columbium alloy fasteners attach various body components.

Coated columbium alloys were evaluated by Aerojet-General Corporation for Apollo and Transtage exit cones.[70] The Apollo cone is 8 ft high and 8 ft in diameter, while the Transtage cone is 4 ft high and 4 ft in diameter. Simple aluminide and silicide coatings exhibited the required capabilities for these applications.

General Dynamics evaluated the use of TRW Cr–Ti–Si

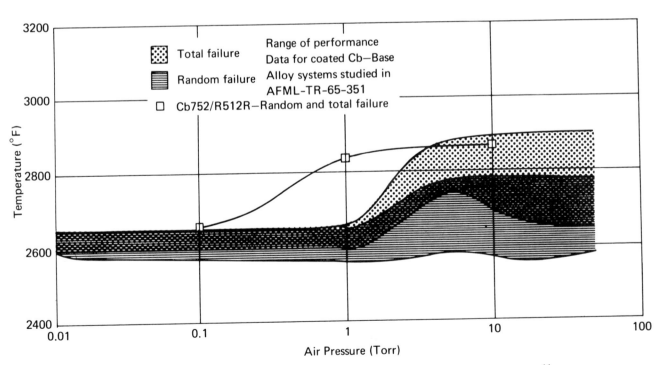

FIGURE 34   Maximum temperature for 4-hr lifetime for Cb 752/R512E and R512R system.[61]

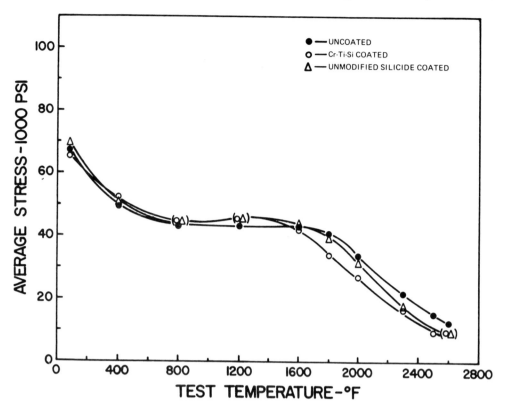

FIGURE 35   Yield strength of uncoated and coated 30-mil B-66 alloy sheet.[64]

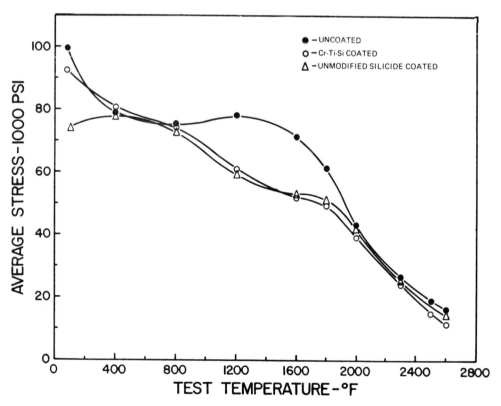

FIGURE 36   Tensile strength of coated and coated 30-mil B-66 alloy sheet.[64]

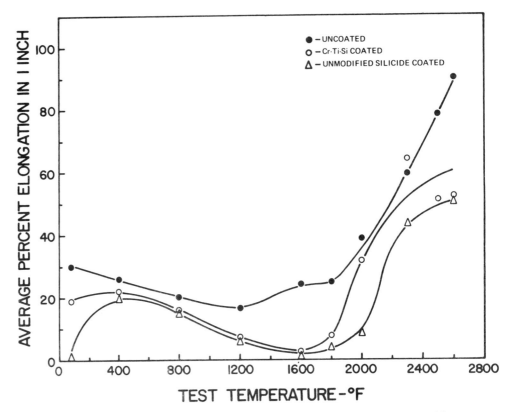

FIGURE 37   Tensile elongation of uncoated and coated 30-mil B-66 alloy sheet.[64]

and Vac Hyd Lunite 3-coated Cb-752 alloy as heat shield materials for a thermally insulated support structure applicable to hypersonic and re-entry vehicles.[71] Corrugated panel sections up to 20 in. × 20 in. were coated, assembled, and evaluated under simulated re-entry conditions.

A major columbium alloy structures program was the ASCEP (Advanced Structural Concepts Experimental Program) conducted by Martin Marietta Corporation.[72] A 1/3-scale model wing–fuselage–cryogenic tank section of a high L/D winged re-entry vehicle was fabricated from D-43 columbium alloy. Brazed-honeycomb heat shield panels up to 16 in. wide and 38 in. high, and other substructural hardware, were Cr–Ti–Si-coated prior to assembly. Incompatibility of the Cr–Ti–Si coating with residual titanium braze alloy and nonoptimized manufacturing-scale coating application conditions produced a high incidence of defects in the Cr–Ti–Si protective coatings. Repair of the Cr–Ti–Si-coated panels with fused silicides proved very successful.

The use of coated columbium alloys for the fabrication of gas-turbine components has been explored for over 10 years as a means of increasing the operating temperatures of turbine hot section parts. Blades, vanes, combustors, and other components subjected to temperatures approaching the turbine inlet gas temperatures are candidates for the use of coated columbium. Pratt and Whitney Aircraft has been most active in evaluating coated columbium alloy turbine materials. In two recent Air Force–funded programs,[57,58] Pratt and Whitney evaluated all potential coating systems for protecting columbium blade and vane alloys. Testing included the assessment of properties such as oxidation, hot-gas erosion, impact, thermal and mechanical fatigue, creep-rupture, tensile, and melting point. In oxidation–erosion, silicides of the Cr–Ti–Si type proved most protective, while in thermal fatigue, the TNV-12 and R-512 fused silicides were superior. An engine test of fused-silicide-coated columbium alloy vanes culminated the program.

DISCUSSION AND RECOMMENDATIONS

A very large number of laboratory and hardware evaluation studies have been performed on coated columbium alloys, both by the coating developers and by independent laboratories and potential users. A tremendous quantity of test data has been generated in these programs, and many of the coating systems discussed in the previous sections have shown excellent potential for utilization in aerospace and aircraft gas-turbine applications. The bulk of these data is noncomparative because of the lack of standardized test procedures; however, the laboratory tests have demonstrated that several coating systems can meet the hypothetical performance goals established for these requirements. Although the high expense of hardware programs is well recognized, flight or test-bed evaluation of actual coated hardware is vitally needed to assess the true capabilities and reliability of these coating systems. Many of the lab tests, particularly those intended to simulate propulsion system environments, are incapable of establishing certain critical conditions encountered in service. Coatings are frequently optimized around these laboratory tests, and there is no assurance that the test results can be related directly to service performance. Careful analysis must certainly be made of the laboratory data that have been generated, and real performance criteria must be established to permit judicious screening of the available systems. Subsequent hardware evaluation of coating systems would then represent a major step forward in assessing the state of the art of coatings for columbium alloys.

The early work on heat-resistant alloys of the iron–nickel–cobalt class, and more recent work on the ductile noble metals, was unsuccessful primarily as a result of interdiffusion between the substrate and coating elements. Degradation of base metal properties, eutectic and intermetallic phase formation, and premature coating failure were the salient consequences. The barrier-layer approach appears to offer the only way to circumvent deleterious interdiffusion, and work in this area should be renewed or continued.

Many of the silicide-type coatings have demonstrated exceptional corrosion wearout capabilities and, in many cases, rather surprising tolerance to mechanical abuse. The basic deterrent to their more extensive use is the lack of reproducibility and reliability of the systems. The Cr–Ti–Si composition has demonstrated an excellent inherent resistance to oxidation. However, none of the many methods investigated for its formation, including vacuum pack, fused slurry, fused salt, and CVD, has been successful in producing Cr–Ti–Si coatings that meet the reliability standards expected of structural materials. This deficiency should be attacked from a more fundamental viewpoint.

Sylvania's relatively recent fused-silicide coatings show excellent capabilities for protecting columbium. The wettability of the braze-type coating makes it particularly adaptable to coating complex hardware with faying surfaces and to utilization as a repair coating method. Again, thickness control, poor resistance to mechanical abuse, and the effect of the coating process on base metal properties hamper its promotion for many applications.

Other relatively untested systems that have demonstrated unique protective characteristics are (a) the Solar TNV-12 porous silicide coating (glass-sealed) in which the porous structure appears to mitigate thermal and mechanical stresses in the coating, while the glass seals the coating surface; (b) the Battelle-Geneva Sn-Al-impregnated silicide coating, which provides self-healing, possibly FOD tolerance, and yet an apparent capability for operation under high gas flow conditions; and (c) the NGTE glass-sealed silicide coating which reportedly shows promise for gas turbine applications.

The several silicide-type coatings discussed in the past few paragraphs are being further optimized and evaluated by their respective developers; but, generally, these programs are broad in nature, and the paths they follow are left to the discretion of the developers. It would seem more appropriate to conduct application-oriented programs aimed at optimizing certain systems for specific applications and specific environmental criteria and culminate such programs with a service or hardware evaluation. Further modification and optimization of existing coating systems for columbium, based on laboratory evaluation, does not appear fruitful.

Another area deserving of attention is the tailoring of coating systems to substrate materials and the use of a bi-metal structural concept whereby a very "coatable" columbium alloy is employed as a surface material on the structural shape, and a strong, less corrosion-resistant columbium alloy is utilized as the core. The performance and reliability of the coating would be greatly enhanced by the coatable substrate. Such a program aimed at gas-turbine applications of columbium is currently in progress.[73]

Standardization of evaluation tests is vitally needed if efficient and effective use of laboratory test data is to be achieved. ASTM committee C-22-VI is currently working on this problem, and the NMAB has made recommendations in this area in the past. The effort should be accelerated and program-sponsoring agencies should recommend the utilization of such test procedures in order to implement the standardization program.

Finally, there is a definite need for basic data in the broad area of high-temperature coatings, alloys and corrosion. Data on diffusion rates, melting points, identification of complex phases and alloys, vapor pressures, and gas-metal reactions are grossly lacking in the literature. The development of coating formation and degradation mechanisms and the establishment or prediction of material limitations are severely hampered by this lack of data. Two areas should be pursued—one aimed at the generation of basic properties for high-temperature elements, alloys, intermetallics, oxides, etc, and a second effort directed specifically to characterizing existing promising coating systems to provide data that would aid in the further optimization of these coatings.

REFERENCES

1. Krier, C. A. Nov. 24, 1961. Coatings for the protection of refractory metals from oxidation. DMIC Report 162.

2. Gibeaut, W. A., and E. S. Bartlett. January 10, 1964. Properties of coated refractory metals. DMIC Report 195.

3. Klopp, W. D. January 15, 1960. Oxidation behavior and protective coatings for columbium and columbium base alloys. DMIC Report 123.

4. Coated refractory metal technology. 1965. MAB-210-M.

5. Hirakis, E. C. Feb. 1959. Research for coatings for protection of niobium against oxidation at elevated temperatures. WADC TR 58-545.

6. Wlodek, S. T. Feb. 1961. Coatings for columbium. Electrochem. Soc.

7. Beach, J. G., and C. L. Faust. May 24, 1955. Electroplated metals on niobium. BMI 1004.

8. Spretnak, J. W., and R. Speiser. June 15, 1957. Protection of niobium against oxidation at elevated temperatures. NRL 467-15.

9. Spretnak, J. W., and R. Speiser. March 15, 1958. NRL 467-16.

10. Neff, C. W., R. G. Frank, and L. Luft. Oct. 1961. Refractory metals structural development program. ASD TR 61-392, Vol. II.

11. Peterson, L. E. Sept. 11, 1963. Oxidation resistant coating program for columbium in Apollo and transtage exit cones. AD 439102.

12. Brown, B. F., *et al.* Nov. 1960. Protection of refractory metals for high temperature service. NRL Report 5550.

13. Brown, B. F., *et al.* Jan. 31, 1961. Protection of refractory metals for high temperature service. NRL Report 5581.

14. Klopp, W. D., and C. A. Krier. March 1961. Zinc coatings for the protection of columbium from oxidation at elevated temperatures. DMIC Memorandum 88.

15. Buchinski, J. J., and E. H. Girard. Oct. 1964. Study of ductile coatings for the oxidation protection of columbium and molybdenum alloys. AD 608135.

16. Girard, E. H., J. D. Clarke, and H. Breit. May 1966. Study of ductile coatings for the protection of columbium and molybdenum alloys. Contract NOw 65-0340-f.

17. Rexer, J. Aug. 1968. High temperature protective coatings for refractory metals. Final Report, Carbon Products Division, Union Carbide Corporation. Contract NAS N-1405.

18. Foldes, S. Nov. 1960. Oxidation resistant fusion coatings on F-48 columbium alloy. FPLD, General Electric. R60-FPD-620.

19. Carlson, R. G. 1960. Oxidation resistance of aluminum dip-coated (Aldico) columbium alloys. Columbium Metallurgy Symposium.

20. Culp, J. D., and B. G. Fitzgerald. Aug. 18, 1964. Recent refractory metal coating studies at McDonnell Aircraft. Presented at the Refractory Components Working Group Meeting.

21. Culp, J. D., B. G. Fitzgerald, and J. C. Sargent. April 12, 1965. Recent refractory metal coating activities at McDonnell Aircraft. Presented at the Refractory Composites Working Group Meeting.

22. Culp, J. D. Oct. 17, 1966. Protective coatings for refractory metal hardware. Presented at the Refractory Composites Working Group Meeting.

23. Wurst, Cherry, Gerdeman, and Gecht. June 1967. The evaluation of materials systems for high temperature aerospace application. AFML-TR-67-165.

24. Cox, J. D. Dec. 31, 1964. An evaluation of Cr-Ti-Si protective coating for columbium alloys B-66 and D-43. GDA-ERR-AN-707. General Dynamics/Astronautics.

25. Aves, W. L., *et al.* June 1964. Diffusion coating process for columbium base alloys. ML-TDR-64-71.

26. Aves, W. L., and G. W. Bourland. Aug. 1965. Investigation and development of techniques to extend the utility of pack processes and compositions for coating-molybdenum and columbium alloys. AFML-TR-65-272.

27. Lorenz, R. H., and A. B. Michael. Oct. 1966. Oxidation resistant silicide coating for columbium and tantalum alloys by vapor phase reaction. Presented at 118th Meeting of Electrochemical Society, Houston, Texas.

28. Jefferys, R. A., and J. D. Gadd. May 31, 1961. Development and evaluation of high temperature protective coatings for columbium alloys. ASD TR 61-66 Pt. I and II.

29. Gadd, J. D., and R. A. Jefferys. Nov. 1962. Advancement of high temperature protective coatings for columbium. ASD-TR-62-934 Pt. I.

30. Warmuth, D. B., J. D. Gadd, and R. A. Jefferys. April 1964. Advancement of high temperature protective coatings for columbium alloys (11). ASD-TDR-62-934 Pt. II.

31. Gadd, J. D. April 1965. Advancement of protective coating systems for columbium and tantalum alloys. AFML-TR-65-203.

32. Fisch, H. A., and J. D. Gadd. May 1968. Manufacturing methods for high temperature coating of large columbium parts. Contract AF-33(615)-2018, AFML-TR-68-134.

33. Zupan, J., *et al.* Sept. 1963. Development of practical techniques for applying protective coatings to columbium alloys. RTD-TDR-63-4063.

34. Zupan, J., *et al.* Dec. 1964. Investigation of practical techniques for coating refractory metals. ML-TDR-64-304.

35. Ortner, M. H., and K. A. Gebler. June 1965–Dec. 1966. Electrophoretic deposition of refractory metal coatings. Contract AF 33(615)-3090, IR 8-344 (I-VI).

36. Wakefield, G. F. Dec. 1966. Final report on "Refractory metal coatings by chemical vapor deposition." AFML-TR-66-397.

37. Batiuk, W., *et al.* April 1967. Fluidized bed techniques for coating refractory metals. AFML-TR-67-127.

38. Dec. 26, 1963. Development of oxidation resistant coatings for columbium alloys–fluidized bed process. D-2 81108-1, Contract AF 33(657)-7132.

39. Dec. 27, 1963. Development of oxidation resistant coatings for columbium alloys, vacuum pack process. D2-81108-2, Contract AF 33(657)-7132.

40. Stetson, A. R., and V. S. Moore. June 1965. Development and evaluation of coating and joining methods for refractory metal foils. RTD-TDR-63-4006, Pt. III.

41. Stetson, A. R., *et al.* July 1965–June 1966. Refractory metal coatings by the fused-salt process. Contract AF 33(615)-3173, RDR-1395-1, 2, 3, 4.

42. Stetson, A. R., and A. G. Melcalfe. Sept. 1967. Development of ductile coatings for columbium-base alloys. AFML-TR-67-139, Pt. I.

43. Nejedlik, J. F., and J. D. Gadd. Oct. 1968. Coatings for long-term intermediate temperature protection of columbium alloys. AFML-TR-68-170.

44. Wimber, R. T., and A. R. Stetson. July 1967. p. 100 *in* Development of coatings for tantalum alloy nozzle vanes. Contract NAS 3-7276, NASA Cr-54529.

45. Holloway, J. E. 1965–1967. Evaluation and improvement of coatings for columbium alloy gas turbine engine components. Contract AF 33(615)-2117, Progress Reports 1–10.

46. Priceman, S., and L. Sama. Sept. 1965. Development of slurry coatings for tantalum, columbium and molybdenum alloys. AFML-TR-65-204.

47. Priceman, S., and L. Sama. Oct. 1965–July 1967. Development of fused slurry silicide coatings for the elevated temperature oxidation protection of columbium and tantalum alloys. STR-66-5501, 14, 17, 20, 25.

48. Restall, J. E. Aug. 1967. Oxidation-resistant coatings for re-fractory metals on aircraft gas-turbine engines. Met. Mat.
49. Stecher, P., and B. Lux. June 1968. Protection contre l'oxy-dation d'alliages de niobium. 6th Plansee Seminar.
50. Wurst, J. C., and J. A. Cherry. Sept. 1964. The evaluation of high-temperature materials. ML-TDR-64-62.
51. Wurst, J. C., J. A. Cherry, P. A. Gerdeman, and N. L. Hecht. July 1966. The evaluation of materials systems for high tem-perature aerospace applications. AFML-TR-65-339, Pt. I.
52. Wurst, J. C., J. A. Cherry, W. E. Berner, and D. A. Gerdeman. June 1967. The evaluation of materials for aerospace applica-tions. AFML-TR-67-165.
53. Gerdeman, D. A., J. C. Wurst, J. A. Cherry, and W. E. Berner. March 1968. The evaluation of aerospace materials. AFML-TR-68-53.
54. Grimm, F. C. July 1962. Evaluation of Pfaudler coating for oxi-dation protection of columbium alloys. Report 8886. McDonnell Aircraft Corporation.
55. Stein, B. A., and W. B. Lisagor. Aug. 1964. Preliminary results of a study of 12 oxidation-resistant coatings for Cb-10Ti-5Zr columbium alloy sheet. 9th Meeting of Refractory Composites Working Group.
56. Cox, J. D. Dec. 1964. An evaluation of a Cr-Ti-Si protective coating for columbium alloys B-66 and D-43. GDA-ERR-AN-707. General Dynamics Corporation.
57. Hauser, H. A., and J. F. Holloway, Jr. July 1966. Evaluation and improvement of coatings for columbium alloy gas-turbine engine components. AFML-TR-66-186, Pt. I.
58. Hauser, H. A., and J. F. Holloway, Jr. May 1968. Evaluation and improvement of coatings for columbium alloy gas-turbine engine components. AFML-TR-66-186, Pt. II.
59. Perkins, R. A., and C. M. Packer. Sept. 1965. Coatings for re-fractory metals in aerospace environments. AFML-TR-65-351.
60. Lavendel, H. W., et al. Aug. 1965. Investigation of modified
61. Culp, J., and B. Fitzgerald. Oct. 1967. Evaluation of fused slurry silicide coating considering component design and reuse. Contract F33615-67-C-1574.
62. Form, G. W., and W. M. Baldwin, Jr. 1956. Effect of brittle skins on the ductility of materials. ASTM Proc. 56.
63. Gadd, J. D. Jan. 1963. Design data study for coated columbium alloys. NOw-62-0098c.
64. Warmuth, D. B., and J. D. Gadd. April 1964. Design data study for coated columbium alloys. Contract NOw 63-0471c.
65. Allen, B. C., et al. Feb. 1967. Elevated temperature ductility minima and creep strengthening of coated and uncoated colum-bium alloys. AFML-TR-66-89.
66. Allen, B. C., and E. S. Bartlett. Sept. 1967. Elevated tempera-ture tensile ductility minimum on silicide coated Cb-10W and Cb-10W-2.5Zr. Transactions quarterly, Vol. 60, No. 3.
67. Klein, M. J. Oct. 1968. Structural and mechanical effects of interstitial sinks. Contract NAS 7-469.
68. Bracco, Sama, and Lublin. May 1966. Identification of micro-structural constituents and chemical concentration profiles in coated refractory metal systems. ML-TR-66-126.
69. Defense Metals Information Center. Feb. 1965. Review of Recent Developments.
70. Peterson, L. E. Sept. 1963. Oxidation resistant coating program for columbium in Apollo and transtage exit cones. AD 439102.
71. Cross, R. I., and W. E. Black. April 1967. Optimization of insu-lation and mechanical supports for hypersonic and entry ve-hicles. AFML-TR-66-414.
72. Wilks, C. R. Nov. 1967. Advanced structural concepts experi-mental program-project ASCEP. AFFDL-TR-67-146, Vol. III.
73. Preliminary work under contract AF 33(615)-67-C-1688. Development of columbium alloy combination for gas-turbine blade applications. TRW, Inc.

MOLYBDENUM*

INTRODUCTION

Development of oxidation-resistant coatings for molybde-num and its alloys has been pursued on an active basis for well over 20 years. During this period, the nature and di-rection of development efforts have shifted dramatically as the major requirements and applications for coated molyb-denum have changed. As a result, more is known today about the fundamental and applied coating technology for molybdenum than for any of the other refractory metals and alloys.

The first major programs on coatings for molybdenum were undertaken in the late 1940's when designers sought materials for use at temperatures beyond the range for Ni- and Co-base superalloys in high-performance ram-jet and aircraft gas-turbine engines. These early efforts largely cen-tered on the use of oxidation-resistant metal plating, or cladding, and intermetallic compounds such as oxides, alu-

minides, and silicides to protect molybdenum substrates. A useful class of coatings based on molybdenum disilicide began to evolve in the early 1950's just about the time that interest in refractory metals for air-breathing propulsion systems declined.

In the late 1950's, a need for refractory metal heat shields on lifting re-entry and hypersonic flight vehicles arose. Applications in this area expanded rapidly, and a major research and development effort on coated molyb-denum alloys was initiated. Work for the most part centered on the promising silicide-base coating systems. Compositions were optimized, large-scale manufacturing processes were developed, coating systems were fully characterized both in the laboratory and in test flight vehicles, and fundamentals governing the performance and failure of the coatings were investigated. The period from 1958 to 1968 was one of high-level activity that provided a sound technological basis for the use of coated molybdenum and its alloys in a wide range of applications.

Unfortunately, the anticipated applications did not ma-terialize, and interest in coated molybdenum had declined

*Prepared by R. A. Perkins.

rapidly by the late 1960's. The technology that had been developed has been used successfully to meet limited requirements in chemical rocket propulsion systems and a number of industrial applications. At present, however, without a large-scale field of application for coated molybdenum, little, if any, effort or consideration is being given to further advancing the state of the art. The purpose of this section is to review the basic coating technology for molybdenum and its alloys that has evolved over the past 20 years, to describe the coating capabilities that exist today, and to indicate the possible direction for future efforts in improving overall capabilities.

FACTORS GOVERNING COATING TECHNOLOGY

Coating technology is governed by four important characteristics of molybdenum and its alloys: oxidation, strengthening mechanism, ductile-to-brittle transition, and fabricability. The solid solubility for oxygen and nitrogen is extremely low. These elements do not tend to cause embrittlement by diffusion into molybdenum and its alloys in the manner characteristic of the columbium and tantalum alloys. Hence, slow leakage through the coating is not necessarily detrimental, and the use of porous (ceramic) coatings can be considered. The oxide of molybdenum (MoO_3) melts at $1460°F$ and has a high vapor pressure. At normal service temperatures ($2000°-2400°F$) the oxide is a gas. When a coating fails, the substrate oxidizes rapidly and is converted to a gaseous phase. Holes develop in the substrate as shown in Figure 38. Coating failures on molybdenum are characterized by hole formation at each failure site. At early stages these failures are difficult to detect, since the coating may appear to be intact while the substrate is oxidized underneath.

Molybdenum trioxide melts at $1460°F$ and the liquid is a good flux for other oxides. Liquid oxide formation can accelerate coating failures in this temperature range. At higher temperatures, the oxide volatizes as fast as it forms and significant volumes of liquid oxide cannot be formed. This aspect of behavior permits coated molybdenum to be used at much higher temperatures than coated Cb or Ta. Coatings on these materials whose oxides are not volatile can fail catastrophically at temperatures where liquid oxides are formed.

Molybdenum sheet alloys (Mo-0.5Ti, TZM, TZC) are strengthened by strain hardening and dispersed carbide phases. Strain hardening also is used to lower the ductile-to-brittle transition temperature. Coatings must be applied at temperatures below the recrystallization temperature to preserve high strength and ductility. Maximum coating deposition temperatures are in the range of $2000°-2200°F$ for

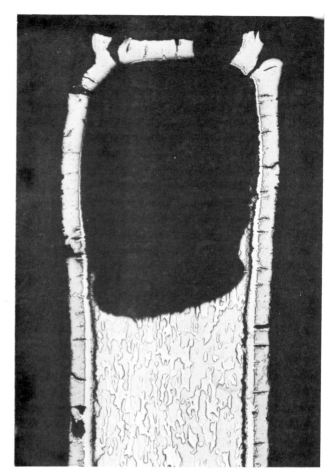

FIGURE 38 Hole in TZM substrate developed at a local defect failure in an $MoSi_2$ coating.

short time processes. The coatings, in general, should not contain strong carbide forming elements. In such cases, an interstitial sink effect can occur wherein the coating removes carbon from the substrate with resultant weakening of the alloy.

Molybdenum and its alloys are notch sensitive. Brittle coatings applied to the surface generate sharp notches and raise the ductile-to-brittle transition temperature. The effect is most pronounced in bending and in impact. Fatigue properties also are reduced. Ductile metallic coatings are required for maximum toughness and ductility of the composite. Parts coated with intermetallic compounds are brittle at room temperature under many loading conditions.

Molybdenum and its alloys are difficult to fabricate. Sheet products tend to delaminate in the plane of the sheet if improperly cut or drilled. The coating of edges is particularly difficult. Many diffusion coatings open up delaminations, and rejects due to edge defects are high. Coatings need to be formulated for good edge protection. Molybdenum normally is not welded, because of recrystallization softening and embrittlement. Parts are joined by mechanical fasteners. Protection of faying surfaces at joints is difficult, and coatings must be formulated to penetrate deeply into contact areas. Coating processes must be applicable to treatment of both detail parts and complex assemblies, since in some cases (e.g., riveted joints) parts must be recoated after assembly.

The foregoing discussion serves to point out that the coating technology for molybdenum has been developed to suit the particular characteristics of this material. The coating is not merely a paint. Instead, it is an integral part of the substrate and has a significant effect on mechanical properties and behavior. The substrate, in turn, has an effect on coating behavior. With molybdenum, this interplay becomes very critical and has been a significant factor in guiding coating development.

METALLIC COATINGS

Ductile metallic overlays were among the first materials investigated as coatings for molybdenum. The materials and processes used are summarized in Table 18. Reviews of past work are presented in References 1–3. Nickel and nickel-base alloy coatings gave useful protection to 2200°F. Coating lives of 100 to 500 hr were realized at 2000°F and below. In general, however, reliability of plated systems was poor. A thermal expansion mismatch between Ni and Mo causes spalling on thermal cycling. Interdiffusion results in formation of brittle intermetallics at the coating–substrate interface, which also contribute to spalling. Nickel and its alloys in general are not suitable for direct application to molybdenum.

Chromium is the most promising of the metallic coatings for molybdenum. It affords excellent oxidation protection and is compatible with the substrate. Chromium, however, is embrittled by nitrogen and will crack and spall on repeated thermal cycling. A nickel or nickel-alloy overlay will protect chromium from nitridation. Thus, a duplex coating of Cr plus Ni or NiCr will give good service on molybdenum. The maximum service temperature is limited to about 2200°F by the oxidation resistance of the Ni alloy overlay. This system has considerable merit for use in the lower temperature range. It is not self-healing, however, and presents a number of design and fabrication restrictions in the use of coated parts.

Noble metal coatings (Pt, Ir) can be used on molybdenum at temperatures to about 2600°F.[4] They provide a significantly longer useful life at all temperatures than the duplex Cr–Ni coatings. The life of these coatings is limited by interdiffusion with the molybdenum substrate, and studies have been made on the use of diffusion barriers to extend life. Hafnium and iridium plus tungsten barriers show some promise. Coating life also is governed by a thermal expansion mismatch, and cyclic use will degrade life. Thick coatings are required (3–5 mils) for good performance. This results in a costly system and limits potential applications. The low emittance (0.2–0.4) of noble metal coatings is another factor that limits utility in many applications.

ALUMINIDE COATINGS

Coatings based on compounds with aluminum received considerable attention during the early 1950's in initial attempts to develop molybdenum parts for aircraft gas turbines. Successful coatings formed from Al–Cr–Si, Al–Si, and Al–Sn alloys were developed by a number of companies, as indicated in Table 19. Background information and general reviews of the technology developed are presented in References 1 and 3–7.

The aluminide coatings are applied as thick overlays using a variety of spray or dip processes (Table 19). Pack-diffusion processes currently used to apply aluminide coatings to Ni- and Co-base alloys apparently were not used in coating molybdenum. The reason for this is not clear, but it may be due to the fact that pack processes in general are used to prepare unalloyed or slightly modified coatings, whereas the aluminide coatings for molybdenum were highly alloyed. It is significant that aluminide coatings for molybdenum were investigated before many of the coating processes in use today were fully developed. Many of the deficiencies of aluminide coatings are found to be due to poor processing practices and lack of control of coating characteristics. The advances in process technology and inspection that ensure high-quality coatings today were not developed until after the interest in coatings for molybdenum had shifted from an aluminide to a silicide base. It is likely that aluminide coatings of greatly improved performance and reliability could be developed for molybdenum with the improved process technology that is available today.

Of all the brittle intermetallic coatings tested on molybdenum, aluminide coatings have the least degrading effect on mechanical properties. This is particularly true for the Sn–Al

TABLE 18   Metallic Coatings for Molybdenum

Process Type	Materials	Thickness Range (mils)
Electrodeposition	Cr, Ni, Au, Ir, Pd, Pt, Rh	0.5–3.0
Flame-sprayed	Ni–Cr–B, Ni–Si–B, Ni–Cr, Ni–Mo	5–10
Clad or bonded	Pt, Ni, Ni–Cr, Pt–Rh	2–20
Molten bath	Cr	0.5–1.0

TABLE 19   Aluminide Coatings

Type	Composition	Deposition Process	Thickness Range (mils)	Developer
Al–Cr–Si	20% Al + 80% (55Cr–40Si–3Fe–1Al)	Flame spray	7–10	Climax
Al–Si	88% Al–12% Si	Flame spray hot-dipped	0.5–7	NRC
Sn–Al	90% (Sn–25 Al)– 10% $MoAl_3$	Slurry- dip or spray	2–8	Sylcor (G. T. & E.)

TABLE 20   Silicide Coatings

Type	Trade Name	Developed by	Deposition Process
$MoSi_2$	Disil	Boeing	Fluidized bed
	PFR-6	Pfaudler	Pack cementation
	L-7	McDonnell-Douglas	Slip pack
	LM-5	Linde	Plasma spray
	$MoSi_2$	Vitro	Electrophoresis
$MoSi_2$+Cr	W-2	Chromalloy	Pack cementation
	Durak-MG	Chromizing	Pack cementation
	PFR-5	Pfaudler	Pack cementation
$MoSi_2$+Cr, B	Durak-B	Chromizing	Pack cementation
	W-3	Chromalloy	Pack cementation
$MoSi_2$+Cr, Al, B	Vought II, IX	Chance Vought	Slip pack
$MoSi_2$+Sn–Al		Battelle-Geneva	Cementation and impregnation

coatings in which the aluminide layer is covered by a ductile metallic Sn–Al alloy. This coating system appears to have superior creep and fatigue behavior. The coating also has excellent self-healing characteristics by virtue of the metal alloy reservoir, which is liquid at service temperatures. Use in low-pressure environments is limited, of course, by the vapor pressure of tin. This coating is the only aluminide coating commercially available today. It finds wide use as a coating for tantalum and its alloys, and current technology can be used to apply the coating to molybdenum.

In general, the aluminide coatings provide good oxidation protection to molybdenum at temperatures up to 2800°F. Life at higher temperatures is very short (less than 1 hr), probably as a result of rapid interdiffusion. The performance at lower temperatures overlaps that of the silicide-base coating systems. Insufficient work has been done to adequately characterize the oxidation behavior of aluminides on molybdenum, and overall performance and reliability, by-and-large, are not known. It is very likely that performance will

be degraded in air at reduced pressure; however, no studies have been made to evaluate environmental effects.

SILICIDE COATINGS

Interest in coatings for molybdenum shifted from an aluminide to a silicide base in the mid-1950's. The evolution of the W-series of coatings by Chromalloy Corporation in 1953–1954 promised a far greater potential for the development of a reliable high-performance coating system. At least a dozen varieties of the basic silicide coating have been developed by nearly as many different companies in the succeeding years, as shown in Table 20. Reviews of oxidation and mechanical behavior are given in References 1, 3, 5, and 8–10.

With few exceptions, most of the silicide coatings are deposited by pack-cementation diffusion processes. The basic nature of the process limits the size and complexity of

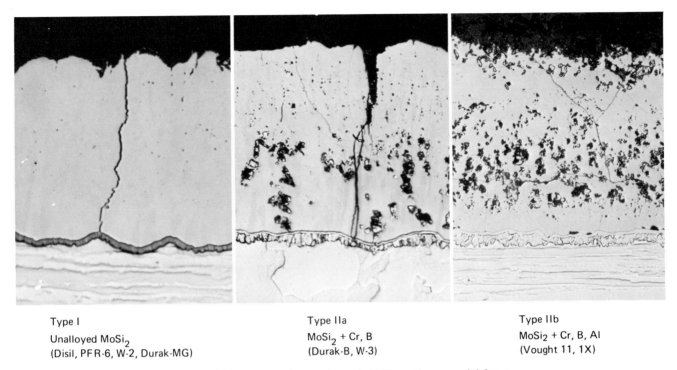

Type I	Type IIa	Type IIb
Unalloyed MoSi$_2$	MoSi$_2$ + Cr, B	MoSi$_2$ + Cr, B, Al
(Disil, PFR-6, W-2, Durak-MG)	(Durak-B, W-3)	(Vought 11, 1X)

FIGURE 39   Representative structures of silicide coatings on molybdenum.

parts that can be coated. The largest parts coated to date are 18 in. in the longest dimension.[11,12] Retorts with two large dimensions (4 × 6 ft) can be processed provided the third dimension is limited to 15–18 in.[13] Large three-dimensional retorts require excessively long heating times and are not practical for the manufacture of uniformly coated parts. The fluidized-bed process overcomes this difficulty. The largest bed designed and operated successfully was 18 in. in diameter.[12,14] This equipment, however, is not operable today. Although limited in size, the technology for silicide coating of molybdenum has been developed to a very advanced state. High-quality reproducible parts can be manufactured on a routine production basis.

The deposition processes also are limited to the manufacture of fairly simple unalloyed or lightly modified silicide coatings. The slurry processes that were developed for depositing highly alloyed coatings on columbium have not been used to any great extent for coating molybdenum. This is due to the fact that columbium and its alloys largely displaced molybdenum in hot-structural and gas-turbine applications due to better fabricability, toughness, and ductility at low temperatures. The slurry processes could be applied to molybdenum if the need arose for more highly alloyed coatings.

The silicide coatings on molybdenum have a fairly simple structure as shown in Figure 39. The majority of coatings are unalloyed or lightly modified and have a Type I structure consisting of 1.5–3.0 mils of MoSi$_2$ over a fractional-

mil interface layer of Mo$_5$Si$_3$. Boron-modified alloys of Type IIa or IIb also have a 1.5–3.0-mil-thick layer of MoSi$_2$. As shown in Figure 39, however, these coatings have a more complex interfacial zone and contain one or more dispersed phases. As a result of this simple structure, the behavior of silicide coatings has been comparatively easy to characterize. More is known of the fundamentals that govern protection and failure for these simple systems than for any other high-temperature coating system.[9,10,15] This basic understanding of behavior has been an important factor in the development of high-performance coated hardware and the effective utilization of coated molybdenum alloys in a wide range of applications.

As discussed above, the useful life of silicide coatings for molybdenum is directly proportional to coating thickness (parabolic function) and inversely proportional to temperature (exponential function). This is due to the fact that life in most applications is governed by interdiffusion with the substrate. As shown in Figure 40, an MoSi$_2$ coating is completely converted to Mo$_5$Si$_3$ in less than 2 hr at 3000°F. Although the coating thickness is doubled, it will fail when all of the MoSi$_2$ is converted to Mo$_5$Si$_3$, since the lower silicide is less oxidation-resistant. Diffusion data and layer growth measurements can be used to predict coating life, as shown in Figure 41. The theoretical curves of life versus temperature were calculated for three different coating thicknesses from layer growth measurement data.[16] It can be seen that the actual life of many different silicide coat-

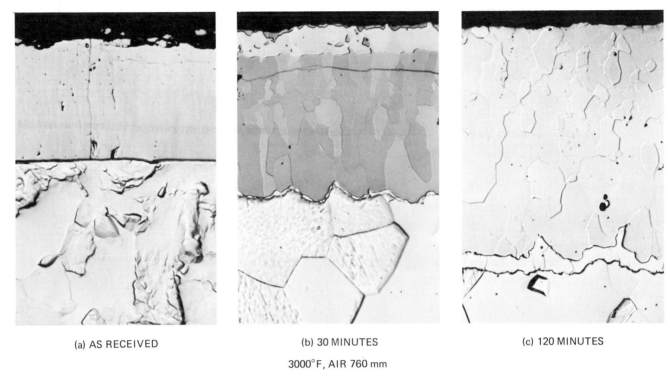

(a) AS RECEIVED   (b) 30 MINUTES   (c) 120 MINUTES

3000°F, AIR 760 mm

FIGURE 40   Changes in coating structure and composition as a result of interdiffusion with the substrate.

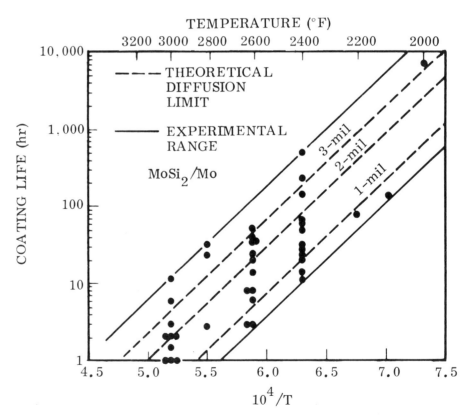

FIGURE 41   Comparison of experimental life data with theoretical life data based on layer growth measurements.

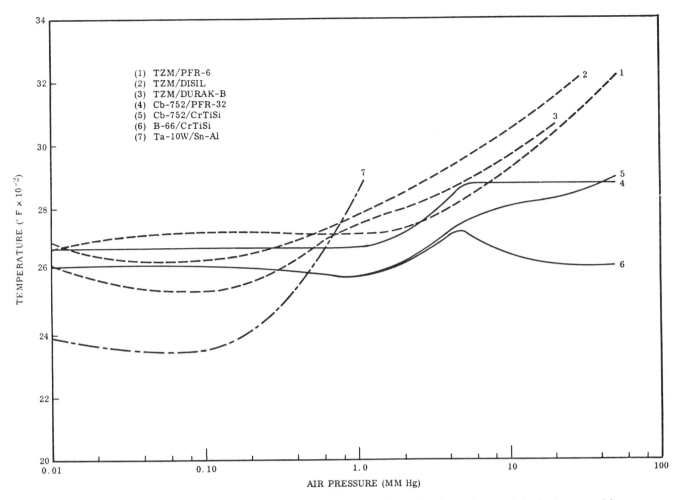

FIGURE 42   The effect of air pressure on the maximum temperature for a 4-hr life of silicide-coated refractory materials.

ings on molybdenum with thickness of 1–3 mils is in excellent agreement with calculated limits. Each plotted point is a published test value for a specific coating system. At least six different systems are represented by these data. This set of curves presents a realistic estimate of life that can be expected from the best available silicide coatings on molybdenum.

These data show a range in coating life of about 1½ cycles at any temperature. For example, at 2400°F, published data show a range of lifetimes from 10 to 150 hr. This, in most cases, is the direct result of variations in effective coating thickness. The effective thickness usually is less than the actual (total) thickness as a result of defects (cracks, fissures, etc.) in the coating. In most silicide coatings on molybdenum, V-shaped cracks penetrate the outer half of the coating in many areas. Failures occur first at the root of these cracks when all the $MoSi_2$ is converted to Mo_5Si_3. The effective coating thickness in many cases is only 50 percent of the total thickness. In one case, performance of a coating system was not changed by removing the outer one half of

the silicide layer before testing.[10] Performance of these coating systems is largely defect-controlled. Significant advances in performance and reliability can be realized if coating defects can be mitigated or removed by improved processes.

A novel approach to the control of defects is provided in a new coating system developed by Battelle-Geneva.[17] In this system, a porous $MoSi_2$ coating is deposited on molybdenum and subsequently infiltrated with a Sn–Al alloy. The coating, in effect, is a combined silicide–aluminide coating. The defect pattern is altered by this process, and the liquid Sn–Al alloy (at temperature) effectively covers coating defects. Coating lifetimes are comparable to those obtained with fully dense $MoSi_2$ coatings at temperatures of 1800° to 3200°F. Reproducibility, however, appears to be better for the porous infiltrated systems.

A major deficiency in the performance of silicide-base coatings appears when the system is used in low-pressure environments.[10,18,19] As shown in Figure 42, silicide coatings that will protect TZM substrates for four hours at

TABLE 21   Oxide Coatings

Type	Designation	Deposition Process	Thickness Range	Developer
ZrO_2–Glass	–	Frit (enamel)	5–30	NBS
Cr–Glass	–	Frit (enamel)	5–10	NBS
Cr–Al_2O_3	GE 300	Flame spray over Cr plate	8–15	General Electric
Al_2O_3	Rockide-A	Flame spray	1–100	Norton
ZrO_2	Rockide-Z	Flame spray	1–100	Norton
ZrO_2	ZP-74	Troweling	100–300	Marquardt

3000° to 3200°F in air at 1 atm cannot be used above 2700°F in air at pressure of 0.1 to 1.0 Torr. This degradation in performance is common to all silicide coatings on all refractory metal substrates and is the result of loss of silicon to the atmosphere in the form of Si, SiO, and SiO_2 vapors. Recent studies with coated Cb alloys indicate that heavy modification of silicide coatings by alloying can improve the performance at low pressure.[20] In no instances, however, has the attack been eliminated. The low-pressure attack of $MoSi_2$ coatings occurs preferentially at hairline cracks and fissures. Thus, the control and mitigation of coating defects is another approach to improved performance in low pressure environment.

In spite of these shortcomings, silicide coatings on molybdenum are still the best systems available for high-temperature service from an oxidation point of view. These systems are in the most advanced state of development and are commercially available today for a wide variety of applications. In general, performance capabilities and reliability satisfy current requirements for the majority of applications in which molybdenum alloys can be used. Problems with fabrication, toughness, and ductility are the major factors that limit the applicability of silicide-coated molybdenum alloys.

OXIDE COATINGS

Refractory oxides (ceramics) are the only materials suitable for the oxidation protection of molybdenum above 3000°F. The types of oxide coatings that have been used on molybdenum are summarized in Table 21. The deposition processes used result in the formation of comparatively thick coatings compared with the intermetallic diffusion coatings. Many of the oxide coatings tend to be porous and are used as thick overlays to provide adequate protection. Also, many were developed initially as thermal insulating coatings with oxidation protection as a secondary consideration.[21]

Ceramic coatings suffer from one common problem: They crack on thermal cycling and tend to spall from the substrate. For short-time (minutes), single-cycle use (e.g.,

rocket motors), they can provide very useful protection from oxidation to 3500°F (Al_2O_3) and to 4000°F (ZrO_2). For long-time (hours) or multiple (cyclic) use, the metallic reinforced and attached systems developed by Marquardt[21] show considerable promise. Although the coatings are not impervious to oxygen, they reduce the partial pressure at the metal surface to values sufficiently low that oxidation rates can be tolerated.

SUMMARY

The performance capabilities of various coatings for molybdenum based on useful life in static air at atmospheric pressure are summarized in Figure 43. The time to failure and temperature limits for metallic coatings and aluminides overlaps the performance range for silicides. The silicides have the broadest range of applicability and provide the most versatile coating systems. Silicide coatings are in the most advanced state of development of all coatings for molybdenum. High-quality, reliable coatings can be applied commercially to large, complex shapes and assemblies.

None of the coatings for molybdenum will provide universal protection. All are subject to the formation of defects, and oxidation behavior is largely defect-controlled. Performance will be degraded under conditions of cyclic use and in low-pressure environments. In spite of their many shortcomings, the coatings for molybdenum will provide adequate protection for the many applications in which molybdenum can be used. The problem is not one of developing an adequate coating system. Instead, it is one of selecting the most suitable coating system on the basis of design, fabrication, and end-use considerations and of characterizing the minimum levels of performance and reliability in simulated and actual service environments. Mechanical limitations with respect to forming and joining, a tendency to brittle fracture, and the dependence of strength and ductility on a strain-hardened state are the major factors that limit the use of coated molybdenum alloys in high-temperature applications today.

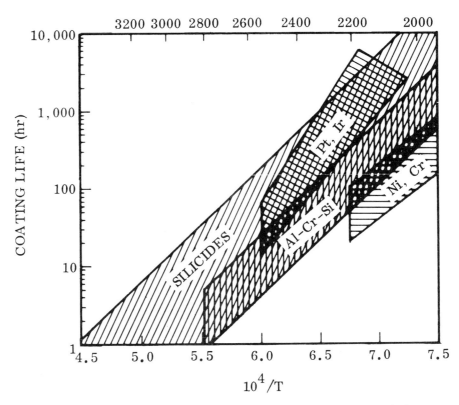

FIGURE 43   Performance capabilities of coatings for molybdenum in air at atmospheric pressure.

REFERENCES

1. Bartlett, E. S., H. R. Ogden, and R. I. Jaffee. March 6, 1959. Coatings for protecting molybdenum from oxidation at elevated temperatures. DMIC Report No. 109.

2. Vaaler, L. E., C. A. Snavely, and C. L. Faust. April 1, 1953. Introductory plating studies on protecting molybdenum from high-temperature oxidation. USAEC BMI-813.

3. Krier, C. A. Nov. 24, 1961. Coatings for the protection of refractory metals from oxidation. DMIC Report No. 162.

4. Girard, E. H., J. F. Clarke, and H. Breit. May 1966. Study of ductile coatings for the oxidation protection of Cb and Mo alloys. Final Report, Bur. of Naval Weapons Contract NOw 65-0340-f.

5. Coated refractory metal technology—1965. Nov. 1965. Materials Advisory Board, Report MAB-210-M.

6. Blanchard, J. R. June 1955. Oxidation resistant coatings for molybdenum. WADC-TR-54-492, December 1954. Part II.

7. Wilks, C. June 1961. Evaluation of coatings for molybdenum. Martin Corp. Engineering Report 11462-6. Contract NOw 60-0321C.

8. Gibeaut, W. A., and E. S. Bartlett. January 1964. Properties of coated refractory metals. DMIC Report 195.

9. Perkins, R. A. 1965. Status of coated refractory metals. J. Spacecraft 2:520–523.

10. Perkins, R. A., and C. M. Packer. Sept. 1965. Coatings for refractory metals in aerospace environments. AFML-TR-65-351.

11. Chao, P. J., G. J. Dormer, and B. S. Payne, Jr. May 1964. Advanced development of PFR-6. AFML-TDR-64-84.

12. Stratton, W. K., et al. March 1965. Advances in the materials technology resulting from the X-20 Program. AFML-TR-64-396.

13. Dormer, G. J., B. S. Payne, Jr., and E. G. Pike. Aug. 1965. Final report on manufacturing methods for high-temperature coating of large molybdenum parts. AFML-TR-65-282.

14. Fluidized bed techniques for coating refractory metals. 1965. IR-8-184 (I, II). The Boeing Co. Contract AF 33(615)-1392.

15. Bartlett, R. W. 1964–1965. Investigation of mechanisms for the oxidation protection and failure of intermetallic coatings for refractory metals. ASD-TDR-63-753, Pt. I, II, III.

16. Perkins, R. A. Feb. 1964. Coatings for refractory metals: Environmental and reliability problems. Proceedings of the ASM Golden Gate Conference on Materials Science and Technology for Advanced Applications. Vol. II.

17. Stecher, P., and B. Lux. 1968. Oxidationsschutz von Molybdän durch eine MoSi$_2$/SnAl Schutzschicht. J. Less-Common Met. 14:407.

18. Perkins, R. A., L. A. Riedinger, and S. Sokolsky. 1963. Oxidation protection of refractories during re-entry. Space/Aeronautics 39(6):107–115.

19. Packer, C. M., and R. A. Perkins. 1967. Performance of coated refractory metals in low-pressure environments. Refractory Metals and Alloys–IV, Vol. 2, Proc. of AIME Metallurgical Society Conference, Vol. 41, New York.

20. Childers, M. G. May 1968. Behavior of Sylcor R512E coated Cb-752 columbium alloys in a low pressure environment. Summary of the Thirteenth Refractory Composites Working Group Meeting. AFML-TR-68-84.

21. Blocker, E. W., C. Kalleys, S. V. Castner, and S. Sklarew. Oct. 1961. Reinforced, refractory, thermally insulating coatings. SAE Technical Paper 417D.

TANTALUM AND TUNGSTEN*

INTRODUCTION

The following represents an up-to-date survey of protective coatings for Ta and W alloys with sufficient comment for reference purposes. The emphasis is on government-sponsored work and reports, since very little work has been financed and published by private industry. For convenience, the coatings are divided into three classes: intermetallics, metallics, and ceramics. There is a small section devoted to evaluation programs, since several large programs have fallen exclusively into this latter category. Practically all significant programs have been initiated since 1960. A DMIC report[1] gives a good survey of the field to that time.

TANTALUM ALLOYS

Much of the work on Ta alloys was the outgrowth of approaches taken with columbium alloys, which in turn followed molybdenum-alloy work. Accordingly, the development of coatings is of relatively recent vintage. Most of the emphasis was on the diffusional growth of intermetallic layers, and this field, accordingly, represents the bulk of the work done to date. A number of commercial coatings have been successfully developed.

INTERMETALLICS

One of the earliest programs was that at Battelle,[2,3] where the application of silicides by pack-cementation techniques was investigated using single- and two-step processes. At least three alloy substrates were used, in addition to pure Ta. The most effective coatings developed were two-step processes, Si + B, Si + Mn, and Si + V. However, "pest" was very prevalent, except for the Ta-30Cb-7.5V substrate. Fairly good lives could be obtained with this substrate even with an unmodified coating. No commercial or semicommercial coatings have survived from this work.

The Battelle program was followed up by Solar,[4] where a two-step pack-cementation process was evolved that has the semicommercial designation Pa-8. This consists of heavily titanizing the surface from a 90Ti-10W pack, followed by straight siliciding.

*Prepared by L. Sama and S. Priceman. Copyright © 1969, American Society for Metals.

In early work at Sylvania,[5] aluminide and beryllide coatings were investigated. Beryllides were applied through the vapor phase at 1800°F. Although offering protection, their major drawbacks were cracking due to expansion mismatch, complexity in coating layer formation, and rapid interaction with the substrate at 2500°F or above. Diffusion was much slower in Ta-10W than in pure Ta. Aluminides were first applied by hot-dipping and pack cementation. Later, a series of Sn-Al-Mo-base slurry coatings were developed[6] and scaled up to commercial size.[7] These coatings are limited primarily by poor resistance to reduced pressure and erosion at very high temperatures.

Kofstad in Norway made a basic study[8] of the behavior of vacuum-hot-dipped samples of Ta and Ta-10W in Al and Al alloys over a range of temperatures and at reduced pressures. Al-Cr-coated samples appeared to give best results. This was not intended as a commercial coating development.

Studies of electrophoretically deposited binary disilicides have recently been performed at Vitro,[9] where combinations of Mo or W with V, Cr, and Ti were covered. Tests were at 1500°F and 2400°F. Only the Mo-V system showed no "pest." W and Re barriers were also studied. This process can be considered commercial for small parts.

At Boeing,[10] a fluidized-bed three-step Si-V-Si process was developed for Cb and Ta-10W alloys. Although tests at 1 atm and reduced pressure were more successful on Cb alloys, the process can be considered semicommercial for tantalum alloys also.

Recent and continuing work at Solar[11] aimed at Ta-alloy turbine vanes has led to the successful development of complex Ti, Mo, W, and V modified silicide coatings applied by a two-step method: a slurry plus high-temperature sinter of an alloy layer followed by a straight silicide pack. Protection for hundreds of hours at 1600°F and 2400°F in furnace tests has been obtained.

Present development work at TRW[12] is concerned with obtaining a scaled-up process for applying W over Ta alloys as a barrier layer for subsequent siliciding. This type of coating is being further evaluated on another program,[13] where development work is proceeding in the investigation of metal oxide composites as future coatings for extremely high temperatures. These are very complex systems with up to five components.

Recent Air Force-sponsored work at Sylvania[14,15] resulted in the development of a fused-silicide coating system for Cb and Ta alloys that appears practical for coating large, complex aerospace sheet-metal components. One composition, Si–20Ti–10Mo (R512C), is under advanced evaluation.[16] The coating is protective to Ta alloys for hours at temperatures to 3000°F and for over 150 hours at 2000°F.

METALLICS

Primary attention in recent years has been given to the development of hafnium-base alloys and of the platinum metals as possible coating or cladding candidates for very-high-temperature protection. At IITRI, this course was pursued, beginning with a study of binary Hf–Ta and Ir-base alloy oxidation behavior[17] and progressing to cladding[18] studies of Hf–Ta on Ta and Ta–10W and finally to the development of Hf–Ta slurry coatings.[19]

Related alloy oxidation studies were made at Solar[20] on Hf–Ta, Hf–Ni, and Hf. However, more emphasis was placed on Ir–Rh and Ir–Pt alloys. Diffusion couple studies were also made. This study progressed to a second program[21] wherein diffusion barriers of W were utilized and Ir–Rh coatings applied by fused salt-bath deposition. HfO_2 overlays were used to enhance emittance, but limitations in the form of thermal cycling cracks in the composites were encountered.

In a recent program at Sylvania,[22] attempts were made to increase the short-time protective life of Hf–Ta slurry coatings through the use of Al, Cr, and Si additives. A more successful method of improving intermediate temperature life was developed by means of a duplex coating consisting of a sintered HfB_2–$MoSi_2$ layer overlaid with a Hf–Ta slurry. However, the maximum temperature capability of Hf–Ta was lowered to 3300°–3400°F.

CERAMICS

In relation to ceramic coating development, which led to the Pa-8, basic studies were concurrently carried out at Solar.[23] These involved measurements of expansion coefficients of bulk oxides modified with various alloying additives. In addition, the compatibility of possible oxide-barrier-reservoir systems was studied: e.g., Hf–HfO_2 reactions.

TUNGSTEN ALLOYS

Most of the coating work on tungsten has been aimed at service temperatures above 3000°F and has utilized pure tungsten as a substrate. Success in practical coating development has been quite limited, and consequently a considerable amount of effort has been devoted to more fundamental studies.

INTERMETALLICS

The major approach has been with pure or slightly modified silicides. The use of tungsten wires as substrates was found convenient in early work at New York University[24] and General Telephone and Electronics Laboratories.[25] Both investigations encountered the "pest" problem over a considerable temperature range. However, the disilicide was found protective to 3300°F.

Later development work at TRW[26,27] was more extensive, and tungsten sheet samples were utilized as substrates. In the first program, pack coatings were investigated in single and multiple cycles. The more successful approaches were to modify the silicides with W, Zr, and Ti, reportedly to generate a modified complex oxide on subsequent oxidation or in special oxidation treatments. The more promising systems were further developed in a second program. A cursory examination was made of the effect of HfB_2 and other compound additives applied by plasma spraying. It appears that a W-modified silicide pack coating of commercial availability was obtained from this work.

METALLICS

Although platinum metal development work discussed previously was aimed at tantalum,[21] the use of tungsten as a diffusion barrier makes the study significantly applicable to tungsten substrates.

Some practical coating work on tungsten has been reported by Consolidated Controls[28] for a "Multi-Element-Umbrella" type, consisting of iridium electrodeposited from a fused-salt bath followed by a plasma-sprayed ZrO_2 layer to prevent volatilization losses during very-high-temperature oxidation. Low-weight losses were obtained in 9 hr at 3270°F. Some cracking of the oxide overlay occurs because of thermal cycling.

Since the major interest in tungsten is for very high temperatures, little attention has been given to the use of more common oxidation-resistant alloys as clads or coatings. However, in a program at General Electric, the practicality of using nickel-base braze materials[29] as protective coating for 1500°F or so was established.

CERAMICS

Early work was done at the University of Illinois,[30] where tungsten wires were coated with oxide-plus-glass mixtures by resistance heating and then torch tested above 3000°F. It was reported that a glass-plus-zirconia, a silicide, and a combination of the two gave protection. No commercial coatings were developed.

In a study at Gulton Industries,[31] complex glass-ceramic frits were applied, but problems in bonding and expansion mismatch were encountered and little success or actual data were reported.

It was recognized early by the Air Force that a comprehensive program would have to be devoted to the development of very-high-temperature coatings for tungsten. Accordingly, a more fundamental study was let to GT&E Laboratories, where, in a series of reports[32-34] released over a 4-year period, the basic problems were attacked from a theoretical standpoint. The major problem was essentially resolved to be limitations in the stability of potential protective oxide films, primarily from a diffusional standpoint. From a more practical point of view, it appeared difficult to generate potentially attractive complex oxides from various types of reservoirs. This work was supplemented by assists from A. D. Little on oxide vaporization studies and from ManLabs, where interdiffusion effects were investigated from a reservoir-substrate standpoint.

The GT&E Laboratories work has been followed up by a program at Battelle.[35] Several classes of complex oxides representing different crystal structures are being studied on the basis of thermal stability related to vaporization and diffusion effects. These oxides have been primarily rare earth-zirconia combinations.

Related oxide studies have been carried out by Tem-Press,[36] in which the stability of complex oxides containing tungsten was investigated.

One of several more practical development programs was recently completed at Solar,[37] where nose-cap materials were under study. These consisted of ThO_2 and HfO_2 overlays on various types of W and Ta substrates. Best results were obtained with ThO_2 over a W wire mesh, to 5300°F, and with HfO_2 to 4900°F. Ta required an intermediate protective coating.

An earlier NASA program was carried out at IITRI[38] in connection with W for use in rocket engines. Various nitride, carbide, and metal oxide combinations were developed and evaluated. The most promising were metal oxide systems consisting of layers of W and the oxides HfO_2, Y_2O_3, and $SnZrO_3$ applied by plasma spray.

EVALUATION PROGRAMS

A limited evaluation was carried out at Marquardt[39] with three commercial silicide packs on Ta-10W, a Marquardt vapor-deposited SiC on Ta-10W and W, and a Marquardt ZrB_2 plasma-sprayed on Ta-10W. Some development work was done with TaN, Bn, and MoC coatings on both types of substrates. In a later program,[40] six commercial coatings were evaluated on Ta-10W to 3000°F.

More extensive evaluations were performed on Sn-Al-coated T-111 foil at Convair,[41] Solar,[42] and Douglas.[43] Electrophoretic coatings of WSi_2 on T-222 and Ta-10W fasteners were tested at Standard Pressed Steel.[44]

Evaluation programs of more limited extent have been in progress at ManLabs[45] for WSi_2 on W rods and Sn-Al on Ta-10W rods for temperatures above 3000°F and at McDonnell,[46] where three Ta alloys with W plus WSi_2 coatings were tested to 3600°F.

In related coating evaluation work, GT&E Laboratories and Sylvania recently completed an extensive analysis of commercially available coatings on both W and Ta substrates.[47]

SUMMARY

Modified silicide coatings represent the most useful class of coatings on tantalum alloys. In general, they require moderate-to-heavy modification by metallic elements other than tantalum. Considerable practical evaluation and testing is still required for most of these, since most development programs have dealt with restricted test types at only a few temperatures. With moderate modification the silicides are capable of protection to about 3000°F for several hours with some increase in life at lower temperatures—on the order of 10 hr at 2800°F and 500 hr at 2000°F. More heavily modified coatings have demonstrated either longer lives at temperatures below 2500°F or some increase in temperature capability above 3000°F, *but not both*. No silicide coatings have proven capable of more than 1 hr at 3500°-3600°F. The practicality and reliability of processing complex parts has not been proven for most systems but is feasible, depending on part complexity, size, and coating process. For protection up to 1 hr at 3800°F, it has been demonstrated that Hf-Ta coatings are feasible and may be practical for certain applications.

A more restrictive situation exists with tungsten. Essentially, only the pure disilicide has emerged as useful, and it appears protective for a short time to 3500°-3600°F. In the 2000°-3000°F range in simple tests, apparent lives of 10-50 hr are attained. The major problem is to prepare samples that do not laminate. Accordingly, the entire problem of protecting tungsten is still academic.

However, tungsten may be potentially attractive as a diffusion-barrier layer on other metals, since its silicide has potentially the highest temperature capability, interdiffusion with other metals is slow, and it is less reactive with the platinum-group metals. Although little practical coating work has been demonstrated, it may be feasible to obtain longer lives and higher temperatures—for example, with iridium overlays on tungsten diffusion barriers. How practical this type of system would be remains to be demonstrated with particular service applications.

REFERENCES

1. DMIC Report 162. November 24, 1961. Coatings for the protection of refractory metals from oxidation. Defense Metals Information Center, Battelle Memorial Inst.
2. Klopp, W. D., *et al.* March 1962. Development of protective coatings for Ta-base alloys. ASD-TR-676. Battelle Report to ASD WPAFB, Contract AF 33(616)-7184.

3. Hallowell, J. B., D. J. Maykuth, and H. R. Ogden. April 1963. Coatings for Ta-base alloys. Battelle Report ASD-TDR-63-232 to ASD WPAFB on Contract AF 33(657)-7909.

4. Stetson, A. R., H. A. Cook, and V. S. Moore. June 1965. Development of protective coatings for Ta-base alloys. Solar Report to AFML No. TR-65-205, Part I on Contract AF 33(657)-11259.

5. Lawthers, D. D., and L. Sama. 1961. High temperature oxidation resistant coatings for tantalum base alloys. Sylvania-Corning Report No. ASD-TR-61-233 to ASD WPAFB on Contract AF 33(616)7462.

6. Sama, L. February 1963. High temperature oxidation resistant coatings for Ta-base alloys. GT&E Labs Report No. ASD-TDR-63-160 to ASD on Contract AF 33(657)-7339.

7. Sama, L., and B. Reznik. August 1966. Development of production methods for high temperature coating of Ta parts. Sylvania Report to AFML No. AFML-TR-66-217 on Contract AF 33(657)-11272.

8. Kofstad, P. April 1966. Mechanisms of protection and failure in aluminide coated Ta and Ta base alloys. Central Inst. for Ind. Research, Oslo, Norway. Report No. AFML-TR-66-60 to AFML on Contract AF 61(052)-834.

9. Ortner, M. H., and S. J. Klach. December 1966. Development of protective coatings for T-222 alloy. Vitro Labs Report No. NASA CR-54578 to NASA on Contract NAS 3-7613.

10. Batuck, S., et al. April 1967. Fluidized bed techniques for coating refractory metals. Boeing Co. Report No. AFML-TR-67-127 to AFML on Contract AF 33(615)-1392.

11. Wimber, R. T., and A. R. Stetson. July 1967. Development of coatings for Ta alloy nozzle vanes. Solar Report No. NASA-CR-54529, RDR-1396-3 to NASA on Contract NAS 3-7276.

12. Kmieciak, H. A., and J. D. Gadd. May 1 to July 31, 1967. Manufacturing techniques for application of duplex W/Si coating on Ta components. Fourth Interim Progress Report No. ER-6970-4, TRW Inc. to AFML on Contract AF 33(615)-5405.

13. Ebihara, W. T., and K. C. Lin. October 1 to December 31, 1966. Development and characterization of high temperature coatings for Ta alloys. TRW Second Interim Progress Report No. ER 6972-2 to AFML on Contract AF 33(615)-5011.

14. Priceman, S., and L. Sama. Sept. 1965. Development of slurry coatings for Ta, Cb, and Mo alloys. AFML-TR-65-204, Sylvania Electric Products to AFML on AF 33(615)-1721.

15. Priceman, S., and L. Sama. Dec. 1966. Development of fused slurry silicide coatings for the elevated temperature oxidation protection of Cb and Ta alloys. Annual Summary Report STR 66-5501.14, Sylvania Electric Products to AFML on AF 33(615)-3272.

16. Yaffe, M. L. January 1, 1968. Tantalum elevon built for 3000°F test. Aviat. Week Space Technol.

17. Rausch, J. J. Nov. 1964. Protective coatings for Ta base alloys. IIT Research Inst., Report No. AFML-TR-64-354 to AFML on AF 33(657)-11258.

18. Hill, V. L., and J. J. Rausch. January 1966. Protective coatings for Ta base alloys. IIT Research Inst., Report AFML-TR-64-354, Pt. II, to AFML on Contract AF 33(657)-11258.

19. Hill, V. L., and J. J. Rausch. September 1966. Protective coatings for Ta base alloys. IIT Research Inst. Report AFML-TR-64-354, Pt. III, to AFML on Contract AF 33(657)-11258.

20. Wimber, R. T. November 1965. Development of protective coatings for Ta base alloys. Solar Report No. ML-TDR-64-294, Part II to AFML on Contract 33(657)-11259.

21. Dickson, D. T., R. T. Wimbler, and A. R. Stetson. October 1966. Very high temperature coatings for Ta alloys. Solar Report No. AFML-TR-66-317 to AFML on Contract AF 33(615)-2852.

22. Lawthers, D. D., and L. Sama. March 1 to May 31, 1967. Development of coating for protection of high strength tantalum alloys in severe high temperature environments. Sylvania Report No. STR-67-5601.12, Progress Report #4, to AFML on Contract AF 33(615)-5086.

23. Ohnysty, B., et al. January 29, 1965. Task II–development of technology applicable to coatings used in the 3000–4000°F temperature range. Solar Report No. ML-TDR-64-294 to AFML on Contract AF 33(657)-11259.

24. Goetzel, C. G., and P. Laudler. March 1961. Refractory coatings for tungsten. N.Y.U. Research Div. to WADD, Report No. WADD TR-60-825 on Contract AF 33(616)-6868.

25. Pranatis, A., C. Whitman, and C. Dickinson. 1961. Reactions and protective systems involving tungsten and refractory compounds–Part II, protection of tungsten against oxidation at elevated temperatures. GT&E Labs Report at Fifth Meeting of Refractory Composites Work Group, Dallas, August 8–10, 1961.

26. Nolting, H. J., and R. A. Jefferys. May 1963. Oxidation resistant high temperature protective coatings for tungsten. TRW Report No. ASD-TDR 63-459 to AFML on Contract AF 33(616)-8188.

27. Nolting, H. J., and R. A. Jefferys. July 1964. Oxidation resistant high temperature protective coatings for tungsten. TRW Report ML-TDR-64-227 to AFML on Contract AF 33(657)-11151.

28. Engdahl, R. E., and J. R. Bedell. 1967. Two protective systems for refractory metals operating in air at high temperatures. Consolidated Controls Corp., presented at 13th Refractory Composite Working Group Meeting, Seattle, Wash., July 1967.

29. Baxter, W. G., and F. H. Welch. July 1961. Physical and mechanical properties and oxidation resistant coatings for a tungsten-base alloy. APEX-623 G. E. NMPO and FPLD to AF and AEC on Contract AF 33(600)-38062 and AT (11-1)-171.

30. Bergeron, C. G., et al. August 1960. Protective coatings for refractory metals. University of Illinois Report to WADD No. WADC TR-59-526, Part II on Contract AF 33(616)-5734.

31. Rabin, W. J., and H. P. Tripp. March 1961. Development of corrosion resistant coatings for use at high temperatures. Gulton Industries Report No. WADD-TR-60-495 to WADD on Contract AF 33(616)-6374.

32. Nicholas, M. G., et al. April–December 1961. The analysis of the basic factors involved in the protection of tungsten against oxidation. GT&E Labs Report ASD-TDR-62-205, Part I, to ASD on Contract AF 33(616)-8175.

33. Dickinson, C. D., and L. L. Seigle. July 15, 1963. Experimental study of factors controlling the effectiveness of high temperature protective coatings for tungsten. GT&E Labs Report ASD-TDR-63-744 to ASD on Contract AF 33(657)-8787.

34. Brett, J., and L. L. Seigle. Aug. 1965. Experimental study of factors controlling the effectiveness of high temperature protective coatings for tungsten. GT&E Labs Report AFML-TR-64-392 to AFML on AF 33(657)-8787.

35. Phalen, D. I., et al. July 14, 1967. Investigation of Oxidic-type materials as high temperature protective coatings for tungsten. Battelle Third Progress Report to AFML on Contract AF 33(516)-5249.

36. Phillips, B., et al. 1965–1966. Research on criteria for selection of alloys and surface treatments for inhibition of tungsten oxidation. Tem-Press Research Corp., State College, Pa., Report ML-TDR-64-230, Part II, August 1965 and Part II, August 1966, to AFML on Contract AF 33(657)-11235.

37. Licciardello, M. R., and B. Ohnysty. April 1967. Development of composite structures for high temperature operation. Solar

Report No. AFFDL-TR-67-54 to AFFDL on Contract AF 33(615)-1831.

38. Bliton, J., *et al.* October 1963. Protective coatings for refractory metals in rocket engines. IIT Research Inst. Report No. B237-12 to NASA on Contract NAS 7-113.

39. Cox, P. B., *et al.* Feb. 15, 1963. Phase I summary technical report of the calendar year 1962, Air Force–Marquardt contributing engineering program. Report 25,065, Vol. VIII, Marquardt Corp. on Contract AF 33(600)-40809.

40. Kallup, C., and S. V. Castner. 1965. Cyclic thermal testing of coated refractory metals. Marquardt Report, presented at 67th Annual Meeting of American Ceramic Society, Philadelphia, Pa., May 1965.

41. Kerr, J. R., and J. D. Cox. February 1965. Effect of environmental exposure on mechanical properties of several foil gage refractory alloys and superalloys. General Dynamics/Convair Report AFML-TR-65-92 to AFML on Contract AF 33(657)-11289.

42. Moore, V. S., and A. R. Stetson. 1964–1965. Evaluation of coated refractory metal foils. Solar Report RTD-TDR-63-4006,

Part II to AFML on AF 33(657)-9443, December 1964; also Part III of same report on AF 33(615) 2049, June 1965.

43. Leggett, H., *et al.* August 1965. Mechanical and physical properties of super alloy and coated refractory alloy foils. Douglas Aircraft Report AFML-TR-65-147 to AFML on AF 33(657)-11207.

44. Roach, T. A., and E. F. Gowen. September 1966. Structural fasteners for extreme elevated temperatures. Standard Pressed Steel Report AFFDL-TR-66-107 to AFFDL on Contract AF 33(657)-11684.

45. Kaufman, L. June 1967. Stability characterization of refractory materials under high velocity atmospheric flight conditions. Manlabs Inc. Progress Report #3 to AFML on Contract AF 33(615)-3859.

46. Jackson, R. E. May 1 to Aug. 1967. Tantalum system evaluation. McDonnell Astronautics Interim Technical Report #5 to AFFDL, Contract AF 33(615)-3935.

47. Bracco, D. J., *et al.* May 1966. Identification of microstructural constituents and chemical concentration profiles in coated refractory metal systems. GT&E Labs Report AFML-TR-66-126 to AFML on Contract AF 33(615)-1685.

GRAPHITE

REVIEW OF WORK UP TO 1964

Several comprehensive reviews on refractory coatings for graphite have been published. In 1963, Chown *et al.*[1] reviewed (73 references) the performance of coatings made of various refractory carbides, borides, and silicides. They concluded that SiC, $MoSi_2$, and SiC in combination with ZrB_2 or TiB_2 are the most oxidation-protective coating systems. In the same paper, the failure mechanisms of SiC and $MoSi_2$ coatings were also analyzed. The status in 1964 was reviewed by Criscione *et al.*[2] (209 references) and by Schulz *et al.*[3] (63 references). In addition to reviewing the then-existing coating systems for graphite, Criscione and Schulz and their colleagues[2,3] considered the actual or anticipated performance limitations of the various classes of refractory materials, e.g., metals, oxides, carbides, borides, and nitrides. It was concluded that while SiC coatings perform satisfactorily to approximately 1650°C, any breakthrough in coating technology for protection at substantially higher temperatures must come in one of two ways: (a) coating with a noble metal, e.g., iridium; or (b) multilayer coatings consisting of an outer layer of a refractory oxide and an inner, or carbon-barrier, layer of a refractory carbide or boride. If all coating components are refractory and stable, the ultimate performance limit of double-layer coatings would be predictable from a knowledge of the diffusion rates (either grain boundary or bulk diffusion) of oxygen through the oxide and/or of carbon through the barrier layer; other mechanisms, however (e.g., spalling due to thermal expansion mismatch), could lead to much earlier failures. The Russian work, again up to 1964, has been reviewed by Samsonov and Epik[4] (149 references almost exclusively to Russian literature). Samsonov's article is divided into several chapters dealing with boride, carbide, nitride, and silicide coatings on metals, a chapter on coatings for graphite, and a chapter on the properties of high-temperature compounds. As far as coatings for graphite are concerned, Samsonov reveals, of course, nothing that is particularly new. One does, however, get the impression that, at least up to 1964, Russian engineers and scientists were following the same reasoning and were struggling with the same problems as their U.S. counterparts.

Due to the extensiveness of the cited reviews (particularly References 1, 2, and 3) only the newer work, mostly aimed at obtaining protection for graphite at temperatures substantially above 1650°C, is discussed further.

RECENT COATING DEVELOPMENTS

LITERATURE SEARCH

The more recent literature (since mid-1964) pertaining to coatings for graphite is listed in References 5 through 36. To make this reference list more useful, a few words summarizing the content of the paper have been added in most cases. Work on relatively low-temperature coatings for commercial applications, nuclear-fuel-element coatings, and other work believed to be outside the interests of the Committee on Coatings has not been included. Progress reports are listed only where appropriate summary reports are not yet available. Papers that present only work discussed in more detail in contract reports have also been omitted.

Recent graphite coating developments are principally

directed towards two applications: (a) re-entry protection for structural parts and (b) graphite rocket-nozzle liners. The Russian work for re-entry protection[5–8,11,20,24,25] still appears centered around the various carbides and nitrides, although this may simply be a reflection of the time lag or secrecy inherent to Russian publications. Most current American work on re-entry protection for graphite involves the use of iridium and iridium alloys.

IRIDIUM COATINGS FOR GRAPHITE

The research on iridium coatings for graphite is discussed in References 26–28. It has been shown that iridium is virtually impervious to oxygen diffusion up to 2100°C[26]; that iridium does not form a carbide and that diffusion at the iridium–graphite interface is negligible[26,27]; that the highest temperature at which iridium will protect graphite is the eutectic temperature of approximately 2280°C[27,28]; that minor amounts of impurities (B, Si, Fe) can lower the eutectic temperature by as much as 200°C; and that the protectiveness of iridium coatings is limited only by the rate of oxidation of iridium, which was determined as a function of temperature, oxygen partial pressure, and gas flow rate.[26–28] It was also shown that the thermal expansion characteristics of iridium require the use of graphite substrates with a high thermal expansion coefficient[26] and that iridium can, in principle, be applied to graphite by slurry-dipping and sintering,[26] cladding,[26] vapor-plating,[26,27] and electroplating.[26]

In an effort to extend the upper temperature limit to which iridium coatings can be used, Harmon[29] investigated selected iridium–osmium–rhodium, iridium–osmium–platinum, and iridium–rhenium–rhodium alloys, determining melting points, carbon eutectic temperatures and compositions, and thermal expansion characteristics of these alloys. He found that additions of rhenium and osmium increase the eutectic temperature with carbon, while iridium–rhodium and iridium–platinum alloys have lower carbon eutectics than pure iridium. An iridium alloy containing 30 mole% rhenium exhibited the highest eutectic temperature with carbon (approximately 2480°C). The thermal expansion characteristics of these alloys were found to be similar to those of pure iridium.

Present follow-up contracts to the work discussed in References 26–28 are concerned with improved methods of depositing iridium on graphite. Wright et al.[30–34] prepared pore- and crack-free iridium coatings that exhibited good adherence by plasma deposition followed by isostatic hot-pressing. The method now preferred[34] for coatings of pure iridium consists of nine steps:

1. Outgas graphite at 2000°C.
2. Plasma-arc-deposit iridium.
3. Outgas at 2000°C.
4. Wrap specimen in graphite foil (previously outgassed at 2000°C).
5. Enclose wrapped specimen in thick-wall, mild-steel container.
6. Electron-beam-weld container vacuum-tight.
7. Gas-pressure-bond at 1090°C for 2 hr at 10,000 psi.
8. Remove specimen from container specimen by machining and grinding.
9. Buff specimen to remove graphite foil.

In addition to optimizing deposition and processing parameters, Wright et al. also determined the influence of various graphite grades on coating performance,[32–34] the emittance of iridium in both air and vacuum to 2000°C,[32,34] and the use of ThO_2 and/or HfO_2 additions to the iridium coating.[32,33] Simultaneously with the work of Wright et al., Macklin and LaMar[35,36] investigated other methods for depositing iridium on graphite, including vapor deposition, electroplating from fused-salt and aqueous solutions, and thermal decomposition of iridium resinate. Certain problems were encountered with each method. A combination coating consisting of a thin inner layer deposited from iridium resinate and an outer layer deposited from a fused-cyanide electrolyte is at present under more detailed evaluation.[36]

REFERENCES

1. Chown, J., R. F. Deacon, N. Singer, and A. E. S. White. 1963. Refractory coatings on graphite, with some comments on the ultimate oxidation resistance of coated graphite. In P. Popper, ed. Proceedings of a Symposium on Special Ceramics 1962. Academic Press, New York.
2. Criscione, J. M., R. A. Mercuri, E. P. Schram, A. W. Smith, and H. F. Volk. June 1964. High temperature protective coatings for graphite. ML-TDR-64-173, Part I, AD 604-463.
3. Schulz, D. A., P. H. Higgs, and J. D. Cannon. July 1964. Research and development on advanced graphite materials. Vol. 34, Oxidation resistant coatings for graphite. WADD TR 61-72.
4. Samsonov, G. V., and E. P. Epik. 1966. Coatings of high temperature materials. Part I, H. H. Hausner, ed., Plenum Press, New York.
5. Burykina, A. L., A. N. Krasnov, and T. M. Yevtushok. 1965. Plasmic and Diffusive Coatings on Graphite, Porosh, Metal, (No. 12). Trans. in Sov. powder met., (December 1965), Application on graphite of Cu, Al, Nichrom, Mo, and W coatings by plasma spraying; Ti, Zr, Nb, and SiC coatings by application of a metal layer on graphite with subsequent diffusion treatment.
6. Burykina, A. L., T. V. Dubovik, T. M. Yevtushok, A. N. Krasnov. 1965. Coatings of aluminum nitride on graphite. High temp., 3:883-85 (Nov.–Dec.). Trans. from Teplofiz, vysokikh temperature, 3:940-42 (Nov.–Dec.). Various nitriding conditions for Al layer on graphite.
7. Burykina, A. L., T. M. Yevtushok. 1965. Coating of niobium carbide and boron carbonitride on graphite. Inorg. mats., 1:914-16 (June). Trans. from Izvest. Akad. Nauk SSSR, Neorg. Mat.,

1:996-98 (No. 6, June). Includes results of application of Ti and Zr to graphite surface.

8. Burykina, A. L., T. M. Yevtushok. Dec. 1965. Titanium and zirconium carbide coatings on graphite. NASA TT F-9861 (N66-16148).

9. Blome, J. C., D. L. Kummer. 1964. Ceramics and graphite for glide reentry vehicles. Paper presented at the 1964 Fall mtg. of the Am. Ceramic Soc. McDonnell. Nosetip-ZrO_2; CrO coated ZrO_2; graphite with SiC over-coated with ZrO_2, heading edges— ATJ graphite coated with SiC; prop. of Al_2O_3, ZrO_2, ThO_2.

10. Berkeley, J. F., A. Brenner, W. E. Reid, Jr. June 1967. Vapor deposition of tungsten by hydrogen reduction of tungsten hexafluoride: Process variables and properties of the deposit. Electrochem. Soc., J., 114:561-8. Developed as protective coating for graphite in rocket nozzles, National Bureau of Standards.

11. Hrbek, A., B. Krotil. 1966. Silicon spray coating in the gas phase. Korose Ochr. Mater. 10:25-7 (Ca, 66:58511) Si layers on Mo_2 on graphite. Coat is heated to 1600° in H atm. to increase strength.

12. Hill, M. L. 1967. Materials for small radius leading edges for hypersonic vehicles. Amer. Inst. Aer. and astr. ATAA/ASME 8th structures . . . . conf. . . . Candidate materials: carbide-graphite compositions, coated graphite, BN, fused SiO_2 composites, and refractory metal alloys.

13. Kalnin, I. L. May–June 1967. Uniform silicon coatings on graphite by chemical vapor deposition. Electrochem. tech., 5:220-3 Au and Ni precoats provided uniform composite Si-SiC coatings.

14. Krier, C. A. 1966. Protective coatings, Powell, F. Carroll, ed. Vapor Deposition . . . c. Includes coatings for graphite rocket-nozzle inserts, coatings for refractory metals, oxidation-resisting coatings for graphite. Boeing Company.

15. Leonard, B. Aug. 1964. Performance of materials in regions of particle impact (11 × 14 rocket motor test program): Final report. IN: U.S. Air Force Matls. Lab., Summ. of 9th Refractory Comp. Working Group Meeting. Various elbow duct material: W, ZTA graphite pyrolytic graphite, Graphite G, CFZ graphite, AHDG-IA graphite, JT-0832 composite, ablative materials, SiC and pyrolytic C coatings, SiN bonded SiC, fused SiO_2, W-coated ATJ, TaC-coated HLM-85 graphite.

16. Wakelyn, N. T., and R. A. Jewell. 1965. Method of protecting carbon base materials from oxidizing environment, NASA flash Sheet 65-0051. Protect C with a nitride coating and the Si on nitride.

17. Nieberlein, V. A. Jan. 1965. Vapor-deposited tungsten coatings on graphite, Am. Ceram. Soc. Bulletin, 44:14–17. On nozzles.

18. Sitzer, D. H. Sept. 1964. Flame sprayed nickel aluminide coatings . . . how and where to use them. Met. Prog., 86:128, Ni-Al coats on graphite; areas of appl.

19. Simmons, Capt. W. C. 1965. High temperature oxidation resistant coatings for graphite, U.S. Air Force Materials Lab., AFML-TR-65-29, Air Force Materials Symp. Review.

20. Sagalovich, V. V., and G. A. Volkova. Aug. 1965. Contact interaction of titanium with niobium carbide on graphite, Akad,
nauk, SSSR, Izvest., Neorg. Mat., 1:1345-48. Trans. in Inorganic materials, 1:1228-31.

21. Hoertel, F. W. 1964. Vapor deposition of tungsten on Merm rocket nozzles. U.S. Bureau of Mines, Rept. of Investigations 6464. On graphite, on Re-coated graphite, on BeO.

22. Nieberlein, V. A. July 1, 1964. Vapor-plating of tungsten for rocket applications. U.S. Redstone Arsenal, RR-TR-64-6, Coat Graphite.

23. Goodman, E., and R. Thompson. July 1965. Research and development of erosion and oxidation resistant coatings. Value Engineering Co., Final Summary Report, Navy Dept. Bureau of Naval Weapons, Contract No. N600 (19) 61804 . . . Cr-Cr_2O_3 coatings on graphite, ATJ.

24. Zazonova, M. V., and A. A. Appen. July 1964. Double-layer coatings to protect graphite from oxidation in an atmosphere of air at 1400°, J. Appl. Chem. USSR, 37:1443-47, Transl. from Zhur. prik. khim., 37:1447-52 (No. 7, July 1964). Internal layer Al_2O_3; external layer a vitreous silicide containing MoSi.

25. Zemskov, G. V., and A. I. Shestakov. 1965. Diffusion metallization of graphite powders, Porosh Metal., 1-5 (No. 9). Trans. in Sov. powder met., (Sept.). Carbides formed on graphite particle surfaces. 1. Coatings, Carbide-Manufacture; 2. Carbon-Protection; 3. Carbides, Chromium – Manufacture.

26. Criscione, J. M., et al. January 1965. High temperature protective coatings for graphite. AFML-TDR-64-173, Part II. AD 608 092.

27. Criscione, J. M., et al. December 1965. High temperature protective coatings for graphite. AFML-TDR-64-173, Part III, AD 479 131.

28. Criscione, J. M., et al. February 1967. High temperature protective coatings for graphite. AFML-TDR-64-173, Part IV. AD 805 438.

29. Harmon, D. P. February 1967. Iridium-base alloys and their behavior in the presence of carbon. AFML-TR-66-290.

30. Wright, T. R., J. D. Weyand, and D. E. Kizer. April 10 to July 10, 1966. The fabrication of iridium and coatings on graphite by plasma-arc deposition and gas pressure bonding. Battelle Memorial Institute, Contract No. AF 33(615)-3706, First Quarterly Progress Report.

31. Same as Reference 30, Second Quarterly Progress Report, July 10 to October 10, 1966.

32. Same as Reference 30, Third Quarterly Progress Report, October 10, 1966 to January 10, 1967.

33. Same as Reference 30, Fourth Quarterly Progress Report, January 10, 1967 to April 10, 1967.

34. Same as Reference 30, Fifth Quarterly Progress Report, April 10, 1967 to July 10, 1967.

35. Macklin, B. A., F. R. Owens, and P. A. LaMar. August 1967. Development of improved methods of depositing iridium coatings on graphite. AFML-TR-67-195.

36. Macklin, B. A., and P. A. LaMar. May 1967. Development of improved methods of depositing iridium coatings on graphite. Progress Report No. 4, AF Contract No. AF 33(615)-3617.

4

Applications for Coating Systems

INTRODUCTION

Oxidation-resistant coatings for superalloys, refractory metals, and graphites have been developed to the point where adequate performance and reliability can be realized in a wide range of high-temperature applications. Processes for the manufacture of complex parts and assemblies have been scaled up to commercial production levels, and high-quality components have been manufactured and tested successfully. A large number of coated parts are in use today in high-temperature propulsion systems and structural applications. The pattern of use for coating systems that has evolved during the past 20 years and the projected use of current and more advanced coating systems are reviewed in this section of the report.

Success in using coated metals requires a careful matching of design features, manufacturing procedures, and service environments to the characteristics of specific coating systems. No one system has universal applicability, and the performance of all coating systems is determined by a complex interrelation of many factors. Even the best coating system has an imposing number of limitations with respect to its use. This chapter presents an objective analysis of the capabilities and limitations of various coating systems in terms of specific applications. Particular attention is devoted to the factors in each application that govern performance.

GAS TURBINES

APPLICATIONS

Gas-turbine engines are being developed for numerous industrial, marine, and vehicular applications. Industrial engines, which already provide some primary electrical power generation and stand-by power for larger electric-generating facilities, will be used more frequently in future years for a variety of purposes. Marine engines are used to power high-speed boats and ships, including relatively large vessels in the submarine-chase and transportation classifications. Efficient and economical engines will be developed for large commercial and military trucks and military tanks. Gas turbines may also be used to power high-speed ground-transportation systems. These newer applications and increased demand for improved performance in commercial and military aircraft engines will result in unprecedented growth in gas-turbine engine technology and production during the next fifteen years.

LAND-BASED

Engines to be used in trucks and tanks will probably be in the 1,000- to 2,000-hp class. They will be exposed to heavy dust and foreign-particle damage. Along seacoasts they will ingest appreciable quantities of corrosive salt-laden air. In colder climates, ingestion of salt from road de-icing will cause serious corrosion problems. It is anticipated that most of the technology required for perfecting these engines will come from knowledge gained in the development of small aircraft gas-turbine engines.

115

MARINE

Marine engines will be used for both electrical power generation (shipboard and shore-based) and ship propulsion. Horsepower ratings of these engines will range up to 30,000 hp. Engines will operate in air containing appreciable quantities of salt. The development of these engines also will depend upon aircraft gas-turbine engine technology.

AIRCRAFT

Lifting-type gas-turbine engines are being developed to power various vertical takeoff and landing (VTOL) aircraft. Three designs are being considered: direct-lift, lift-cruise, and a combination of direct-lift and lift-cruise. In direct-lift powered aircraft, the lifting engines will be used only during takeoff and landing operations. During normal flight, additional gas-turbine engines will power these aircraft. Lift–cruise powered aircraft differ in that the same engines power the aircraft during its entire flight profile. Utilization of both direct-lift and lift–cruise engines possibly will be most advantageous. Both types of engines will ingest reasonably large quantities of dust and other foreign matter during takeoff and landing operations from unprepared landing sites.

Direct-lift engines will most likely be mounted vertically inside the aircraft (perhaps in the fuselage). They will be used to raise and lower the aircraft above the landing site. At altitude, other engines will take over and power the aircraft (and the lift engines will be shut down). Lift engines, therefore, must be short in length, light in weight, and capable of high performance. These requirements may be achieved at a sacrifice in efficiency. Engines will be used for about 3 min during each takeoff or landing. Nearly 80 percent of their total operating time will be of a cyclic nature. It is anticipated that the first-generation direct-lift engines will have a time between overhaul (TBO) of about 50 hr; eventually, this will be increased to 100 hr or more. The compression ratio of these engines will be about 10:1 for fighter aircraft and 15:1 to 20:1 for transport-type aircraft. Design engineers want to use a minimum of air cooling in the turbine section of these engines. They also desire to use as high a turbine-inlet temperature as possible. (The latter will depend upon availability of materials.) It is anticipated that the turbine-inlet temperatures will be somewhere between 2400° and 2700°F. Without cooling, metal temperatures could approach the turbine-inlet temperatures. The use of cooled superalloys is considered to be the most practical approach for solving the materials problem at the present time. It is planned to cool turbine blades to about 1800°F. Vanes will be cooled to about 2000°F, although these components may have local hot-spots that approach the gas temperature. Cooling techniques being considered include: simple convection, impingement cooling (i.e., the coolant gas impinging against the inside of the leading edge), and film cooling (transpiration). It is anticipated that engine-performance requirements will dictate film cooling, although the problems associated with using this cooling technique are severe.

Lift-cruise engines will not only provide the power for vertical takeoff and landing, but they will be capable of being maneuvered into position for horizontal flight. Their design will be somewhat similar to advanced versions of currently used (cruise-type) gas-turbine engines. Time between overhaul for these engines will be up to 2,000 hr. Initial compressor ratios of 25:1 will eventually approach 40:1. Turbine-inlet temperatures will increase from about 2400°-2500°F in 1970 to about 2750°-2950°F by 1980. The 2400°F gas temperature will result in a metal temperature of about 2200°F. Because of the long life-expectancy requirement, lift-cruise engine designs probably will employ cooled superalloys.

Rapid performance improvements are also programmed for conventional cruise-type aircraft gas-turbine engines. The trend in future engines will be toward higher pressures and higher bypass ratios. Increases in turbine-inlet temperatures are required for optimum cycle efficiency. For long-time service (in excess of 10,000 hr required for commercial aircraft) the maximum permissible turbine-inlet temperature with uncooled blades and vanes is approximately 1800°-1900°F (Figure 44). With air cooling, turbine-inlet temperatures as high as 2200° to 2300°F are permissible, even for long-term service. More sophisticated cooling systems (film cooling) may permit turbine-inlet temperatures to be increased to 2700°F. It is expected that cruise-type engines having increasingly higher turbine-inlet temperatures will be developed. Various cooling techniques will be used to lower the metal temperature sufficiently to permit the use of advanced superalloys.

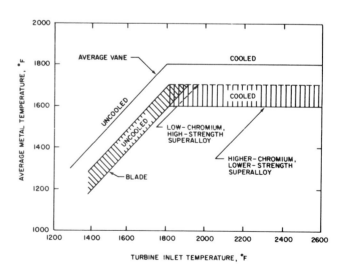

FIGURE 44   Design criteria for first-stage blades and vanes.

ENVIRONMENTS

The phenomena that contribute to the deterioration and failure of hot-section components in gas-turbine engines are: oxidation and sulfidation (hot corrosion), creep, thermal fatigue, erosion, foreign-object damage, and overtemperature damage.

Sulfidation (sometimes referred to as hot corrosion or "black plague") is an accelerated form of high-temperature corrosion associated with the presence of Na_2SO_4. Sodium chloride (ingested by the engine from air) at temperatures above $1100°F$ is converted by SO_2 and SO_3 (from combustion of sulfur impurities in fuel) to Na_2SO_4. Condensed Na_2SO_4 (sodium sulfate vapor–air mixtures are innocuous to superalloys, in contrast to solid or liquid sodium sulfate) is the cause of sulfidation.[1] Disruption or dissolution of protective oxides followed by sulfide formation, chromium depletion in the substrate, and accelerated oxidation describes the phenomenon in general terms. On airfoils, sulfidation is characterized by heavy blistering or swelling in the leading-edge area (heaviest attack generally occurs at locations where the temperature is greatest), accompanied by preferential cracking of the oxide layer at the leading edge in a spanwise direction parallel to the blade axis.[2] Because (for a given NaCl content in the air) Na_2SO_4 can condense at higher temperatures as the pressure increases, it is expected that sulfidation attack will become more serious as engine compression ratios increase. The increased occurence of sulfidation attack in service is also related to the lower effective chromium contents of recently developed high-strength, creep-resistant nickel-base superalloys. Optimum high-temperature mechanical properties for these nickel-base alloys are obtained by compositional modifications that include lowering the chromium content. This reduction in chromium content reduces the hot-corrosion resistance of these alloys. Although cobalt alloy development is not as advanced as development of the nickel-base alloys, cobalt alloys are far more resistant to hot corrosion than the high-strength nickel-base alloys.

Creep is another factor that sometimes limits the useful life of a turbine blade or vane. Centrifugal loading at high temperatures can cause extension of (rotating) blades; pressure differentials between the concave and convex surfaces can cause bowing of (stationary) vanes. This problem prompted metallurgists to develop a family of superalloys of lower chromium content that have excellent high-temperature strength and creep properties, but at a sacrifice of hot-corrosion resistance.

Thermal fatigue (a form of low-cycle fatigue) is frequently responsible for the cracking observed in hot components in gas-turbine engines. Flameholders, turbine blades and vanes, transition ducts, afterburner leaves, and combustion liners are typical of the components that can fail by thermal fatigue. It is reasonable to believe that oxidation, particularly at grain boundaries, can produce sites for the initiation of thermal-fatigue cracks.

Erosion by carbon particles and ingested particulate matter removes protective oxides at airfoil leading edges and promotes premature failure of engine blades and vanes. Maximum erosion of blades and vanes generally occurs at a point near the midspan, where the highest temperatures occur. It appears that short-run aircraft experience more erosion damage than do long-run aircraft. This is logical since more frequent takeoffs and landings generate a larger quantity of carbon particles.

Foreign-object damage (FOD) from debris ingested by the engine (or segments from the breakdown of components upstream in the engine) is regarded by many manufacturers as a major reason for replacing turbine blades and vanes in current gas-turbine engines. As aircraft operation from unprepared runways increases, a corresponding increase in the frequency of FOD is expected.

Over-temperature damage (e.g., resolution of the alloy constituents and incipient melting) can occur during a momentary temperature overshoot. Frequently associated with "hot-starts," such an overshoot can seriously affect the structural integrity of many turbine components.

Since superalloys are being used in engines where metal temperatures sometimes exceed 85 percent of their melting points, cooling techniques will be necessary to lower the metal temperatures of many components as turbine-inlet temperatures exceed about $1900°F$. It is not believed that refractory metal alloys of Mo, Cb, W, or Ta will be used to any large extent in near-future aircraft engines because available oxidation-resistant coatings for these materials are not sufficiently reliable. The same would be true of chromium-base alloys if an otherwise suitable alloy were available. Since superalloys will continue to be used in aircraft gas-turbine engines for many years, it is necessary to develop alloys and techniques for minimizing the deterioration and failure of hot-section components. To achieve the long life expectancies required for future engines, it is mandatory that sulfidation and oxidation be minimized. Although future alloy development may assist in reducing these deleterious effects, protective coatings appear to offer the most promising approach. In addition to minimizing the hot-corrosion problem, advanced coating systems may also be beneficial in reducing thermal and mechanical fatigue and improving erosion and FOD resistance.

PERFORMANCE

CURRENT STATUS

Life expectancies for hot-section components in aircraft gas-turbine engines depend to a large extent upon the mission profile of the aircraft. Nonaggressive environments and periodic descaling frequently permit uncoated blades in commercial aircraft engines to function for as long as 12,000 hr without degradation by oxidation or hot corrosion. In engines used to power antisubmarine-type aircraft, sulfidation can reduce blade longevity to about 1,200 hr. Uncoated blades in helicopter engines operating in Southeast Asia can experience severe sulfidation after only 800 hr. The uncoated vanes in commercial helicopter engines are sometimes nearly destroyed by sulfidation after only 300 hr of operation from offshore oil drilling sites. High-temperature protective coatings for superalloys in gas-turbine engines can significantly extend their life expectancies.

The basic requirements of a coating for a superalloy substrate in gas-turbine engines dictate that: (a) it must resist the thermal environment; (b) it must be metallurgically bonded to the substrate; (c) it should be thin, uniform, light in weight, reasonably low in cost, and relatively easy to apply; (d) it should have some self-healing characteristics; (e) it should be ductile; (f) it should not adversely affect the mechanical properties of the substrate; and (g) it should exhibit diffusional stability with the substrate.

A number of organizations have developed high-temperature protective coatings for superalloys. Many of these coatings have been used effectively in gas-turbine engines to extend the life expectancies of hot-section components.[3-5] The important coatings available at the present time are described in Table 11, p. 68. Inspection of Table 11 reveals that (excluding the vitreous or glassy-refractory coatings) all of the currently important coatings for superalloy components in gas-turbine engines use aluminum as the primary coating constituent. Processes used to obtain these aluminum-rich coatings are: pack cementation, hot-dipping, slurry, and electrophoresis. (In addition, some experimental coatings are being applied by physical vapor deposition and cladding.) There probably is some redundancy among the 38 aluminum-rich coatings available from the 14 vendors. Such redundancy cannot be determined, however, because processing details are generally regarded as proprietary information by the coating developers.

Regardless of the process, all of the aluminum-rich coatings are based upon the interdiffusion of aluminum (without or in conjunction with added elements) and the substrate. For nickel-base alloys, processing parameters are adjusted so that the coating consists primarily of a relatively thin (0.0005 to 0.004 in.) layer of nickel aluminide (NiAl).[6] Resistant to both sulfidation and oxidation, this high-melting-point intermetallic phase exists over the range 23.5 to 36 wt% aluminum. It is important that the aluminum content of the coating be within these compositional limits because aluminum contents greater than 36 wt% result in the formation of a brittle, low-melting-point Ni_2Al_3 (delta) phase; aluminum contents less than 23.5 wt% favor the formation of Ni_3Al (gamma prime) phase, which has relatively poor hot-corrosion resistance.

Nearly all currently available coatings are modified by the addition of small amounts of an alloying element or elements to the NiAl. This may be accomplished by codeposition of the alloying element or elements with the aluminum or by prealloying the substrate surface prior to aluminizing. Although various coating-performance improvements reportedly result from the use of these alloying elements, no definite conclusions regarding the role of a particular additive have been made at this time.

Similar coatings also are available for cobalt-base alloys. Compared to nickel-base alloys, however, higher coating temperatures and longer diffusion times are required to form equivalent coating thicknesses on cobalt-base alloys.

Extensive testing of currently available coatings for superalloys has been conducted. The majority of this evaluation work has consisted of testing isothermal and cyclic (static or dynamic) oxidation (weight-gain or weight-loss), mechanical properties, low- and high-cycle fatigue, thermal shock, hot-gas erosion, sulfidation and impact. A significant amount of these data are summarized in a report by Jackson and Hall.[7] Comparing them is virtually impossible, unfortunately, because the various investigators used different testing techniques, and rarely reported how their coatings compared with others without coding the names of competitive coatings.[4,7] Also, comparisons frequently are unrealistic because a coating developer may optimize his coating process based upon a given test technique and then evaluate competitive coatings using the same test. Regardless of these factors, aluminide coatings definitely improve the oxidation-erosion resistance of superalloys.[8]

The work of Hamilton et al.[2] shows that an aluminide coating can have no deleterious effect on the stress-rupture behavior of IN-100 and Alloy-713C nickel-base alloys (Figure 45). However, the effect of coatings on stress-rupture behavior is controversial. Recent testing at Lycoming Division of AVCO Corporation[9] shows that the stress-rupture properties of uncoated thin sections are about 50 percent of those obtained by using thick-section test bars; coatings can reduce the stress-rupture properties of the thin sections by another 25 percent. The stress-rupture testing of thin sections (which will simulate thin trailing-edge situations and cooled-blade and vane designs) is also favored by other engine manufacturers. Additional work has shown that the thermal fatigue and mechanical properties of various superalloy substrates are not adversely affected by aluminizing.[6] Bartocci,[10] however, reports that diffusion coatings (or the thermal cycle used during their application) have a detrimental effect on high-temperature fatigue properties and may also affect the stress-rupture properties of some nickel-base alloys. Thus, the exact effect of protective coatings on various substrate properties is not well known. It is known that coatings can be "tailored" to perform well in a given test, and many of the discrepancies reported in the literature probably result from this effect. Lack of stan-

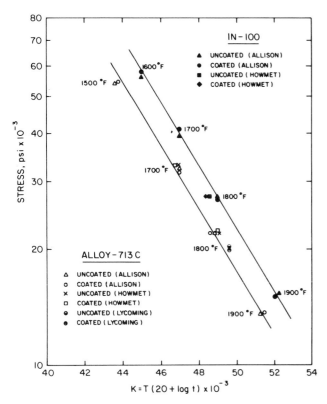

FIGURE 45 Stress-rupture of uncoated and Alpak-coated IN-100 and Alloy-713C.[2]

dardized testing practices, particularly in regard to the effective cross-sectional area of coated samples, is another significant factor contributing to variable results.

It is agreed by various investigators[2,6,10–12] that aluminum-rich diffusion coatings can significantly improve the sulfidation resistance of superalloys used in aircraft and marine gas-turbine engines. The same conclusion is obtained from in-service engine performance data. Allison Division of General Motors Corporation[13] reports that coatings at least double the time at which sulfidation becomes a serious problem for blades and vanes used in engines powering fixed-wing antisubmarine aircraft. Lycoming Division of AVCO Corporation[9] reports that the life expectancies for super-alloy blades and vanes in helicopter engines operating in environments conducive to hot corrosion are doubled to tripled by the use of aluminum-rich coatings. Similar results are forecast from preliminary data obtained at General Electric Company.[14] For over ten years, the Pratt and Whitney Aircraft Company has successfully used protective coatings on turbine components to extend the life of hardware.[15]

Even when sulfidation is not a problem, aluminum-rich coatings are used to extend the life expectancies of super-alloy blades and vanes. (At least one manufacturer uses a coating to ensure the successful performance of a first-stage vane alloy.) For commercial engines, coatings can at least

double the life expectancies of blades and vanes used in gas-turbine engines. In addition, for suitable performance, advanced superalloys require coatings.

None of the current coatings can satisfy all of the basic performance requirements established earlier for protective coatings for superalloys. Existing coatings are limited to use temperatures of less than 2000°F. At 2175° to 2200°F, melting may occur in the diffusion zone of aluminum-rich coatings applied to nickel-base substrates. Fusion in the diffusion zone occurs at about 2300°F for aluminide coatings applied to cobalt-base alloys. Thermal stability is also a problem for current aluminum-rich coatings at temperatures exceeding 2000°F. Aluminide coatings deteriorate with time when they are exposed at elevated temperatures. It is suspected that the breakdown mechanism occurs as follows: In the temperature range 1560° to 1830°F, the mechanism appears to be the outward diffusion of aluminum, while beyond 1830°F the additional inward diffusion of aluminum becomes an important factor.[6] Conversion of aluminum to aluminum oxide and spallation of this oxide, which thus requires continued use of the aluminum reservoir, is the major cause of coating failure below 2000°F. Another factor that limits the performance of aluminide coatings is inadequate ductility.[10] Aluminide-type coatings exhibit low ductility and therefore are highly susceptible to foreign-object damage. Coating thickness is another factor affecting the performance of currently available aluminide coatings. Although thin coatings (0.001 to 0.003 in.) are necessary to minimize thermal spalling, thicker coatings would provide a larger reservoir of aluminum and would be expected to perform for longer times before coating breakdown.

In spite of these limitations, aluminide-type coatings successfully extend the life expectancies of gas-turbine engine blades and vanes. Because the advantages obtained by using coatings outweigh the disadvantages, several hundred thousand superalloy blades and vanes for military and commercial aircraft gas-turbine engines are coated each month by the engine manufacturers and coating vendors in the United States.

FUTURE NEEDS

Compared to the currently available aluminide coatings, future coatings for superalloy hot-section components in gas-turbine engines will have to exhibit superior performance capabilities. With the development of these significantly improved coatings, it is hoped that blade and vane life expectancies in commercial aircraft engines can be increased to more than 12,000 hr. (Average temperatures associated with the blades and vanes in these engines will be about 1700° and 1850°F, respectively.) Even for higher temperature hot-section components in military gas-turbine engines, a twofold to threefold improvement in life expectancy should result from the use of these new coatings.

There is reason to believe that superalloys with improved temperature capability will be developed; this would create the need for coatings having a life expectancy in excess of 500 hr at 2100°F. TD–Ni and TD–Ni–Cr alloys are available, and these materials possibly could be used for flameholders, spray bars, combustion chambers, transition ducts, and turbine vanes in gas-turbine engines if oxidation-resistant coatings were available that could function reliably for long times at 2200°F. Therefore, it is desired that the maximum use temperature of coatings be increased from 2000° to 2200°F. Even a 100°F improvement would be beneficial. However, if aluminide coatings are to be used at these higher operating temperatures, there is a need to increase the melting point of the nickel- or cobalt-poor, aluminum-rich diffusion zone.

Improved diffusional stability is a necessity if advanced coatings are to achieve these longer life expectancies and increased use-temperature capabilities. For example, with aluminide coatings, diffusion barriers (or alloying elements added to the coating that will decrease the thermodynamic activity of aluminum) may be necessary to prevent excessive inward diffusion of aluminum or outward diffusion of nickel or cobalt. Inward diffusion of certain necessary additive elements in future coatings also must be avoided if they promote the formation of undesirable quantities of sigma phase in the substrate. It may also be necessary to prevent the outward diffusion of certain elements that are needed in the substrate but undesirable in the coating.

Improved coatings must possess increased ductility. This will greatly improve their resistance to foreign-object damage and reduce unintentional coating damage during component handling. A ductile coating, or a coating with a relatively ductile intermediate diffusion zone, would be effective in retarding the initiation or propagation of surface cracks. Such a coating could significantly reduce damage and failure related to both mechanical and thermal fatigue.

Future coatings must be reasonably thin. The coating thickness for blades and vanes should be within the range 0.001–0.005 in. Coating thickness is most important at the trailing surfaces of blade and vane airfoils (especially the miniblades and vanes used in small gas-turbine engines) because excessive coating buildup at this location reduces the effective gas-path cross-sectional area. Care must be exercised that excessive interdiffusion does not result in a serious reduction in the load-carrying capability of the substrate. In addition, thin coatings will be required, if it is found necessary to coat the internal passages of complex cooled components in order to prevent oxidation and sulfidation.

In summary, it is desired that future coatings for superalloys used in gas-turbine engines possess the following characteristics: improved diffusional stability; improved resistance to oxidation, sulfidation, and hot-gas erosion; improved ductility; a 200°F increase in operating-temperature capability; improved resistance to thermal and mechanical fatigue; improved resistance to thermal shock; minimum adverse effect on the mechanical properties of the substrate; and some self-healing capability.

MANUFACTURING TECHNOLOGY

CURRENT STATUS

Large scale coating of superalloy substrates has been limited primarily to individual gas-turbine engine blades and vanes. These vary in size from about 1.5 to 6 in. in length. One engine manufacturer does coat vane sections (X-40 vanes with L-605 rails) that are about 2.5 in. × 4 in. × 10 in. in size. Other hot-section components that are coated on a more or less routine basis are thermocouple tubes (used to measure turbine-inlet temperatures), and Hastelloy X as cast stainless-steel fuel nozzles.

The aluminide coatings for these production items are applied using conventional pack cementation, vacuum-pack cementation, slip-pack cementation, Cr-plate followed by hot-dipping, and slurry techniques. Certain inherent manufacturing advantages and limitations are associated with each of the coating processes, and these are thoroughly reviewed in Chapter 5.

Briefly, conventional pack cementation is advantageous in that the pack supports the component being coated and eliminates the need for special holding fixtures that can cause "blind spots" in the coating. A large number of blades and/or vanes can be coated simultaneously. (One manufacturer loads ten 4 in. × 12 in. × 20 in. retorts into a single furnace.) If the retort is evacuated before being placed in the furnace, the process is referred to as vacuum-pack cementation. The vacuum process appears to improve coating "throwing power" and may be advantageous if the inside walls of long, small-diameter holes in cooled blades and vanes are to be protected from oxidation and sulfidation. Other advantages associated with properly conducted conventional and vacuum-pack cementation processes include uniformity of coating thickness and structure, economy (based upon pack reusability), reproducibility, simplicity of processing procedure and ease of control, and simplicity and low cost of processing equipment and facilities (excluding vacuum). Both cementation processes do have the disadvantage of long heating-up and cooling-down times that can adversely affect the mechanical properties of the substrate. (Post-coating heat treatments have been used successfully to improve the substrate mechanical properties of coated components.) In the slip-pack cementation process, the thermal-cycle time can be reduced if the component to be coated is not supported in a retort containing inert refractory oxide powder that must be heated and cooled during the coating process; this process, however, necessitates the use of special supporting fixtures. The slip-pack cementation process is especially advantageous for the economical coating of large objects. Slurry-coating processes also are advantageous for coating relatively large components and complex shapes. Melting slurries have the added advantages of high fluidity and the capability to coat areas of limited accessibility. The slurry process is quite versatile: It utilizes simple equipment, heating-up time is relatively short, and masking of areas where no coating is desired is very simple. In addition, the process is adaptable for coating repair by local slurry application. The main disadvantages associated with slurry-coating processes are difficulties in coating the inside walls of long, small-diameter holes and in the control of coating thickness.

At present, turbine-blade alloys being coated include Waspaloy, IN 100, Udimet-700 (cast and wrought), B-1900, Alloy-713C, and Udimet-500. Turbine-vane alloys being coated include René 41, Alloy-713C, X-40, WI-52, MAR-M302, B-1900, and Udimet-700.

Although blade shrouds are coated, the root sections generally are not, because the presence of a coating can lead to fit-up problems. In addition, several engine manufacturers believe that the relatively brittle aluminide coating can promote fatigue cracking in the root section. The root section generally is masked prior to coating using mechanical devices or sprayed-on (CaO-base) temporary coatings. At least one engine manufacturer coats the entire blade and then removes the coating from the root section by grinding. The question as to whether fatigue cracks are promoted by the presence of a coating at the root section is controversial; one engine manufacturer reports that although the coating may become cracked, the cracks do not propagate into the substrate.

Most of the engine manufacturers and coating vendors do not intentionally coat the inside walls of holes in cooled blades and vanes. One engine manufacturer using a vacuum-pack cementation technique reports that coatings can be reliably deposited inside small-diameter holes to a thickness approximately one-half that deposited on the outside. Another manufacturer states that holes having a diameter of 0.030 in. can be uniformly coated to a depth of 2.5 in.; 0.010-in.-diam holes can be coated to a depth of 0.5 to 1 in.

Gas-turbine engine vanes are routinely recoated and placed back into service. Many vanes are still in use after being recoated 8 to 12 times. The technique of recoating blades is not as well advanced, but it appears that these components may be capable of being recoated more than twice before excessive substrate metal (especially at the trailing edges) is lost by grinding and/or interdiffusion. In addition, blades are sometimes recoated without removing the used coating.

Larger hot-section components, generally for marine engines, have been successfully coated on an experimental basis. Some of these items have included 32-in.-diameter Hastelloy X transition ducts, Hastelloy X combustion chambers, and Hastelloy X cross-fire tubes. Components as large as 36 to 48 in. in diameter and 1 to 10 in. deep have been successfully coated experimentally.

FUTURE NEEDS

It is anticipated that, in addition to increased production of coated blades and vanes, protective coatings for superalloys will be applied to other gas-turbine engine components. This practice will be mandatory for marine engines. Larger hot-section components that may require coatings include combustion chambers, transition ducts (if they are used), afterburner liners, nozzle flaps, flameholders, spray bars, fuel-nozzle shrouds, cross-fire tubes, and flameholder beams (support for the flameholder). Several manufacturers use integrally cast turbine wheels (i.e., the disk and blades for each stage are cast as one section) and integrally cast nozzles (i.e., the vanes and rails for each stage are cast as one section). The integrally cast complex sections are about 15 in. in diameter and 4–5 in. deep. Another component that also may require coating in the near future is a cooled, 14-in.-diam, 4-in.-deep bearing support for a developmental tank engine.

Increasing component size is not the only manufacturing problem that must be solved. Miniature blades and vanes for smaller gas-turbine engines may require protection, and the successful coating of these components may prove even more difficult. In addition to the problems associated with extremes in component size, coating technology must be extended to include a growing list of newer gas-turbine engine alloys. Some of the additional alloys that have to be coated include B-1910, MAR-M421, Nimonic 108, Nimonic 105, Nimonic 80A, C-1023, René 77 (Udimet-700 with phase-composition control), René 80, TD-Ni, and TD-Ni-Cr. Coating process variables generally vary from alloy to alloy, and there may be a need for a manufacturing specification for each specific coating–alloy combination.

An increasing amount of recoating work is desired, and improved stripping techniques will be required. In many repair cases, it may not be necessary to completely remove the old coating. For example, low-activity pack-cementation packs are known to be beneficial for many coating-repair operations.

If it is established that the long, small-diameter walls of internal passages in cooled gas-turbine engine components must be coated, significant improvements in currently available coating processes will be required. Although slurry, conventional-pack cementation, and slip-pack cementation coating techniques probably will not be satisfactory, improvements in the vacuum-pack cementation process may provide for the reliable deposition of thin coatings inside these holes. The thin-walled sections associated with cooled components create another manufacturing problem, in that coating repair work will be even more difficult because of the unavailability of adequate material for reworking, especially at trailing surfaces.

In general, improved process control for the various coating techniques will be required. This will be difficult because: (a) Concentration requirements that give adequate diffusion rates for current coating processes lead to diffusion imbalance and Kirkendall voids, (b) the inherent limitations of the currently used coating processes make it difficult to control the coating composition perfectly, (c) thermodynamic and kinetic factors limit freedom in the codeposition of many multiple-element combinations, (d) the substrate composition must be included as part of the coating, and (e) all of the current coating processes are diffusion-controlled.

It may be possible to obtain improved coatings by using the fluidized-bed vapor-deposition technique. Essentially a pack process (where the pack is fluidized by a high flow of inert gas), this technique provides high rates of uniform heating. Unfortunately, limitations associated with this process are numerous, and include: (a) the need for large volumes of inert gas when coating large components (or high-production items) that necessitates a gas-recycling capability, (b) difficulty in coating limited-access areas, (c) the need for an adequate heating source to maintain the required temperature profile in the fluidized bed, (d) difficulty in maintaining a uniform bed composition, (e) the need for techniques to prevent distortion and damage to components from the bed action, and (f) relatively costly scale-up.

It should be clear that better techniques than conventional pack cementation are needed for applying high-temperature coatings to superalloy substrates. Some encouraging results are being obtained by using claddings and the physical-vapor-deposition (PVD) coating technique. Newer processes, such as pyrolytic deposition from organometallic vapors or solutions and electrolytic deposition from fused salts, are too experimental to be evaluated at this time; they should not be discounted, however, as possibly effective means of depositing protective coatings on superalloy substrates.

For the near future, it is believed that most coating improvements will be realized by the addition of modifiers to basic aluminide coatings using established (but probably multiple) coating processes. Encouraging data are reported already for coatings composed of $Cr-Al-Cb-Y_2O_3$ and $Fe-Cr-Al-Y$.[15] The concept of adding oxide particles (e.g., Al_2O_3) to aluminide coatings is relatively new, but it is believed that titanium (or other metals) in the coating may be able to reduce the Al_2O_3 chemically and release additional aluminum in order to prolong the integrity of the surface oxide layer of the coatings.[14]

Other future manufacturing requirements will include automated coating processes to reduce coating costs and the use of nondestructive testing (NDT) methods capable of predicting the useful remaining life of hot-section components.

One additional future manufacturing requirement can be mentioned, but the possibility of its solution is considered remote compared to the others. A need exists for a simplified procedure to coat turbine-blade tips (e.g., a "brush-on" coating that could be "cured" at about $800°F$).[9] This need occurs because when blade tips are coated new, the blade tip must be finished–i.e., machined before it is coated. Thus, it is not possible to make minor adjustments in blade length during engine assembly (fit-up). If a simplified coating procedure were available, coated blade lengths could be changed and the tips recoated to provide the necessary oxidation or sulfidation resistance.

REFERENCES

1. DeCrescente, M. A., and N. S. Bornstein. 1968. Formation and reactivity thermodynamics of sodium sulfate with gas turbine alloys. Corrosion, 24:127–133.
2. Hamilton, P. E., K. H. Ryan, and E. S. Nichols. 1967. Nickel-base alloys and their relationship to hot corrosive environments. Hot Corrosion Problems Associated with Gas Turbines, ASTM STP 421, Am. Soc. Testing Mat., pp. 188–205.
3. Puyear, R. B. July 12, 1962. High-temperature metallic coatings, Machine Design, 34:176–184.
4. Lane, P., and N. M. Geyer. Feb. 1966. Superalloy coatings for gas turbine components, Journal Metals, 18:186–191.
5. Long, J. V. March 1961. Refractory coatings for high-temperature protection. Metal Progress, 79:114–120.
6. Llewelyn, G. 1967. Protection of nickel-base alloys against sulfur corrosion by pack aluminizing. Hot Corrosion Problems Associated With Gas Turbines, ASTM STP 421, Am. Soc. Testing Mat., pp. 3–20.
7. Jackson, C. M., and A. M. Hall. April 20, 1966. Surface Treatments for Nickel and Nickel-Base Alloys, NASA Technical Memorandum No. NASA TM X-53448 (AD-634076), George C. Marshall Flight Center, Huntsville, Alabama. pp. 5–74.
8. Sippel, G. R., and J. R. Kildsig. 1966. Laboratory tests for evaluation of surface deterioration in gas turbine engine materials, Allison Research and Engineering, 2nd Quarter, pp. 20–35, Allison Energy Conversion Division of General Motors Corporation, Indianapolis, Indiana.
9. Freeman, W. R., Jr. April 16, 1968. Private communication, to J. R. Myers and J. D. Gadd, AVCO Corporation, Stratford, Connecticut.
10. Bartocci, R. S. 1967. Behavior of high-temperature coatings for gas turbine engines, Hot Corrosion Problems Associated with Gas Turbines, ASTM STP 421, Am. Soc. Testing Mat., pp. 169–187.

11. Graham, L. D., J. D. Gadd, and R. J. Quigg. 1967. Hot corrosion behavior of coated and uncoated superalloys, Hot Corrosion Problems Associated with Gas Turbines, ASTM STP 421, Am. Soc. Testing Mat., pp. 105–122.

12. Betts, R. K. Oct. 2, 1967. Selection of coatings for hot-section components of the LM1500 marine-environment engine, General Electric Company Report No. R67FPD357, Marine and Industrial Department, Lynn, Massachusetts/Cincinnati, Ohio. (Research conducted under Contract No. NObs 94437.)

13. Hamilton, P. E., D. Scott, K. H. Ryan, and E. S. Nichols. March 27, 1968. Private communication to J. R. Myers and J. D. Gadd, Allison Division of General Motors Corporation, Indianapolis, Indiana.

14. Levine, D., and M. Levinstein. March 26, 1968. Private communication to J. R. Myers and J. D. Gadd, General Electric Company, Cincinnati, Ohio.

15. Talboom, F. P., J. D. Varin, E. Hovan, and I. Linask. April 17, 1968. Private communication to J. R. Myers and J. D. Gadd, East Hartford, Connecticut.

CHEMICAL PROPULSION

INTRODUCTION

This section reviews the use of coating systems in missile and spacecraft propulsion engines. The review covers liquid- and solid-fueled rocket engines with missions ranging from launch to orbital or flight-path control. Nuclear space propulsion systems are also reviewed. The requirements envisioned for future coatings applications in all of the above areas are considered, with most emphasis directed toward rocket engines.

Selection of rocket engine materials presents a difficult problem since, upon ignition, combustion products instantaneously fill a rocket chamber and nozzle. Gas temperatures rise to values as high as 9000°F (approximately 5000°C), while chamber pressures increase by from tens to hundreds of atmospheres. The engine experiences both thermal and mechanical shock. During operation, which lasts from a few seconds to hundreds of seconds, the highly reactive combustion products exhaust at very high velocities. Multiple reuse requirements further complicate the problem. Currently, two general approaches are employed to allow rocket engines to survive long enough for mission completion. The first involves cooling the nozzle to allow the use of easily fabricable structural materials, such as stainless steel or aluminum. This approach is most commonly used in liquid-propellant rockets, especially in launch applications. Oxidation-resistant coatings generally are not used in these engines. The second approach involves insulated or radiation-cooled hot structures. These must be coated, in many cases, to meet performance requirements.

Nozzle cooling in liquid-fueled rockets is most often achieved by pumping the liquid fuel through longitudinal tubes that comprise its internal surface. The cold fuel removes sufficient heat that the tube surfaces in contact with the hot exhaust are kept at temperatures below that at which strength is lost and chemical reactions become serious. This approach is called regenerative cooling. It can be supplemented by depositing a thermally insulative coating over the tube surfaces. The coating produces a further thermal

drop across the gas-nozzle interface and so either minimizes the coolant pumping requirements, makes possible a higher combustion temperature, or allows longer firing times. Another technique used to cool nozzles in liquid-fueled rockets involves film-cooling. Here, fuel is injected directly into the chamber in such a way as to form a cool, gaseous layer along the engine's surfaces. In solid-fueled rockets, some internal surface cooling is achieved by ablation, which is the controlled thermal decomposition of organic composites. The decomposition absorbs heat endothermically, providing the surface cooling. Finally, thermal lag coatings have been used to keep both liquid- and solid-fueled rocket engines cool. Thermal lag coatings suffice if the firing times are short enough for the coating to slow down conductive heat transfer to the point that the material beneath does not overheat during firing.

In the second rocket engine construction approach, commonly used for attitude-control rockets, the nozzles are made of superalloys (e.g., L-605), refractory metals (Mo, Ta, or Cb), or graphite and are allowed to run hot. With the refractory metals, corrosion is a major concern, and coatings based on compounds of silicon or aluminum have been found to be adequately protective. Because of the relatively short operation times, rocket engines of the other materials are generally coated with materials of high emittance so that structural temperatures are controlled by enhanced radiative heat loss.

LAUNCH (BOOSTER) ENGINES

CURRENT STATUS

In almost every case, the nozzles used on large, regeneratively cooled booster-rocket engines, which range from the Atlas to the F-1 (the first stage engine for the Saturn V), are not coated. These engines generally have relatively low expansion ratios and, for the most part, are not very long. With proper design, shorter lengths ordinarily can be adequately cooled regeneratively by the fuel, with or without supplemental film-cooling. With adequate cooling providing 100%

design reliability, there is no need to consider coatings, with their additional problems of cost and variable reliability. The general procedure is to shorten the nozzle and then attach an extension skirt that is either ablatively or radiatively cooled.

The engines used in the second and third stages of launch systems are also generally uncoated. Most of these engines have high expansion ratios, making regenerative cooling more difficult. However, the RL-10 and the J-2 engines, both being considered for second- and third-stage use on the Saturn V, are uncoated. This is possible because liquid hydrogen is the fuel, and it has adequate heat capacity to cool the longer lengths. One exception is the Agena D engine used with Atlas to launch the Lunar Orbiters, Nimbus weather satellites, etc. This engine has a 10-mil alumina flame-sprayed coating on its titanium thrust cone, to provide a thermal lag, and its aluminum alloy combustion chamber is sprayed with tungsten carbide. Another exception is the trans-stage of the Titan II, which employs a thermal insulative coating consisting of a proprietary metaloxide mixture that is plasma-sprayed onto the nozzle tubes. The second stage of the Titan II also uses an ablative protection system for its skirt extension.

Another of the few rocket engines currently employing regenerative cooling supplemented by an insulative coating is the XLR-99. This engine powered the X-15 experimental aircraft. In it, the stainless-steel nozzle tubes, cooled with ammonia fuel, are further protected by plasma-spraying with a graded zirconia–nickel chromium alloy mixture deposited on a plasma-sprayed molybdenum undercoating. Coatings extend the number of firings that are possible for the XLR-99 engine.

Most of the large solid-fueled rocket engines employ ablative protection for their nozzles. In most cases this ablator consists of a fiber-reinforced (silica or asbestos) phenolic resin. Such a technique is used for the large 260-in.-diameter engine. It is also generally employed on a variety of launch systems for military applications. These include not only the long-range, long-firing-time systems but also some of the short-range systems if their firing times are great enough.

When the firing times are very short in liquid- and solid-fueled rocket engines for short-range missiles, thermal lag coatings may be used. These coatings slow down the conductive heat transfer during the short firing time (frequently less than 5 sec), so that the heat from combustion does not degrade the mechanical properties of the light-weight structural alloys. They also provide short-time corrosion protection. For example, the Bullpup air-to-ground missile employs flame-sprayed zirconia on both the aluminum combustion chamber and the nickel–chromium alloy precoated copper nozzle in order to combat thermally induced mechanical degradation and corrosion–erosion from the exhaust gases of over 4000°F (2200°C). The solid-fueled Scout missile

has a flame-sprayed zirconia coating to protect its ZTA graphite nozzle from corrosive and erosive attack. In the Lance surface-to-surface missile, a metal-graded zirconia thermal lag coating is employed.

FUTURE NEEDS

In contrast to current practice, in many of the advanced regeneratively cooled rocket-engine concepts, thermally insulative coatings are envisioned as an integral part of the engine. In such concepts, gas temperatures, chamber pressures, and heat transfer rates are higher than those presently encountered because of either higher performance engine designs or the use of higher energy fuels such as LH_2/LF_2. Thus, higher heat fluxes to the structure and to the coolant can be expected. The advanced concepts will probably not use refractory metals in regeneratively cooled engine structures. Nickel alloys, Type 347 stainless steel, and Inconel-X will probably continue to be used as structural materials as they have been with the lower energy propellants (LO_2/LH_2 or N_2O_4 hydrazine). Continued use of these metals, however, will make it necessary that the heat fluxes be reduced. The problem of reducing them may be compounded by designs that incorporate fuels with low specific heats, by designs that provide insufficient cooling for the engine, or by designs in which film-cooling is impractical.

The several possible approaches to this problem of heat flux control include transpiration cooling, improved film-cooling, flame liners, thermal barriers, and various combinations of these. At this time, however, thermal barrier coatings appear most promising. This approach has already had some success in the X-15 engine. Thermal barrier coatings may include refractory oxides, refractory metals, or mixtures or graded layers of both. Also being considered are the refractory metal carbides, nitrides, and borides. These coatings may be applied by plasma-spraying or slurry application procedures. Undercoatings of intermediate melting point may also be incorporated into the coating design to relieve thermal shock and improve coating adherence to the engine structure. If coatings are the means of achieving the mandatory reduction in the heat flux to the engine structure, then any coating loss will bring about local overheating and early, if not catastrophic, engine failure. Furthermore, since the combustion products of the high-energy propellants are highly corrosive to the anticipated structural materials, the coatings must also provide corrosion resistance. Thus, both environmental and coating design factors are of critical concern.

The anticipated rocket engine environments in the many advanced engine concepts are as varied as the concepts themselves. As in the past, thermal shock, gas erosion, mechanical stress transients, thermal cycling, nonuniform coating thickness and combustion that lead to hot spots, and mechanical vibration are major problems to be faced. Degradation by chemical reaction with the combustion products may also become of serious concern. Such reactions could change the thermal properties of a coating and move it off the design value, as occurs if zirconia is partially reduced by hydrogen. The substoichiometric oxide has higher thermal conductivity than the starting material. Reactions could also induce metal embrittlement.

For nozzles ranging from 3 in. in diameter to very large boosters, H_2-O_2 combustion will develop environments containing H, H_2, O, O_2, and OH with gas temperatures of about 5900°F (about 3250°C). Chamber pressures will range from 1,000 to 2,000 psia. Single firing times greater than 2 min with multiple reuse of the regeneratively cooled nozzle are desired. Under these conditions, coating surfaces may reach 4000°F (2200°C). Coatings having thermal resistances of from 120 to 250 in.^{2} sec-°R/Btu will be expected to drop heat fluxes to the nozzle from 60 Btu/sec-in.^{2} (calculated for an uncoated nozzle cooled with liquid hydrogen) down to around 20 Btu/sec-in.^{2} for a coated nozzle similarly cooled. The latter heat flux will allow the liquid hydrogen to keep the tube surfaces under the coating cooler than 1600°F (approximately 870°C) and thus structurally sound.

For small second-stage engines, 3 to 5 in. in diameter at the throat, the eventual use of the high-energy combination of Flox (liquid oxygen containing 76 wt% fluorine) and liquefied gases, such as propane, is attractive. The highly reactive combustion products will include H, HF, F, F_2, and O_2. Gas temperatures will be about 7000°F (about 3900°C), while coating surface temperatures will reach 4000°F (approximately 2200°C). Chamber pressures will be around 100 psia. For such applications, firing times of 50 sec with a maximum nozzle life of 35 firings are sought. Here, coatings with thermal resistances to 1400 in.^{2} sec-°R/Btu, will be needed to decrease heat fluxes from 6 to 3 Btu/sec-in.^{2}

The regeneratively cooled booster engines all have turbopumps to deliver the propellants to the injector. The superalloy wrought turbine blading experiences heat shock on start-up, but while peak start temperature excursions may exceed 2000°F (1100°C), operating temperatures are around 1600°F (870°C). For future engines, it does not appear that improved turbine blading will require coatings to minimize corrosion and thermal fatigue, mainly because of the short operating times involved.

In the selection of coatings, many design trade-offs are necessary to achieve all of the desired coatings properties. The coating systems may become quite complex. Multicomponent coatings for the advanced systems may be graded in composition or layered and so may be difficult to deposit. Plasma-spray, slurry, and other deposition techniques capable of handling multicomponents and metering them into the coating in the proper ratios must be improved. Also, the

problems involved in the characterization of a graded, multicomponent system and in the development of nondestructive testing techniques to insure 100 percent conformity to design thickness and compositions need more attention.

On large advanced solid rocket boosters, there is no apparent necessity to use coatings on either the graphite nozzles or the carbon–phenolic or silica–phenolic exit cones. This is true because on large boosters (such as the 260-in. booster), any given loss of nozzle material results in a relatively small percentage increase in nozzle diameter; thus, little effect on performance is observed. For tactical rockets and others with smaller nozzle diameters, the same loss of graphite nozzle material has a much larger effect on performance. Here, tungsten or carbide coatings are needed. For example, the Polaris A3 (first stage) has tungsten in the throat and a tungsten liner below.

Re-use of huge booster nozzles for the 260-in.-diam solid propellant rockets could be achieved if a water-cooled, alumina-coated nozzle design under study by NASA is perfected. This nozzle and coating will experience thermal conditions similar to those listed for the H_2-O_2 liquid fuel systems, but the exhaust will contain HCl, Cl_2, O, H_2O, OH, CO_2, etc., as gaseous species. In addition, the presence of liquid Al_2O_3 in the exhaust products further threatens coating life.

MANEUVERING AND CONTROL ENGINES

CURRENT STATUS

Rocket engines for space propulsion use a wide variety of materials for chambers, throats, and nozzle extensions, with the specific choices dictated by mission requirements, fuel system, and operating temperatures and characteristics. Materials now in actual or contemplated use include ablatives; stainless steel and superalloys; pyrolytic graphite and pyrolytic graphite composites; JTA and similar graphite composites; carbides and hypereutectic carbides; borides; molybdenum, columbium, and tantalum alloys; and polycrystalline graphite. Of these materials, only the last four mentioned are customarily coated for oxidation protection.

Rocket motors of up to 300-lb thrust incorporating Cb alloy nozzles or nozzle extensions are now under advanced evaluation. The lunar excursion module will utilize a Cb-103 alloy nozzle with a TRW aluminide coating and ablative throat nozzles. The Apollo service module has ablative throat engines with a Cb-103 extension nozzle coated with a NAA 85 modified aluminide slurry. A 100-lb-thrust Cb-103 nozzle, coated with Sylvania's Sylcor R 512 A (Ti, Cr, Si) slurry coating, forms the backup system to a similar molybdenum engine for the Apollo command module. The coating systems for columbium nozzles were originally developed for the protection of re-entry hardware and thus exhibit the problems and limitations discussed in the section entitled "Hypersonic Vehicles." However, the relative importance of the various problems is not necessarily the same. Low-temperature degradation is not a serious problem with nozzle coatings, while degradation of the alloy, particularly a raising of the ductile–brittle transition temperature (due to coating–substrate interaction), is a very serious consideration. Silicide coatings degrade the ductile–brittle transition temperature of all columbium-base alloys. Cb-103 is one of the least affected by coatings and is widely used for this reason.

Development of improved coatings for columbium rocket engines appears to be feasible. Vac-Hyd Processing Corporation states that a diffusion-bonded slurry-applied silicide coating (VH No. 19) has protected Cb-103 alloy for 6 hr at 3100°F (approximately 1700°C) in noncyclic operation.[1] Sylvania states that their Hf-Ta coating (R515) has a temperature capability to 3600°F (about 1980°C), provided the columbium alloy is first precoated by a vapor-deposited layer of tantalum.[2] Rexer[3] recently determined the rate of depletion of iridium coatings on columbium through intermetallic compound formation and found that a 0.005-in.-thick iridium coating will last at least 200 hr at 1780°C (3236°F). Iridium coatings on columbium should be useful, therefore, to the eutectic temperature of $1890 \pm 20°C$ $(3346 \pm 36°F)$,[4] and their performance would essentially be determined by the oxidation and corrosion resistance of iridium, which is known to be very good.[5,6] A substantial amount of fabrication development and testing, with emphasis on NDT during fabrication and after firing, still remains to be done before these recent developments can be translated into improved coatings for columbium rocket nozzles. The work, however, does point out that the performance characteristics, particularly the operating temperature, of columbium engines could be substantially increased if this should become necessary.

Small $MoSi_2$-coated molybdenum engines are used in present space programs. The Apollo program utilizes molybdenum engines of 50–100-lb thrust for attitude control on the command module and for the LEM module. Similar engines of 100 and 22 lb thrust are envisioned for the Manned Orbiting Laboratory program. The larger (100-lb) engines are fabricated by pressing, sintering, and forging and have an uncoated L605 (cobalt-base alloy) extension bolted to the nozzle. Maximum operating temperature for the nozzle extension is 2100°F (1150°C). The smaller (22-lb) engines are fabricated from bar stock (pressed, sintered, and extruded). Coating application and performance are reportedly identical for both nozzles. The coating consists of pure $MoSi_2$ and is applied by Chromizing Corporation via a pack-cementation process. Both the 100-lb and the 22-lb engines generally use monomethylhydrazine and N_2O_4,

with a maximum operating temperature of 2700°F (approximately 1430°C). The 100-lb engine can be operated with Aerozine 50. Coating life is limited more by abrasion (during handling) than by oxidation. Earlier spalling problems have been solved by improvements in coating technique. Molybdenum engines have so far performed well despite apprehensions about them. Future engines will use the same pure $MoSi_2$ coating, and there appears to be little need or incentive for further coating development. The principal reservation to the use of molybdenum nozzles centers on the relative brittleness of the metal, not on coating performance.

For many rocket engine applications, graphite in its various forms is generally used uncoated, but in a few cases coatings have been applied. The Scout and Lance missiles use a polycrystalline graphite nozzle with a flame-sprayed coating of zirconia. Tests have been performed on attitude-control nozzles of polycrystalline graphite coated with silicon carbide. The very limited use of coated graphites is due to the fact that there are at present no operational coating systems that can protect graphite from oxidation in the 4200°–4700°F (2300°–2600°C) region, where graphite reaches its highest strength.[7] Operational coating systems, principally based on silicon carbide, give reliable protection only to approximately 2900°–3000°F (1600°–1650°C).[8,9] Plasma-sprayed coatings of Al_2O_3 or ZrO_2 are capable of protection at higher temperatures, but only for very short time periods. These oxides are thermodynamically unstable with respect to graphite, and protection is in part due to the low thermal conductivity of the oxides.

Iridium coatings have the potential to protect graphite up to the eutectic temperature of 4140°F (2280°C).[10-12] It is known that iridium is virtually impervious to oxygen,[10] that it does not form a carbide, and that diffusion at the graphite–iridium interface is negligible.[10,11] Pore- and crack-free iridium coatings that adhere well on graphite have been prepared by plasma deposition followed by isostatic hot-pressing.[13] Iridium alloys may offer protection at even higher temperatures. The eutectic temperature of Ir–30 percent Re with carbon was found to be 4500°F (2480°C).[14] The high price, very limited availability, and high density of iridium, however, will probably prevent the use of iridium or iridium alloy coatings for any but the most critical applications.

Substantial efforts have been made to develop tungsten coatings for graphite for use in solid-fuel engines, although none of these coatings appears to be in use yet. Tungsten is normally applied by plasma-spraying or vapor deposition.[15-19] The latter method provides denser and stronger deposits. Most vapor deposition work involves reduction of WF_6 by hydrogen. In a recent study, Berkeley et al.[15] found that the addition of carbon monoxide to the H_2–WF_6 gas mixture resulted in co-deposits of tungsten and W_2C, with

Knoop hardness values (100-g load) ranging from 1,600 to 3,300, while the hardness of pure tungsten deposited under otherwise comparable conditions ranged from 500 to 670. The hardest deposit had a carbon content of 1.1 wt%, corresponding to a W_2C content of approximately 35 percent. While higher room-temperature hardness does not necessarily indicate improved performance in a rocket nozzle environment, the observation is interesting and should be followed up.

FUTURE NEEDS

Future coating requirements for maneuvering and attitude-control engines are likely to be limited. Some use of coatings could possibly be made with hybrid engines and tactical system engines. High-energy final stages are needed for kick engines, orbital transfer engines, and descent or ascent engines for extraterrestial exploration. Typical engines are the Agena, Transtage, Apollo Service Module, and the Lunar Excursion Module. The skirt extensions on all the engines are of the radiative type and are coated with oxidation-resistant coatings.

Because of their high specific impulse, restart capability, and throttleability, hybrid propulsion systems have some appeal for future upper stages. For example, the high flame temperature of Flox/lithium fuel hybrid of approximately 8500°F (4700°C) imposes a severe heating environment. The nozzle is submerged in the motorcase to reduce stage length and to improve combustion efficiency. The nozzle is ablative with a pyrolytic graphite throat. Since there are no solids in the exhaust products, and fluorine compounds do not react with carbon, erosion is reduced in comparison to the problem encountered with solids. For very high temperatures in hybrids (7000°–9000°F, 3400°–5000°C), it is surprising that the pyrolytic graphite nozzles have not experienced difficulties in such regimes. It is unlikely that coatings could be found for adequate protection at these temperatures.

Monopropellant hydrazine rocket engines are attractive for vehicular and satellite attitude control and will probably supersede bipropellent systems. Decomposition of hydrazine with Shell-405 catalyst occurs spontaneously and at low temperature. Temperatures can be controlled and for thrust optimization will be around 1800°F (980°C). Superalloys such as L-605 cobalt-base are being successfully used without coatings for such operation.

For tactical missile propulsion in the past, primary consideration has been given to compactness, storability, and low cost. But in the future, maneuverability by means of throttling and start–restart capability will be required. To attain these ends, storable liquids, hybrids, and start–restart solids will probably be considered for tactical missile propulsion.

Graphite is today's leading rocket nozzle material, but it does have some drawbacks. Its ultimate tensile strength is low. Brittleness makes graphite mechanically inadequate even under moderate loads. Promise is held out for pyrolytic graphite deposited on polycrystalline graphite. For high-performance missiles there appears to be a requirement to use "something better" than graphite. Some designers favor a "hard" throat. For instance, tungsten is a candidate metal, especially if it could form a protective coating with some metal that had been infiltrated into it—infiltrated possibly with zirconium or hafnium metals. Such a "hard" throat should be useful with fluorine or Flox oxidizers, perhaps with hybrid rockets.

REFERENCES

1. According to data sheet supplied by Mr. S. J. Gerardi of Vac-Hyd Processing Corp., June 7, 1968.
2. Sylvania Bulletin on R 500 Series Coatings, August 1967.
3. Rexer, J. June 1968. High temperature protective coatings for refractory metals. Final Technical Report, Contract No. NASw-1405.
4. Giessen, B. C., U. Jaehningen, and N. J. Grant. 1965. J. Less Common Metals 10:147.
5. Criscione, J. M., et al. 1965. High temperature protective coatings for graphite, ML-TDR-64-173, Part II, Jan.; Part III, Dec.
6. Jaffee, R. I., and D. J. Maykuth. Feb. 1960. Refractory materials. DMIC Memorandum 44.
7. The Industrial Graphite Engineering Handbook, published by Union Carbide Corporation, Carbon Products Division, pp. 5A. 02.02 and 5A. 02.03.
8. Chown, J., R. F. Deacon, N. Singer, and A. E. S. White. 1963. Refractory coatings on graphite, with some comments on the ultimate oxidation resistance of coated graphite. P. Popper, "Proceedings of a Symposium on Special Ceramics 1962," Academic Press, New York, pp. 81–115.
9. Schulz, D. A., P. H. Higgs, J. D. Cannon. July 1964. Research and development on advanced graphite materials, Vol. 34, "Oxidation Resistant Coatings for Graphite," WADD TR 61–72, Vol. 34.
10. Criscione, J. M., et al. Jan. 1965. High temperature protective coatings for graphite. AFML-TDR-64-173, Part II, AD 608 092.
11. Criscione, J. M., et al. December 1965. High temperature protective coatings for graphite. AFML-TDR-64-173, Part III, AD 479 131.
12. Criscione, J. M., et al. February 1967. High temperature protective coatings for graphite. AFML-TDR-64-17, Part IV, AD 805 438.
13. Wright, T. R., T. R. Braeckel, and D. E. Kizer. February 1968. The fabrication of iridium and iridium alloy coatings by plasma arc deposition and gas pressure bonding. AFML-TR-68-6.
14. Harmon, D. P. February 1967. Iridium-base alloys and their behavior in the presence of carbon. AFML-TR-66-290.
15. Berkeley, J. F., A. Brenner, and W. E. Reid, Jr. June 1967. Vapor deposition of tungsten by hydrogen reduction of tungsten hexafluoride. Process Variables and Properties of the Deposit. Electrochem. Soc., J., 114:561–8.
16. Hoertel, F. W. 1964. Vapor deposition of tungsten on Merm rocket nozzles. U.S. Bureau of Mines, Rept. of Investigations 6464.

17. Nieberlein, V. A. July 1, 1964. Vapor-plating of tungsten for rocket applications. U.S. Redstone Arsenal, RR-TR 64-6.
18. Beidler, E. A., and J. M. Blocher, Jr. 1961. Experimental coating of graphite rocket nozzle inserts with tungsten by hydrogen reduction of tungsten hexafluoride. Battelle Memorial Institute Final Report, U.S. Navy Bureau of Naval Weapon.
19. Beller, W. 1960. Tungsten nozzles by vapor deposition. Missiles and Rockets 7:23–29.

HYPERSONIC VEHICLES

APPLICATIONS AND ENVIRONMENTS

In this section, the important aspects of coatings for hypersonic vehicles are discussed. These include vehicle characteristics and missions that delineate the requirements and structural design and structural materials that provide constraints on coating usage, characteristics, capabilities, and application methods. Many different types of hypersonic vehicles with significantly different characteristics and missions were available for detailed examination; however, the discussion presented here was narrowed to focus on three essentially different types of vehicle that encompass most of the important variations associated with hypersonic vehicles. These are the manned hypersonic aircraft, manned or unmanned lifting re-entry vehicle, and unmanned single-flight hypersonic missiles. The attendant air-breathing propulsion systems common to all three types of vehicles are also considered.

STRUCTURES

Manned Aircraft

Considerable interest has existed in manned hypersonic aircraft for the past few years, and studies have been made recently by both governmental agencies and industry to define major problem areas and to estimate feasibility. Because of the severe environmental conditions that will be experienced by hypersonic aircraft, the availability of suitable protective coatings appears to be an important consideration in determining feasibility. Factors that have an important bearing on the need and the type of protective coating selected include possible configurations and missions, flight environment and structural surface temperatures, and examples of typical structural designs. Structural-materials requirements and materials selection are reviewed to provide a background for the discussion of protective coatings for hypersonic aircraft. Many of the details presented relative to configuration, environment, and structural design are taken from Reference 1 and updated. Additional information on structures and materials problems appropriate to manned hypersonic aircraft is given in References 2–8.

At present, certain efforts are under way to develop technology that, it is hoped, will make manned hypersonic aircraft feasible at some future date. No approved program for the development of such aircraft is known to exist. Major technological advances in aerodynamics, propulsion, structures, and materials appear necessary before such aircraft become practicable, reliable, and economical.

Three types of missions and configurations are illustrated for hypersonic aircraft in Figure 46. The vehicle shown in the upper portion of the figure applies to commercial transportation with a flight speed of Mach (M) 6 to 8. For spacecraft launching, a Mach 3 to 12 vehicle is shown. For military missions, a vehicle with a speed from Mach 8 to 12 is indicated. The vehicles for all of these missions are large and would probably utilize liquid hydrogen for fuel. Mission times would range from several minutes to more than an hour. A typical hypersonic aircraft configuration is shown in Figure 47. The length of such a vehicle may be on the order of 300 ft with a gross takeoff weight in excess of 500,000 lb. The lower surface heat shields would protect the underlying primary structure from the severe thermal environment. The temperatures that these vehicles would experience would depend strongly on the flight trajectory, configuration, and performance, as is discussed later.

A typical flight corridor for hypersonic aircraft is indicated in Figure 48. The aircraft operates within some dynamic pressure limits at the lower flight velocities; for higher velocities, temperature limits would further restrict the flight corridor. The higher value of dynamic pressure (1,500 lb/ft^2) defines the lower boundary of the flight corridor. The 4000°F temperature line applies to a hemispherical nose of 1 ft radius with an emittance of 0.8. Figure 49 indicates equilibrium surface temperatures achieved during sustained flight at Mach 8 at 88,000 ft. The temperatures shown are the steady-state values expected in a flight of several minutes' duration and are those at which the radiation cooling of the surface is in equilibrium with the aerodynamic heating. Note that only small portions of the structure such as the nose and engine inlet are heated to temperatures in excess of 3000°F and that most of the wing, fuselage, and tail portions experiences temperatures below 2000°F. The temperatures indicated in Figure 56 can be varied significantly by changing the aircraft configuration and performance. Blunting the nose or leading edges or reducing the cruising speed or the dynamic pressure will reduce the temperatures and make the coating problems easier to solve (see Table 22).

The successful utilization of coatings on sheet metal

COMMERCIAL TRANSPORTATION
M = 6 - 8

SPACECRAFT LAUNCHING
M = 3 - 12

MILITARY OPERATIONS
M = 8 - 12

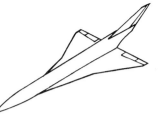

FIGURE 46   Missions and configurations of hypersonic aircraft.

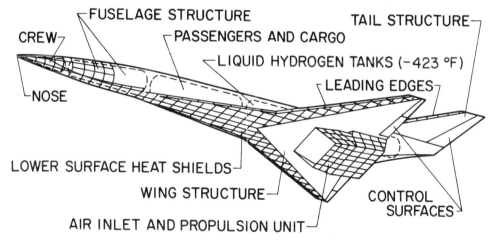

FIGURE 47   Total hypersonic aircraft configuration.

structures is strongly influenced by the design details of the structure. As indicated in Figure 54, large sections of the structure may require the use of surface heat shields (that may be coated) to protect the underlying structure. The remainder of the structure may operate successfully without heat shields. The general arrangement and structural details of a typical wing structure without heat shields are shown in Figure 50. The entire wing structure is shown schematically on the left; wing details are shown in the other views. The structure consists of an array of ribs

and spars. A waffle configuration skin panel is indicated.

In areas of the vehicle where the temperature limits of the load-carrying structure are exceeded, or where fuel tanks are used, the structure must be protected by insulation and heat shields as shown in Figure 51. In this view the load-carrying structure is protected by insulation, and the insulation in turn is protected from the airstream by heat shields. Details of a possible heat shield are shown in Figure 52. The corrugations are oriented in the streamwise direction and are attached to the load-carrying structure by clips. Special

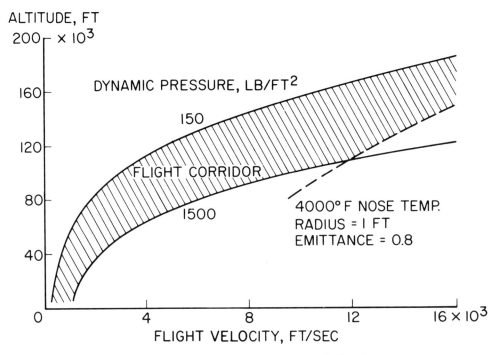

FIGURE 48   Typical flight path for hypersonic aircraft.

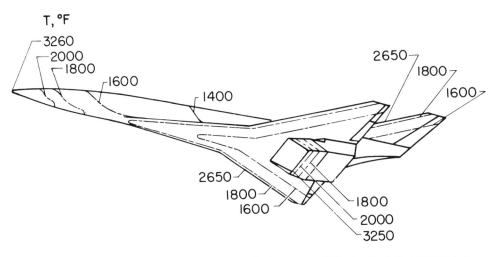

FIGURE 49   Equilibrium surface temperatures during sustained flight at Mach 8 at 88,000 feet.

attention has been given to the details of the attachment clips to permit application of coatings for protection against oxidation or embrittlement and to minimize thermal stresses and distortions.

In view of the wide range of missions that hypersonic aircraft may perform, detailed materials requirements that are applicable to all vehicles cannot be specified; however, it is assumed that certain common requirements such as long life and multiple reuse capability are mandatory for this class of vehicles. Minimum capabilities for different types

of materials that are expected to be utilized in hypersonic aircraft are suggested in Table 23. The materials are specified in terms of the application or location on the vehicle. The maximum temperatures and the desired life or exposure time without refurbishment are also indicated.

Superalloys are indicated for the primary structure with the maximum temperature below 1600°F. Structural heat shields that operate in the 1600° to 2000°F range are also noted to be superalloys. The superalloy sheet thickness for both primary structure and heat shields is minimum gauge,

TABLE 22   Representative Surface Equilibrium Temperatures on Hypersonic Aircraft[a]

| | Equilibrium Temperature, °F | | | | | | | |
| | Nose (q = 2,200 psf) | | Inlet Leading Edge (R = 1/4 in., Λ = 0°) | | Wing and Tail Leading Edge (q = 2,200 psf, Λ = 70°) | | Lower Wing Surface (X = 20 ft) | |
Mach No.	R = 1/4 in.	R = 6 in.	q = 2,200 psf	q = 500 psf	R = 1/4 in.	R = 1 in.	q = 2,200 psf	q = 500 psf
6	2120	1860	2230	2100	1880	1700	1330	1020
8	3260	2570	3250	2980	2650	2320	1600	1180
10	4000	3050	3990	3560	3200	2790	1700	1230
12	4390	3220	4280	4020	3410	2940	1730	1260

[a]Emissivity = 0.8, angle of attack = 4°
q = dynamic pressure
R = radius of curvature
Λ = sweep angle
X = streamwise distance from leading edge

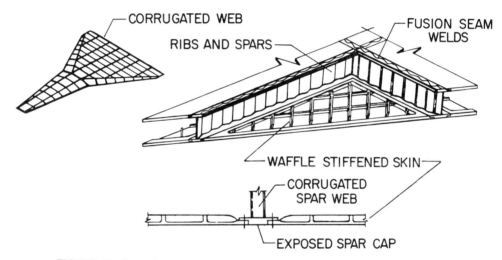

FIGURE 50   General arrangement and details of wing structure for hypersonic aircraft.

generally on the order of 0.008 in. to 0.020 in. The leading edges may be fabricated from refractory metals or cooled superalloys. Several materials are listed for the nose cap, including ceramics, graphite, refractory metals, and cooled superalloys. Titanium alloys are suggested for the fuel tanks (hydrogen), and cooled superalloys with a maximum temperature of 1600°F are indicated for the propulsion system.

The various types of materials suggested indicate that protective coatings will be required for certain areas of this type of vehicle. For much of the load-carrying structure, temperatures below 1600°F are noted, and the need for a coating is open to question. As operating temperatures for superalloys approach 2000°F, coatings may be mandatory, particularly for thin sheet material, in order to prevent destructive loss of metal thickness from oxidation and em-

brittlement. Also, a high surface emittance must be maintained on the external surfaces to aid in structural cooling by radiation. Most superalloy materials will exhibit high surface emittance at high temperatures, provided that care is taken in preparation of the oxidized surface. The application of certain coatings may lower the effective emittance and produce a hotter surface.

In the design process, materials are selected that will meet the life and reliability requirements and at the same time provide minimum weight. To accomplish this, a comparison is made between the weight of the uncoated material with proper allowances for loss of effective thickness due to oxidation and corrosion against the weight of coated metal required to meet the specified performance. The lower weight should dictate the approach selected. The

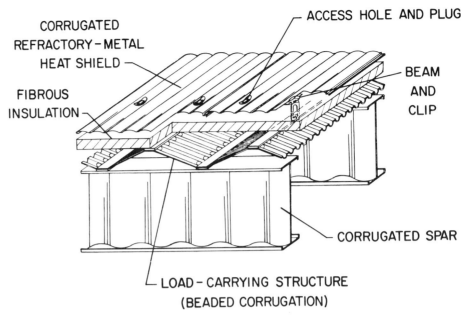

FIGURE 51   Insulated wing structure with corrugated refractory-metal heat shield.

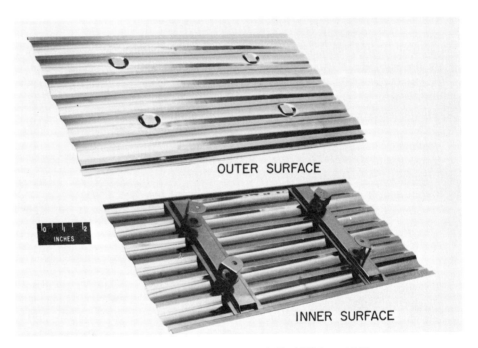

FIGURE 52   Corrugated refractory-metal (Ta–10W) heat shield.

coating thickness for the thin superalloy sheet must be the smallest practicable value to avoid excessive weight penalties. As an example, for an aluminide-type coating, the weight of 1 mil of coating on each side of the sheet of the hypersonic aircraft wetted surface exceeds 1 ton. Each mil of coating applied would thus decrease the payload on the order of 5 percent.

One additional factor needs to be considered: namely,

design approaches that require frequent refurbishment are not feasible for the thin superalloy sheet that comprises the major portion of the surface area of the vehicle. Conversely, for the most severely heated portions, which comprise only a very small fraction of the vehicle surface, various design approaches are potentially possible, and frequent refurbishment may be tolerated. Because of these considerations, the coating requirements for the superalloys that experience

TABLE 23 Suggested Minimum Temperature–Time Capabilities

Application	Material	Temp., °F	Exposure Time, hr.
Primary structure	Superalloys	Below 1600	5,000
Heat shields	Superalloys	1600 to 2000	5,000
Leading edges	Refractory metals Superalloys (cooled)	2000 to 3000	50
Nose cap	Ceramics Graphite Refractory metals Superalloys (cooled)	3000 to 4000	5
Hydrogen tank	Titanium alloys	−423 to 600	5,000
Propulsion system	Superalloys (cooled)	1600	2,000 (cycles)

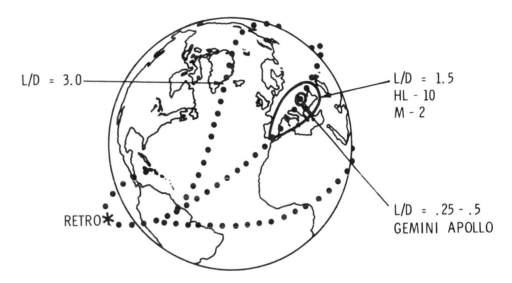

FIGURE 53 Maneuver performance of lifting reentry vehicles.

relatively moderate temperatures pose a challenge of comparable magnitude to that indicated for the more severely heated areas, such as the nose, leading edges, and underside of the vehicle.

Lifting Re-entry

Flexible approaches for returning manned and unmanned space vehicles from orbital or suborbital missions are required to reduce recovery delays and to minimize the expense associated with the deployment of expensive tracking and recovery forces. Lifting re-entry vehicles, in utilizing aerodynamic maneuvering to achieve a wide latitude in re-entry flight path, make it possible to significantly reduce logistics problems and costs associated with recovery. They also provide the benefits of a conventional horizontal land-

ing and a more or less conventional approach to maintaining the vehicle in a condition suitable for reflight (Figure 53). Vehicle size, performance, and exact mission vary widely. However, high-performance lifting re-entry vehicles have numerous similarities. A hypersonic lift over drag of 2.0 to 3.0 will be required to provide the necessary performance. Manned vehicle landing will probably be accomplished with the aid of variable-geometry wings and/or auxiliary jet engines. Rocket propulsion will normally be carried to perform space and re-entry maneuvers. Except for details, materials applications will be common to all designs for this class of vehicle. Discussion of recent programs investigating the concept of high-performance lifting re-entry vehicles appears in References 9 through 11.

The time–temperature characteristics of high-performance

LOWER SURFACE CENTERLINE TEMPERATURES

50 PSF AREA LOADING
.8 SURFACE EMITTANCE

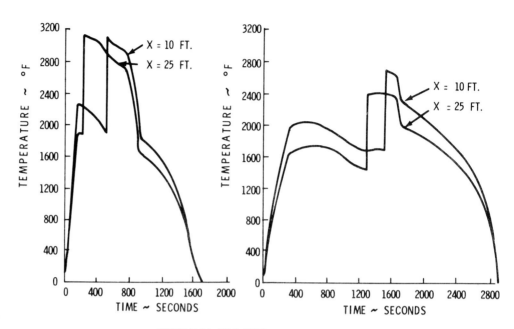

FIGURE 54   FDL-7MC operational re-entry.

lifting re-entry vehicles vary widely, since they are dependent upon the re-entry path chosen. The development of materials for such applications must consider a broad envelope of environmental conditions until specifically identified with a vehicle and mission of defined performance characteristics.

As an example, consider a 35-ft manned re-entry vehicle having a reradiative thermal protection system to illustrate time and temperature characteristics. Figure 54 presents centerline thermal profiles at two locations on such a vehicle for two extreme conditions of re-entry performance. The sharp rise in temperature is caused by transition of the boundary layer from laminar to turbulent flow. It should be noted that the left-hand curve indicates a higher overall temperature requirement for structural materials, while the right hand curve indicates a longer time at temperature for the material and a higher total heat input. A specific vehicle may be required to perform either of these re-entries or, if mission flexibility demands, both of them. The maximum temperatures occurring on this vehicle during a nominal re-entry are shown in Figure 55. These temperatures are not reached simultaneously on all portions of the vehicle, as can be seen in Figure 54, nor are they constant for any long period of time; they are, therefore, representative of a non-equilibrium condition during re-entry.

Economic considerations dictate that a major portion of the vehicle must be reusable. Such basic subsystems as load-bearing structures, avionics, propulsion, thermal protection, and mission equipment must be utilized for as many as 50 to 100 times. The thermal protection system requires the major refurbishment, accounting for 70 to 80 percent of the between-flight maintenance cost. Development of an inexpensive, easily maintained, and reliable method of protecting the internal portions of the vehicle from the heat of re-entry is, therefore, a major requirement in vehicle design.

The majority of vehicles under consideration at the present time utilize a main load-carrying structure separate from the thermal protection system. The primary structure is maintained at a relatively low temperature by insulation or by insulation and active cooling. Thus, conventional aluminum airframe construction techniques can be utilized for the main load-carrying structure. Figure 56 shows the layout of a research version of the FDL-7MC configuration designed to grow in speed capability from an initial Mach 2 to Mach 7, and then to Mach 15 or greater. The use of a conventional skin and frame structure allows growth by merely replacing the outer thermal protection system. In addition to being readily designed and fabricated, a cooled load-carrying structure is the lowest in weight (Figure 57).

Nonintegral tankage will be utilized for carrying internal

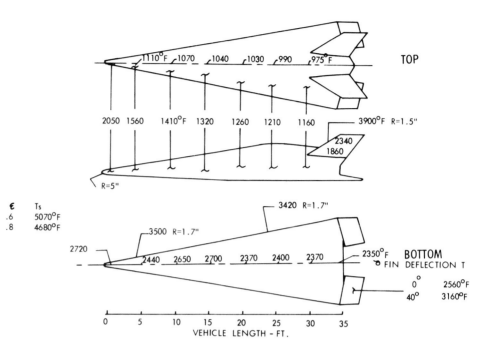

50 PSF AREA LOADING   11° ANGLE OF ATTACK

45° BANK ANGLE   0.8 SURFACE EMITTANCE

FIGURE 55   FDL-7MC maximum surface temperatures.

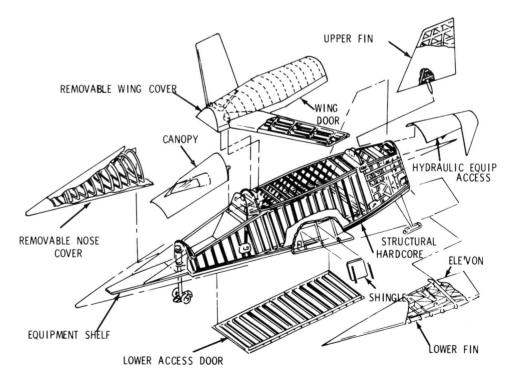

FIGURE 56   GTV structure.

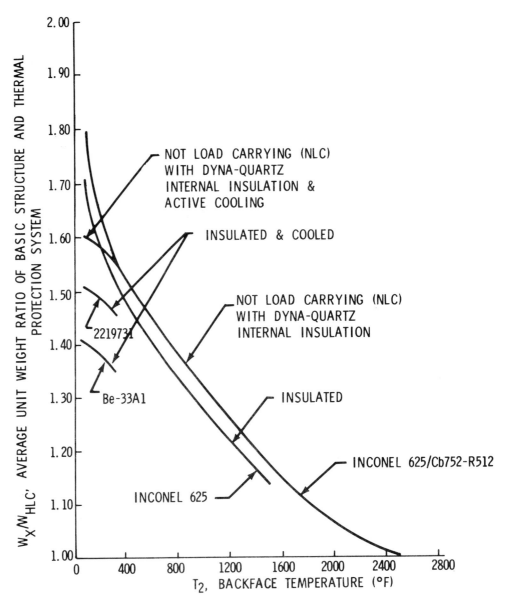

FIGURE 57   Thermostructural concept selection effect on average unit weight of basic structure and thermal protection.

fuel for most vehicles in the near future. Additional design and demonstration work will have to be done before integral tankage will be able to be considered for use in re-entry vehicles. Fin and wing structure such as that shown on the vehicle illustrated will probably be designed to be hot-load-carrying. Integration of hot and cool load-carrying structures creates a design challenge but will generally provide the lightest weight structure. Cutouts in the structure will provide stowage of variable-geometry wings and the landing gear and allow equipment access. These nonoptimum considerations must be taken into account when designing the structure and the thermal protection system.

The greatest potential application for coatings is to protect the sheet-metal radiative-heat shielding from oxidation at elevated temperatures during re-entry. Since the total mission life of this type of vehicle at high temperatures is less than 100 hr, the requirement to coat superalloys does not seem likely. Refractory metal portions of the heat shield must be coated.

The use of radiative thermal protection systems on re-entry vehicles faces considerable competition from ablative thermal protection systems. Cost, reliability, and re-entry conditions are major factors that enter into the trade-off. Figure 58 indicates that a radiative system must be reusable

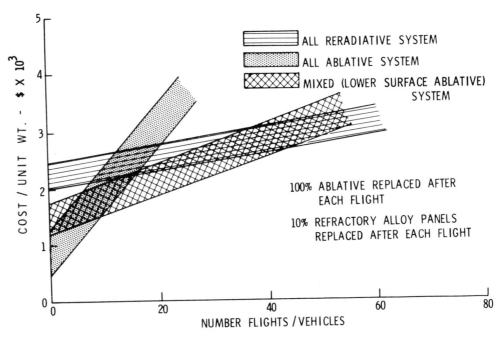

FIGURE 58   Heat shield cumulative cost trends for reusable high L/D vehicles.

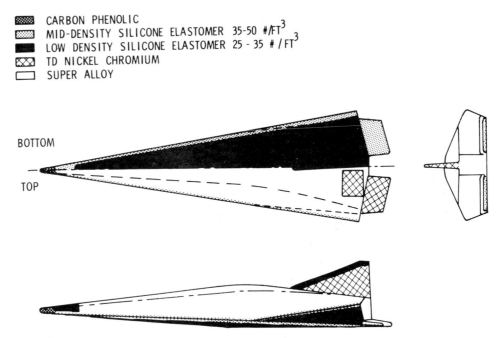

FIGURE 59   Radiative heat shield materials distribution for FDL-5A.

to be cost-effective. For low traffic rates, an ablative-radiative thermal protection system appears to be attractive because if an ablator replaces the refractory metal, cost effectiveness, through reliability and mission flexibility, is enhanced. The distribution of radiative and radiative/ablative heat shield materials on a proposed configuration is shown in Figures 59 and 60, respectively. If mission requirements dictate re-entry at temperatures above the capability of coated refractory metals, alternative materials will have to be used. If, however, competition is on a straight cost-effective basis, ways must be developed to decrease the manufacturing and installation costs of coated refractory

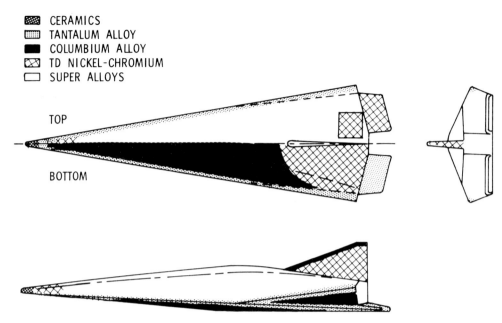

FIGURE 60   Ablative/Radiative heat shield materials for FDL-5A.

metals and to increase sharply the life and reliability of each part. The additional consideration of overlaying a radiative thermal protection system with an ablative material has been generally dismissed because of difficulty in determining the proper ablative thickness and because of excessive ablative bondline temperatures and ablation at the bondline.

Missiles

The environments and materials requirements associated with hypersonic missiles are defined and predicated on the basis that these vehicles are intended for single flights. The environment they encounter covers broad ranges of pressure, altitude, and temperature because the missiles may operate at both high and low altitude in a single flight. The most severe conditions are encountered in the low-altitude, hypersonic-velocity portions of the flight, and these conditions lead to the most stringent requirements for coatings. Although the environment is considerably more hostile than that encountered by manned hypersonic aircraft, the flight time is much shorter, being measured in seconds up to a few minutes. Because of the expendable nature of hypersonic missiles, cost-effectiveness is an important factor in materials selection and requirements. This factor requires careful consideration of such items as materials availability, fabrication ease, and the limited life requirement for structural and propulsion systems. Often, the short-life characteristic permits the use of materials at higher temperatures than allowable for long-time use. Also, insulating or ablative coating can be used more readily.

Temperatures, pressures, and aerodynamic loads depend upon design configuration, speed, and altitude. One of the most challenging problems involves materials used on air inlets, wings, and control surfaces. In low-altitude flights, severe performance penalties result from unnecessarily blunt stagnation areas. Cold-wall heat fluxes to 7000 Btu/ft^2-sec with stagnation pressures over 800 psi and aerodynamic shear loads of 300 psi could be encountered under extreme conditions, generating temperatures over 4000°F. Possible candidate materials for temperatures beyond the capabilities of superalloys include silicide-coated columbium alloys and silicide-coated tantalum alloy claddings or coatings on refractory metals or graphite.

AIR-BREATHING PROPULSION (RAMJET)

In this section, primary attention is directed toward general material requirements for air-breathing propulsion systems applicable to hypersonic vehicles. For hypersonic aircraft, the transition from subsonic to supersonic combustion in a ramjet engine is from Mach 6 to Mach 7. Because of high dynamic pressure considerations, supersonic combustion ramjet engines (Scramjet), rather than subsonic combustion ramjet engines, are most generally considered for use in the hypersonic speed range from Mach 5 to Mach 12. The following discussion will, therefore, be concerned primarily with the material requirements and thermal protection criteria for Scramjet engines.

The internal structure of a Scramjet engine may be shielded from a hostile thermal environment by two fundamental means—passive and active thermal protection. Passive thermal protection is based on the heat-sink, insulative,

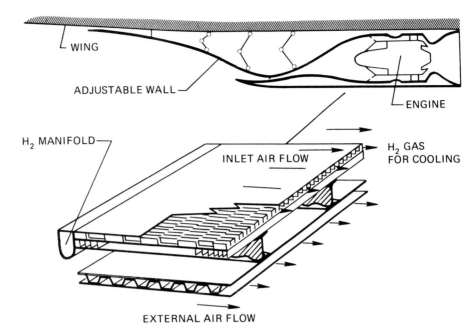

FIGURE 61   Regeneratively cooled propulsion system.

and radiative properties of materials. Active thermal protection utilizes the heat-sink capacity of engine fuel as a coolant for regeneratively cooled or mass-addition-cooled engine surfaces. The regeneratively cooled ramjet engine is usually preferred because of its multi-reuse capability. The ramjet propulsion system poses a severe materials problem because of high pressures and temperatures.

The regeneratively cooled ramjet is shown schematically in Figure 61. In the upper portion of this figure a two-dimensional air inlet propulsion system is shown. The view in the lower portion shows a typical wall cross section. The air entering this type of propulsion system is compressed to high pressures and temperature. The useful net thrust of a Scramjet engine is produced by the contour. Captured airflow is compressed by the inlet contour to a relatively high pressure, but the airflow is maintained at supersonic speeds. While the air is at this high pressure, heat is added by the supersonic combustion of fuel, which increases the *local* volume requirement for continuous expansion of the gas to ambient pressure. The engine contour confines the gas, resulting in local compression and a reduced expansion rate, to give a net forward component in wall pressure (net thrust) over the complete engine. Consequently, the propulsion aerothermodynamic requirements generally establish the engine configuration, which, in turn, establishes the materials design and utilization. The regeneratively cooled propulsion system applicable to a manned hypersonic aircraft shown in Figure 61 is subjected to high heat fluxes and temperatures. The air entering the propulsion system is compressed to as high a pressure as 10 atm, and heat fluxes in

the combustor area may be of the order of 500 Btu/ft^2-sec. These severe thermal environments produce large temperature differences in the structural walls. High thermal stresses that may result can produce thermal fatigue. The metal temperatures in the propulsion system can be controlled, usually by local sizing of the regenerative-cooling system, to accommodate the heat fluxes. The composition of the atmosphere in the propulsion system may be categorized in the following way: In the forward region, oxidizing conditions prevail because of the presence of oxygen in the airstream. In the combustor and aft regions, the atmosphere as a rule consists primarily of combustion products with possible traces of oxygen or fuel.

One of the most challenging requirements of a Scramjet active cooling system is to provide adequate cooling, with the total coolant flow rate being maintained equal to, or less than, the engine combustion fuel-flow requirement. Since the available fuel heat sink is usually constant under a particular design condition, the best approach is to reduce the hot-gas-side heat flux. One technique for reducing the heat flux to the engine structure, without decreasing the coolant capacity from the regenerative system or compromising thrust performance (as in mass addition), is to provide a high-thermal-resistance coating on the hot-gas side wall. Ceramic coatings are very effective as thermal-resistance materials at high heat fluxes. However, the effectiveness of insulating coatings decreases as heat flux decreases.

When a Scramjet flight trajectory includes significant time at high Mach numbers, the inlet leading edge must be actively cooled because its small radius causes the heating rate to be very high. One basic method of accomplishing this is with mass-addition cooling. As opposed to the more generous radii of lifting re-entry vehicles, the radii of leading edges attendant with propulsion systems is dictated by aerodynamic efficiency of the powered vehicle and aerodynamic efficiency of the engine cycle.

On interior engine surfaces, the best available high-temperature structural materials would fail in a few seconds if left unprotected or uncooled even at moderate velocities. Certain exterior components, free to radiate to the atmosphere, may operate at somewhat higher velocities, but nearly all "buried" structure requires some degree of thermal protection. Passive thermal protection systems may be divided into two general classes: sacrificial and nonsacrificial. Sacrificial systems are those in which chemical change or mass loss occurs during operation. Materials used in sacrificial systems include charring ablators, ablating refractories, and infiltrated refractories. Sacrificial systems are generally used for short-term operation in high heat-flux environments—for example, on high-velocity–high-acceleration missiles on nonrecoverable missions that are accomplished in a few seconds or minutes. Nonsacrificial systems utilize a materials heat-sink capacity and thermal insulating properties to maintain underlying structure within desired temperature limits. The primary advantage of nonsacrificial systems is that the materials undergo no physical or chemical degradation during use. Such systems, therefore, may be reusable and, in the case of radiation cooling, can operate continuously or for long periods.

The mission requirements of a Scramjet engine strongly influence the selection of the thermal protection method. It is convenient to relate the thermal protection capabilities to the mission requirements by the flight trajectory variables of Mach number, altitude, and time. Generally, if the flight Mach number is low enough or the flight duration time short enough, a passive thermal protection system may prove adequate for the complete engine. For long-duration missions, active thermal protection for the Scramjet engine appears necessary.

In summary, it can be seen that the propulsion system, integral with a hypersonic vehicle, will face an aerodynamic pressure and temperature environment more rigorous than the vehicle because of the addition of heat (combustion) to an already hostile environment. The two major criteria, consequently, that determine the material requirements, are altitude–Mach number and geometry considerations. Thermal protection techniques, such as those shown in Figure 62, may be used in various sections of the propulsion system. Although approaches from superalloys to insulated regenerative-cooling techniques will be used, depending on the altitude and Mach number, most missions indicate the requirements of a regenerative system with coated or uncoated substrates (superalloy or refractory alloy) or an in-

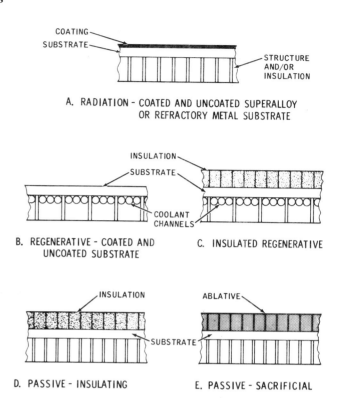

FIGURE 62 Thermal protection techniques.

sulated and regeneratively cooled wall to reduce coolant requirements. Table 24 lists the candidate materials for these thermal protection techniques. Generally, the major classes of materials will be used as follows[12-16]: coated refractory alloys for structure, walls, and leading edges; coated and uncoated superalloys for substructure, regeneratively cooled walls, and actively cooled leading edges; steels and titanium for substructure; and graphites (coated), ceramics, and ablatives for thermal protection.

Coating requirements for the propulsion system are not well established. It appears doubtful that present coatings can survive the high temperatures, thermal cycling and straining, erosion, and corrosion associated with the propulsion system. Continual improvement in coating capability is expected with further development; the most feasible approach, however, may be to use fuel cooling coupled with the selection of structural metals resistant to heat and thermal fatigue. Some refurbishment of the propulsion system materials may be required periodically. If coatings emerge that will improve the thermal fatigue characteristics and resistance to oxidation and corrosion, substantial improvement in propulsion-system life and performance would be obtained. Emittance values for the coatings that may be considered for the propulsion system are not particularly significant because the internal surfaces cannot be cooled by radiation.

TABLE 24 Candidate Materials for Hypersonic Air-Breathing Propulsion Systems

Class – Typical Candidates	Designation
Refractory metals	
Tantalum alloys	Ta-10W, T-111, T-222
Columbium alloys	D-43, Cb-752, C-103, C-132M, etc.
Molybdenum alloys	½Ti-Mo, TZM
Tungsten alloys	Tungsten-3Re
Hafnium–tantalum	
Superalloys	
Nickel-based	Hastelloy X, Inconel 718, Udimet 700, IN-100, TD-Ni, TD-NiC
Cobalt-based	L-605, MAR-M302
Steels and titanium	
Stainless steel	321, 310
High-strength steel	AM 350, 17-7Ph, etc.
Titanium	Ti-6Al-4V, Ti-7Al-4Mo
Graphites	
Commercial bulk	ATJ, JTA, AXF-5Q, etc.
Pyrolytic graphite	
Graphite composites	Carbitex, Pyrocarb, RPG, RPP
Ceramics	
Zirconia	
Magnesia	
Alumina	
Honeycomb composites	
Ablatives	
Silica phenolic	
Asbestos phenolic	

PERFORMANCE

CURRENT STATUS

Silicides–Mo Alloys

The earliest development of protective coatings for refractory metals was for molybdenum, and a significant coating capability now exists. All of the better coatings of interest for hypersonic vehicle application are based on the silicide system. Some of the available coatings are "straight silicides" (essentially $MoSi_2$), although the majority are modified with metallic elements such as boron or chromium. It is generally agreed that such modifiers will improve coating performance; however, modifiers are not mandatory nor are they incorporated in as large amounts as for silicide-coated columbium.

In hypersonic vehicle applications, molybdenum alloys are generally considered for use at maximum "flight" temperatures of 2600° to 3500°F. At lower temperatures, columbium alloys compete with molybdenum. Tantalum alloys are the competitors for higher temperatures. Compared to columbium and tantalum alloys, molybdenum alloys are more difficult to form and join. Welding is seldom used because of embrittlement due to grain growth; however, aerospace structures have been successfully electron-beam-welded. These limitations restrict the size and complexity of coated-molybdenum hypersonic structures. Coated molybdenum has a ductile-to-brittle transition (DBTT) that can vary between wide extremes, necessitating careful handling after coating. Recrystallization resulting from exposure to elevated temperature (~2600° to 2800°F) also causes reduced ductility and strength and will raise the DBTT. In spite of the shortcomings, there are applications for molybdenum on hypersonic vehicles. Its elevated-temperature creep resistance is much better than that for columbium, and its strength-to-weight ratio is superior to that of tantalum over an appreciable temperature range.

Because of the shortcomings mentioned earlier, less emphasis was placed on the development and evaluation of coatings for molybdenum alloys during the past few years compared with the work on coatings for columbium alloys. Consequently, fewer data are available for estimating the performance of silicide-coated molybdenum hypersonic vehicle structures than are available for coated columbium, even though the coating technology for molybdenum is well established. The behavior of molybdenum disilicide has been studied extensively, and it is considered to be one of the most oxidation-resistant coating materials. When used as a coating for molybdenum, its performance is inferior to that of the bulk material. This is partly because microcracks and other such defects are created by the coating–base-metal thermal expansion mismatch, and because lower-order silicides are formed via diffusion of silicon into the substrate.

Improperly processed molybdenum alloys tend to delaminate, and coating defects (fissures or crevices) are likely to occur at edges where delamination is readily produced, particularly on the thin sheet used for hypersonic vehicle components. Delamination-induced coating defects can be virtually eliminated by careful manufacturing and inspection. Also the "pedigreed" TZM alloy produced under the Refractory Metal Sheet Rolling Program appears less subject to delamination than other commercial sheet produced at that time. However, most of the coating performance data available are for the older "first generation" commercial TZM. The commercial TZM available today closely resembles the "pedigree" type and is considerably easier to form than the earlier materials.

The cyclic oxidation protective life at atmospheric pressure and at 2600°F for coated commercial and pedigreed TZM sheet are given in Table 25. In comparing lives, the differences in coating thickness should be noted because lifetime is generally proportional to coating thickness. The high incidence of edge failures shown by the data in Table 25 is typical of the state of the art and points to an area requiring attention to improve reliability and effective life. Edges are prepared for coating by radiusing, but this only

TABLE 25 Life of Coated Commercial and Pedigreed TZM Sheet in Cyclic Oxidation at 2600°F

Coating	Supplier	Nominal Coating Thickness, mil	Percent of Tested Specimens Failing by Indicated Mode[a]				Probable Life at Indicated Levels of Reliability, hr[b]			Observed Number of 1-hr Cycles to Failure	
			H	E	S	C	70 percent	90 percent	95 percent	Minimum	Maximum
20-mil commercial TZM[c,d]											
Disil	Boeing[e]	2.4	42	33	25	–	16	10	7	8	40
W-3	Chromalloy	1.2	–	85	15	–	9	4	3	6	35
Durak B	Chromizing	1.6	–	100	–	–	25	18	15	3	47
Modified silicide	General Technologies	1.6	–	90	10	–	<1	<1	<1	1	9
Sn–Al	General Telephone[f]	3.5	10	45	20	25	8	5	4	7	19
Modified silicide	Ling-Temco-Vought	1.9	–	70	30	–	22	18	16	10	30
PFR-6	Pfaudler	3.0	–	100	–	–	5	<1	<1	2	14
40-mil pedigreed TZM[c]											
Modified silicide											
#1	Ling-Temco-Vought	2.8	–	–	–	–	29	25	24	–	37
Modified silicide											
#2	Ling-Temco-Vought	2.0	–	–	–	–	17	14	12	–	29
L-7	McDonnell	2.0	–	–	–	–	25	18	16	–	42
PFR-6	Pfaudler	5.2	–	–	–	–	26	12	5	–	70

[a] Failure mode: H = suspension hole, E = edge, S = surface, C = catastrophic failure.
[b] Estimated from Weibull plots. Ten to twenty specimens were tested per coating.
[c] Test specimens were ½ in. × 1¼ in. coupons with edges and corners finished by University of Dayton.
[d] Examination indicated that sheet was of poor quality from the standpoint of delamination and that edge and corner finishing was inadequate.
[e] Substrate was 30 mils thick.
[f] All coatings are silicide-base, with exception of the General Telephone aluminide coating.

142

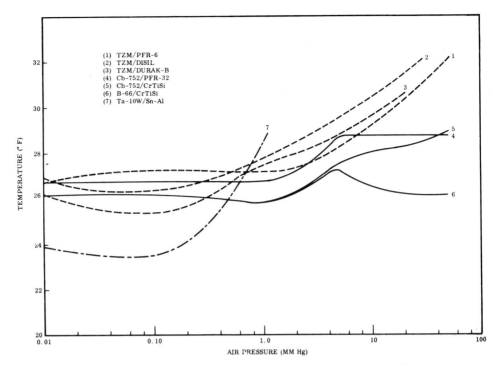

FIGURE 63   Maximum temperature for a 4-hr life of silicide coatings in air at reduced pressures.

minimizes the coating edge crevice formed, rather than eliminating it. Achieving an optimum radius is difficult, particularly on large and complex parts, and only a small radius can be produced on thin-gauge material. Also, a thick silicide coating will generally produce larger crevices that lead to early edge failures. The difference between the maximum and minimum life given in Table 25 is also significant. The early failures were due to defects generally located at edges. On hypersonic structures, the first coating failures observed are likely to correspond to the minimum test coupon lives because of part versus test coupon size considerations. However, the first coating failure, particularly if at an edge, is not likely to lead to immediate structural failure of the part.

The lifetime of silicide-coated molybdenum is substantially reduced when the oxygen partial pressure is below about 5 Torr as shown in Figure 63.[17] Pressures in this range are common for hypersonic vehicles, and the low-pressure degradation substantially reduces the effective coating life. While most systems can survive one re-entry, this behavior severely limits reuse capabilities. Also, since lower pressures may exist on internal surfaces, coating failures may be initiated in limited-access areas that cannot be readily inspected.

The thermal emittance of silicide-coated molybdenum and its alloys has been the subject of considerable controversy and uncertainty. One of the main reasons for this uncertainty is that emittance is a function of temperature, pressure, and exposure time. It appears that the total normal emittance of silicide-coated TZM alloy at 2500°F and at atmospheric pressure is normally between 0.6 and 0.7. Emittance-improvement topcoatings have been developed[18,19] and show promise for increasing the emittance by 0.1 to 0.2 emittance units.

Silicides–Cb Alloys

Because of the earlier work on protective coatings for molybdenum, the initial effort on columbium tended to follow along the lines of pack-siliciding. This was not strictly logical in view of the substantial differences in properties between these two metals and their silicide compounds. Inherent oxidation resistance of $CbSi_2$ is relatively poor. In addition, the molar volume ratio of $CbSi_2$ to Cb and the large mismatch in thermal expansion between this compound and its parent metal promote extensive cracking of pack-silicide coatings on columbium alloys. Nonetheless, widespread effort was expended in developing "modified" pack-silicide coatings. While a broad variation of behavior was noted among the several proprietary pack-silicide coatings, their relative performances were not consistent. The spread in performance is attributed to variations in processing, inasmuch as chemical, x-ray, and electron microprobe analyses have established that the one-step modified pack silicides were basically $CbSi_2$. ($CbSi_2$ has negligible solubility for any of the commonly tried pack additive "modifier" elements.) None of the basic pack-silicide coatings are commercially produced today, although any number of organizations have the capability for such application.

TABLE 26   Representative Test Results for Vacuum-Pack Cr–Ti–Si-Coated Columbium Alloys

Average oxidation test lives at various temperatures

Temperature (°F)	Mean life (hr)	
	D-43 alloy	B-66 alloy
1800	No coating failures observed in >150 hr	No coating failures observed in >150 hr
2500	109	139
2700	36	32

Protective reliability at 2500°F (99 percent confidence)

Probability (%)	Life (hr)	
	D-43 alloy	B-66 alloy
50 (Mean)	109	139
75	91	120
90	74	102
95	64	91
97.5	59	82

Tensile strength properties of 30-mil D-43 alloy

Test Temperature (°F)	Uncoated			As CrTiSi Coated			After Creep Oxidation[a]		
	UTS (psi)	YS (psi)	Percent Elong.	UTS (psi)	YS (psi)	Percent Elong.	UTS (psi)	YS (psi)	Percent Elong.
RT	81,000	62,000	24	72,300	51,400	25	71,800	51,000	25
2000	41,000	37,700	19	32,400	28,800	11	–	–	–
2500	19,300	17,500	43	17,500	15,300	88	–	–	–

[a] 100-hr creep exposure at 2500 psi, 2000°F resulted in approximately ½ percent elongation before tensile tests.

The first reasonably reliable coating developed for columbium was a chromium–titanium alloyed silicide (Cr–Ti–Si) applied by a two-step processing sequence. The original process involved vacuum-pack deposition: First, codeposition of Cr and Ti from a master alloy pack to form a surface layer of $(CbTi)Cr_2$ with an underlying Ti-enriched diffusion zone, followed by vacuum-pack siliciding, during which a columbium-to-silicon ratio gradient is produced throughout a final coating best described as mixed (Cr, Ti, Cb) silicides.

This coating has been repeatedly shown to provide protection for times as long as 150 to 200 hr at 2500°F in static air furnace oxidation tests (with hourly air-cooling). Much longer protective lives are normally realized at lower temperatures. However, slow thermal cycling significantly reduces coating life at all temperatures. Representative oxidation test results are shown in Table 26. Also included in the table are typical as-coated D-43 alloy strength properties. Effects of the thermal processing associated with the coating application, together with the physical and thermodynamic effects of the coating's presence, combine to reduce the effective strength of carbide precipitation strengthened alloys such as D-43 or duplex-annealed Cb-752 by as much as 25 percent. Solid solution-strengthened alloys are not affected in the same way, although there can be adverse changes in ductility and fatigue strength.

The Cr–Ti–Si coating has several limitations that require thorough consideration before it is used. The need for a vacuum process and the long processing time combine to make the coating expensive. For high temperature and long-time applications, such as in multimission re-entry vehicles, diffusion may degrade mechanical properties below a tolerable limit, especially for substrates less than 0.010 in. thick. Low pressure (below 5 mm Hg), combined with high temperatures (above 2500°F), appreciably shorten coating life by chromium diffusion to the surface and subsequent vaporization.[17]

Probably the most successful coating approach for columbium has been the fused-slurry silicides. A number of elements and intermetallics have been studied as modifiers to the basic Si–20Ti and Si–20Cr eutectic slurry compositions. Fusion of the sprayed or dipped slurry in vacuum results in mixed silicide phases that provide high-temperature capability. The compositions Si–20Cr–5Ti and Si–20Cr–20Fe have received the most investigation and testing, but over forty coatings have been developed. Among the slurry coatings studied, the Si–Ti–Cr and Si–Cr–Fe have received the most effort. Typical oxidation and mechanical-property data are shown in Tables 27 and 28. In general, it can be said that these coatings furnish protective lives as long as, or longer than, any coating formerly known, and do not degrade mechanical properties of the substrate alloy.

Extensive characterization of the fused-silicide type of

TABLE 27 Oxidation Properties of Fused-Slurry Silicide-Coated Columbium

Furnace oxidation life of Si–Cr–Fe-coated Cb752

Test Conditions

Temp. (°F)	Cycles	Protective Life (hr)[a]
1600	Isothermal	1300+
1800	16 cycles	595+
2000	16 cycles to RT	595+
2300	8 cycles to RT	189
2400	8 cycles to RT	100 to 189
2500	8 cycles to RT	120 to 189
2500	Hourly cycle to RT	97
2600	Hourly cycle to RT	33 to 44
2700	Hourly cycle to RT	19
2800	Hourly cycle to RT	15
800 to 2500	Slow cycles each hour	96 to 120

Oxidation and re-entry simulation tests for fused-slurry-coated alloys

Coating	Substrate Alloy	Slow Cyclic Oxidation[b] Life (1-hr cycles to 2500°F Maximum Temp.)	Re-entry Simulation Cyclic Life[a] (1-hr cycles to 2500°F Maximum temp.) Low (internal) Pressure	High (external) Pressure
Si–20Cr–5Ti	D43	24	200+	200+
	Cb752	15	177 to 200	161 to 200
Si–20Cr–20Fe	D43	63	200+, 553+	200+
	Cb752	71 to 120	200+	200+

[a] + indicates test stopped without failure; still protective.
[b] Range of lives observed with only a few specimens tested at each condition.

TABLE 28 Mechanical Properties of Fused-Slurry Silicide-Coated Columbium Alloys

Tensile Properties of 18-mil Si–Cr–Ti-Coated D-43 Columbium Alloy (2500°F test temperature)

	UTS (psi)	YS (psi)	Elong. (%)
Uncoated, as received	17,800	10,400	48
Coated with Si–20Cr–5Ti, as-coated	23,000	12,600	15
Coated with Si–20Cr–5Ti, oxidized 8 hr at 2500°F	21,500	12,400	19

Tensile Properties of 30-mil Cr–Ti–Si-Coated Cb752 Alloy

Test Temperature (°F)	Uncoated UTS (psi)	Elong. (%)	As-Coated UTS (psi)	Elong. (%)
RT	79,000	22	79,000	17
1500	54,000	14	51,000	10
1850	43,000	12	44,000	10
2250	35,000	28	30,000	5

coatings has demonstrated the inherent superiority of this type in relation to the other coatings for columbium. It is believed that the slow cyclic and re-entry simulation tests, given in Table 27, establish the basic protectiveness of the fused-silicide coatings for multimission re-entry vehicle applications, except that the effect of stress is not considered. Wettability and flowability experiments and demonstrations of the ability of this type of coating to be applied to practically any size or shape of structure have indicated that the full potential of this coating can be achieved in practice. As far as the combined effects of low pressure and high temperature are concerned, re-entry simulation testing combined with low-pressure studies show the fused-slurry silicide coating to be superior to all other coatings. Statistical reliability tests, defect tolerance tests, and repair studies also provide evidence of the practicality of this type of system.

The reliability of the fused silicides can probably be attributed to several factors. First, as demonstrated in wettability tests, it is quite unlikely that uncoated areas may inadvertently result. It is hard to imagine a grossly inhomogeneous coating resulting from the reaction of a molten alloy, since a substantial part of the reaction takes place before solidification. In addition, it is not possible to have unreacted or partially reacted coating materials adhering to the part after processing—something that can readily happen with the pack-cementation process. In the fused-silicide process, all coating materials are reacted almost instantly; there is little danger of incompletely fusing the coating. However, this rapid reaction can cause the formation of overly thick coatings on low-access surfaces where control of the dip-applied-slurry thickness is very difficult.

There is little question that the fused-silicide coatings will penetrate the faying surfaces of any joint better than any coating applied by other processes and will thereby afford greater protection to these joints. However, the extent to which the protective life of the coated joint approaches the basic protective life of the coating will depend upon such factors as the stiffness of the joint, the length of cantilevered faying sheet, the weld or rivet spacing, the faying surface gap, the flatness of faying sheets, and the thicknesses and thickness ratios of mating parts. These factors will determine whether the joint will be effectively sealed or if the faying surfaces will each have a separate, tapered coating. It has been demonstrated that short stiff joints can be effectively coated and protected, but design limitations are undefined. Fused silicide coatings can now be applied to many re-entry vehicle components with confidence that they will satisfactorily protect the structures through several reuse cycles.

Silicides—Ta Alloys

Most of the early work on coatings for tantalum paralleled previous columbium coating efforts. The same factors previously discussed for columbium militate against successful use of simple pack-silicide coatings. It should be recognized that the problem of identifying promising coatings for Ta is more difficult because interest in the structural use of Ta occurs only for service temperatures greater than 2500° to 2800°F. Over a period of several years, a series of investigations explored the feasibility of two-step, modified-pack silicides. Boron, manganese, or vanadium additions were found to result in occasionally long coating lives, but results were never consistently reproducible. The most effective pack process devised was a two-step coating, consisting of a heavy titanium precoat applied from a 90Ti-10W pack, followed by pack-siliconizing. Test lives as long as 28 hr at 2700°F were achieved. None of these developments resulted in commercially available coatings.

The standard vacuum-pack Cr-Ti-Si coating described for Cb alloys has been applied to Ta with reasonable lives at low temperatures but with an insufficient maximum-temperature capability. A fluidized-bed, three-step process has been demonstrated in which tantalum is silicided, then vanadium is deposited, and finally the coating is resiliconized. Again, maximum service temperature is marginal, and the overly complex process is not considered practical.

Studies of electrophoretically deposited binary silicides have been conducted in which combinations of W or Mo with V, Cr, and Ti disilicides were included. The most interesting combinations are either $MoSi_2$-$TiSi_2$ or $MoSi_2$-$CrSi_2$ electrophoretically codeposited and sintered over a VSi_2 layer. The final step was pack-siliconizing. These coatings have a potential life of several hundred hours, based upon both 1500° and 2400°F testing. Maximum test lives of up to 800 hr at 2400°F were observed, but many short-lived tests also occurred; very poor reproducibility was indicated, and no higher temperature oxidation results were reported.

Current work, aimed at Ta-alloy turbine vanes, has led to the successful development of a series of complex multicomponent (e.g., W-Mo-Ti-V) silicides. The concept for these systems involves applying a surface alloy layer by slurry plus vacuum sintering, to form a somewhat porous precoat; this layer is then silicided by the conventional pack technique, during which most porosity is filled or closed by volume increase within the layer. The thermal expansions of these complex silicides are fairly well matched to that of the substrate alloy, and some residual (closed) porosity serves to arrest cracks, so that exceptional thermal-fatigue resistance results. Protective lives in excess of 600 hr at both 1600° and 2400°F have been consistently observed for several coating compositions within this family of coatings. The W-Mo-Ti-V modifier alloys have demonstrated the greatest potential for reliable long-life performance. Promising coatings have resulted from surface impregnation with a slurry of barium borosilicates glass to improve low-temperature behavior. Work reported to date has not included oxidation testing at very high temperatures (i.e., 2800°F and

TABLE 29 Cyclic Oxidation Protectiveness of Silicide Coatings on Tantalum Alloys

Test temperature (°F)	Maximum cyclic oxidation life (hr)	
	Ti-modified pack-silicide coating on Ta–10W alloy	Si–20Ti–10Mo slurry on T-222 Ta alloy[a]
1000	60	150+
1600	50	150+
1800	50	150+
2000	22	150+
2600	8	27
2700	28	50
2800	3–4	20
3000	1	4–7
3200	0.28 (17 min)	1[b]

[a]+ Indicates test stopped, samples not failed.
[b]Some 3200° samples failed immediately.

above). No mechanical-property data are available for these coatings. Currently, these silicide coatings are more suited to use as turbine materials than to the hypersonic vehicle applications being considered in this section.

Of particular interest is the recent effort to apply high-temperature WSi_2 coatings to Ta alloys by sequential deposition of W and Si. Original efforts to embody this concept involved chemical vapor deposition of tungsten, followed by vacuum-pack siliciding. Such coatings, when applied on small coupons, had lives longer than 16 hr at 3000°F, 4 hr at 3300°F, and 1 hr at 3500°F when furnace-tested. These promising early results have never been successfully reproduced. The innate difficulty has been the application of a uniform, dense, well-bonded tungsten precoat. Recently, effort has been undertaken to apply tungsten coatings to Ta by slurry spray or electrophoresis plus sintering, and to date, consistent results have not been attained. Thus the potential of WSi_2 coatings remains promising but unrealized. It is tentatively planned to test a slurry-sinter W version of this system applied to a large experimental test component.

The fused-slurry silicide coating technology developed for columbium has also been successfully applied to tantalum. A SiTiMo slurry coating has proven the most reproducible. The cyclic oxidation protection afforded tantalum alloys by the SiTiMo fused slurry is shown in Table 29. Maximum temperature capability is 3200°F or less at 1 atm. The slurry SiTiMo has been successfully applied to large experimental components and is under continuing evaluation. Despite considerable activity, coatings for tantalum are still at a rather preliminary stage of development.

Silicides–W Alloys

Most of the coating work on tungsten has been aimed at service temperatures beyond 3000°F and has involved un-alloyed tungsten as the substrate. Almost all development

activity has evolved around pack-silicide coatings, with a little work on ceramic coatings and significant fundamental efforts on the investigation of oxide properties directed toward the identification of promising coating concepts for use at very high temperatures. Success in terms of practical coatings has been quite limited. Unlike silicide diffusion coatings on other refractory alloy substrates, tungsten disilicide (WSi_2) offers some promise for protection above 3000°F, perhaps as high as 3500°F. Several investigators have obtained significant oxidation protection by pack- or vapor-deposited WSi_2 coatings to at least 3300°F. However, a major problem has been recurrence of "pest oxidation" at lower temperatures (i.e., 1500° to 2500°F) where very short protective life is observed.

In an investigation devoted to the study of "metal-bonded metal-modified oxide coatings," coatings were formed by sequentially vacuum-pack-depositing and diffusing a series of modifier metals, including Ti, Zr, W, Si, and B, and then converting the surface to oxide in wet hydrogen. Subsequent oxidation formed the dense protective oxides. The two most promising systems identified were a codeposited silicon-plus-tungsten layer (apparently WSi_2) and a modification that included successive deposits of Ti and Zr followed by W–Si. Extensive furnace oxidation testing showed these coatings to be protective for times on the order of 75 to 100 hr at 3000°F, 16 hr at 3300°, 10 hr at 3400°, and 2 to 5 hr at 3500° to 3600°F. Plasma-flame exposure for up to 2 hr at 3500°F and low-pressure tests for as long as 1 hr at 3500°F were also conducted. The original investigators reported x-ray diffraction identification of a tungsten ortho-silicate as the protective scale on these coatings; in several later studies, however, only WSi was found in the coating and SiO_2 in the oxide scale. In any event, vacuum-pack silicide coatings that provide a reasonable degree of protection to tungsten have been developed. The same coating has also been successfully applied to rocket-nozzle inserts for firings, and a modified coating has been used to protect molybdenum nozzle-throat inserts of attitude-control thrusters for the S-IVB upper-stage Saturn launch vehicle. A laboratory study using only small, coupon-type specimens revealed the W–Si coating to have a 96 percent probability of surviving 5 hr at 3500°F in air at 1 atm.

Limitations on forming of tungsten components and high ductile-to-brittle transition temperatures have combined to minimize the extent of practical work on coatings for tungsten that would have any applicability to hypersonic vehicles. Therefore, the foregoing narrative describes a very preliminary state of the art, and no real performance capability or reliability can be said to exist for coated tungsten structures.

Aluminides–Refractory Metals

Aluminide coatings are the second most widely developed and utilized type of coating for columbium and tantalum alloys. Aluminide coatings may be divided into two types: (a) primarily intermetallic layers such as $CbAl_3$ or $TaAl_3$ and (b) liquid-phase coatings such as tin-aluminum and silver–silicon–aluminum. Compared with the silicides, the aluminide coatings have lower temperature capabilities and shorter protective lives. Aluminide coatings have been used extensively for past and current aerospace vehicle applications chiefly because of processing advantages. Vehicles such as ASSET and BGRV were designed for one flight only and did not require long-life coatings. A processing advantage was gained because most aluminide coatings are applied by the molten (fused) slurry process. This process permits the coating of large and complex assemblies, common to aerospace construction, that have faying surfaces and very limited access areas.

The intermetallic aluminide coatings are not competitive in terms of service life with currently developed silicide coating systems. For example, the upper temperature limit for aluminide-coated columbium is 2800°F for very short times, and the average life at 2600°F in 1-atm air is about 6 hr. Recent developments make it possible for silicide coatings to be applied by a molten-slurry process; thus the aluminides no longer offer an exclusive application advantage.

The liquid-phase coatings have long and reproducible protective lives when used in relatively thick films. At 2600°F in 1-atm air, the Ag–Si–Al coating has an average life of about 115 hr (1 hr cyclic) and a life of 95 hr for a 95 percent level of reliability. Such performance is superior to the silicides; however, the Ag–Si–Al coating was 7.2 to 13.6 mils thick, which is about four times greater than for the silicides. After correcting for thickness, the aluminide performance would be generally comparable to the better silicides; however, the reliability perhaps remains superior. The liquid phase that is present above about 2000°F provides the high reliability of the liquid-phase aluminides; however, it also makes the aerospace use of this type of coating questionable, in view of high volatility and low resistance to aerodynamic shear.

For future hypersonic applications involving a long service life and many reflights, the aluminide coatings have less promise than the silicides. Therefore, the silicide coatings have been discussed in greater detail because they appear closer to meeting the future requirements. Where future requirements include only one or two flights and temperatures below 2800°F, the present aluminide coating state of the art is satisfactory.

Aluminides–Superalloys

Current capabilities and manufacturing technology for diffusion coatings (aluminides) for superalloys are thoroughly discussed in Chapter 3, p. 60. That discussion suggests that it is doubtful that current aluminide coatings can satisfy the long-life and low-weight requirements predicted for future hypersonic flight vehicles where temperatures will exceed about 1800°F. For vehicles designed for relatively short life expectancy (e.g., up to 100 hr), it is questionable whether aluminide coatings for superalloys will offer any advantages with regard to oxidation and/or corrosion resistance for temperatures less than about 2000°F. For all applications, an oxidation-protection coating should be used only when the coating weight is less than the weight of the metal lost by oxidation. Diffusion coatings are not practical for substrate thicknesses of 0.003 in. or less.

Thick insulating oxide-type coatings for superalloys may be useful in shorter-life hypersonic engines; these have been discussed in this chapter under Chemical Propulsion. (p. 123). These coatings could be used for providing a ΔT where gas temperatures are in the 2000° to 3000°F range and could thus permit superalloys to function at about 2100° to 2000°F.

Noble Metals

In any discussion of the potential use of platinum-group metals (Pt, Pd, Ir, Os, Ru, and Rh) for high-temperature protective coatings, it is important to remember that they are, in general, very expensive materials. It is doubtful, therefore, that any of these six metals will ever be used to coat large surface areas. Other factors that must be considered include (a) the physical properties of the platinum-group metals, (b) inherent oxidation resistance, (c) availability, and (d) interaction with substrate materials at elevated temperatures.

Examination of the data in Table 30 suggests that the melting points of both Pt (1769°C) and Pd (1552°C) would somewhat limit their use as high-temperature protective coatings. On the basis of the melting-point criteria, only Ir, Os, and Ru could be used at temperatures above about

TABLE 30 Properties of Platinum-Group Metals Compared with Selected Other Metals

Metal	Melting point °C	°F	Density (g/cm³)	Thermal expansion (%) from 25°C to indicated temp.	
Pt	1769	3216	21.4	1.02 (1000°C)	1.67 (1500°C)
Pd	1552	2826	12.0	1.36 (1000°C)	—
Ir	2454	4449	22.5	0.77 (1000°C)	1.80 (2000°C)
Os	2700	4892	22.6	0.32 (600°C)	—
Ru	2500	4532	12.2	0.46 (600°C)	—
Rh	1966	3571	12.4	1.05 (1000°C)	1.79 (1500°C)
Cb	2415	4379	8.6	0.81 (1000°C)	1.78 (2000°C)
W	3410	6170	19.3	0.47 (1000°C)	1.07 (2000°C)
Mo	2610	4730	10.2	0.56 (1000°C)	1.47 (2000°C)
Ta	2996	5425	16.6	0.69 (1000°C)	1.53 (2000°C)
Cr	1875	3407	7.2	1.07 (1000°C)	2.18 (1800°C)
Co	1495	2723	8.9	1.80 (1000°C)	—
Ni	1453	2647	8.9	1.67 (1000°C)	—
Fe	1536	2797	7.9	1.17 (1000°C)	2.09 (1390°C)

1975°C. The high densities of Pt, Ir, and Os would be of concern, especially if they were used to coat large surface areas of a metal with significantly lower density. Such a "weight penalty" might preclude the use of these metals as coatings for many space-vehicle applications. (Weight also affects the cost since a given volume of high-density platinum-group metal would cost significantly more than an equal volume of a lower-density metal.) The importance of the density factor is indicated by the following data on the surface area that can be coated to a thickness of 0.002 in. by 1 oz of platinum-group metal:

Metal	Surface Area Coated to a Thickness of 0.002 in. by 1 oz of metal (ft²)
Pt	0.31
Pd	0.55
Ir	0.29
Os	0.29
Ru	0.54
Rh	0.53

The thermal expansion data in Table 30 suggest that many platinum-group metals would not be compatible with the metallic substrates on which they might be used as coatings. For example, there is a reasonably large thermal expansion mismatch between Rh (one of the more oxidation-resistant platinum-group metals) and Cb, W, Mo, and Ta. Such a mismatch would probably present problems in maintaining coating integrity during cyclic heating. The validity of this prediction is supported by the work of Goetzel et al.,[20] who encountered a thermal expansion mismatch problem when attempting to use electroplated Rh as a protective coating for W. The low emittance values generally associated

with platinum-group metals may also prevent their use for those applications where radiative cooling is required.[21]

At least two of the platinum-group metals (Os and Ru) are not sufficiently oxidation resistant to protect other materials for long periods. This fact is evident from the data in Figure 64.[22] In an oxidizing atmosphere, osmium oxidizes at a rate similar to that observed for Mo or Re. The oxidation rate of Ru is only slightly lower than that reported for W. Ir and Rh are sufficiently oxidation-resistant to be considered for use as protective coatings. The same is generally true for Pt (which oxidizes at a rate between 4×10^{-2} and 4×10^{-4} mg/cm²/hr at 1425°C) and Pd.

The data in Table 31 show that the United States is dependent upon foreign sources for its major requirements of Pt, Pd, Ir, Rh, and Ru. Sufficient quantities of Os are produced domestically to satisfy the nation's needs for this Pt-group metal. Assuming that only the U.S. Government's inventory would be available, sufficient quantities of Pt group metals would be available to cover the following surface areas with protective coating to a thickness of 0.002 in.

Metal	Surface Area Coated to a Thickness of 0.002 in. by U.S. Government's Inventory (ft²)
Ir	4,050
Pd	405,000
Pt	238,000
Rh	530
Ru	8,100

Additional amounts of Pt-group metal might, of course, become available to supplement that in the U.S. Government's

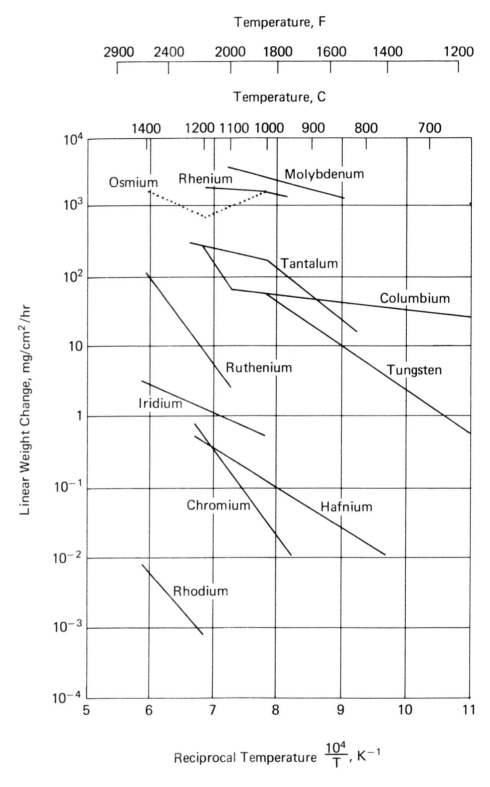

FIGURE 64 Effect of temperature on the oxidation rate of Os, Ru, Ir, and Rh compared to other refractory metals. (Data for Mo, Rh, Os, Ru, Ir, and Re are weight losses.)[22]

TABLE 31 Availability of Platinum-Group Metals

Production in U.S. (troy oz)[a]

Year	Platinum Pri	Sec	Total	Palladium Pri	Sec	Total	Iridium Pri	Sec	Total	Osmium Pri	Sec	Total	Rhodium Pri	Sec	Total	Ruthenium Pri	Sec	Total
1956–1960 (av)	39,092	48,834	87,926	6,229	42,339	48,568	2,538	1,296	3,834	871	364	1,235	1,207	3,097	4,304	920	1,701	2,621
1961	46,113	51,218	97,331	28,988	32,451	61,439	1,903	193	2,096	148	6	154	1,993	1,836	3,829	308	267	575
1962	36,462	71,817	108,279	16,144	56,273	72,417	905	767	1,672	100	99	199	1,016	2,570	3,586	148	576	724
1963	40,290	54,084	94,374	32,799	59,993	92,792	2,270	440	2,710	189	273	462	3,421	1,990	5,411	1,239	319	1,558
1964	30,539	66,043	96,582	27,301	49,879	77,180	3,981	764	4,745	515	928	1,443	6,274	2,338	8,612	2,480	195	2,675
1965	25,247	53,562	78,809	26,339	50,025	76,364	2,628	960	3,588	1,199	763	1,962	4,858	2,590	7,448	1,452	625	2,077

Imports to U.S. (troy oz)

Year	Platinum	Palladium	Iridium	Osmium	Rhodium	Ruthenium
1964	281,922	<83,018	6,615	—	55,804	10,356
1965	349,280	734,881	10,839	—	39,769	8,198

Sold to Consuming Industries in U.S. (troy oz)

Year	Platinum	Palladium	Iridium	Osmium	Rhodium	Ruthenium
1956–1960 (av)	346,076	412,935	6,060	819	21,805	5,154
1961	283,088	508,040	6,547	805	19,174	5,572
1962	304,272	519,860	9,251	1,125	26,063	5,888
1963	424,344	526,527	9,832	1,056	37,068	4,367
1964	451,350	591,432	9,652	1,379	55,426	8,441
1965	411,435	717,085	9,554	1,634	38,910	8,083

End-of-year refiners, importers, and dealers' stocks (troy oz)

Year	Platinum	Palladium	Iridium	Osmium	Rhodium	Ruthenium
1961	255,654	244,910	12,250	3,058	29,258	10,315
1962	256,755	285,173	13,871	2,762	30,692	8,849
1963	320,601	315,756	18,907	1,531	32,900	9,880
1964	378,896	317,691	20,022	1,936	38,388	10,331
1965	422,804	427,450	18,374	1,502	44,531	11,712

U.S. Government Inventory as of 31 December 1965 (troy oz)

Metal	National stockpile	Supplemental stockpile	Total
Iridium	14,000	—	14,000
Palladium	90,000	648,000	738,000
Platinum	716,000	50,000	766,000
Rhodium	1,000	—	1,000
Ruthenium	15,000	—	15,000

Source: Ware.[23]
[a]Pri (primary) is metal refined in U.S.; Sec (secondary) is metal recovered in U.S.

inventory; however, these supplemental quantities could be limited, especially in the event of a national emergency. Furthermore, it must be remembered that the abundance of Au in the earth's crust is five times that of Rh, Ir, Ru, or Os.[24]

Another factor that must be considered when evaluating the potential use of Pt-group metals as high-temperature protective coatings is interdiffusion between the coating and the substrate. An investigation by Buchinski and Girard[25] confirmed the need for diffusion barriers between a 0.002-in.-thick coating and the substrate when Pt or a Pt-10Rh alloy is used to protect FS-82 (Cb-28Ta-10W-1Zr) or TZM (Mo-0.5Ti-0.1Zr) refractory metal alloys from oxidation for 100 hr at temperatures up to 1650°C. The need for diffusion barriers between Pt or Rh and Mo or W had been predicted earlier by Pentecost.[21] In the absence of a diffusion barrier, rapid interdiffusion resulted in early failure of the material systems. Attempts by Buchinski and his co-workers to obtain a satisfactory diffusion barrier resulted in only marginal success. Research by Criscione *et al.*[26] involved the use of electrodeposited (0.0025 to 0.003 in. thickness) and clad (0.005 in. thickness) Ir on Mo, W, Ta, and Cb. Oxidation tests at 1850°C revealed that the loss of Ir was controlled by the rate of oxidation and the rate of interaction with the substrate. The integrity of a coated specimen was greatest for those substrates that re-acted slowly with Ir (an Ir-coated W did not fail in 117 hr at 1850°C). This work further established that interdiffusion is an important consideration when one attempts to protect various substrates with Pt-group metals. Research by Dickson *et al.*[27] revealed that a diffusion barrier of 0.003 in. of W (or Re) is satisfactory for minimizing interdiffusion between Ta and Ir for times up to 1 hr at 2200°C. In the absence of such a diffusion barrier, Ir-Ta eutectic melting (1950°C) would be expected after sufficient interdiffusion had oc-curred.

Oxidation-Resistant Alloys

Hafnium alloyed with 15–30 percent Ta reacts with oxygen at elevated temperatures and forms protective scales that are tenacious and thermal-shock-resistant. Three distinct layers are formed: (a) an outer oxide, (b) a subscale that is a com-posite oxide plus metallic alloy, and (c) a substrate zone that is interstitially hardened. The scales are protective to about 4000°F, although oxygen transport continues at rapid rates such that the scale grows to substantial thicknesses in short times (i.e., hundredths of an inch in 10 min). A mini-mum in the oxidation rate occurs at 20–25 Ta. The Hf-(10-15)Cb system also shows analogous oxidation resistance, although inferior to Hf-20Ta, with compensating indication of superior ductility and fabricability. Oxidation behavior of Hf-Ta alloys is substantially modified by addition of a few percent of elements such as Al, Cr, Si, B, Ir, and Pt. In

static-furnace tests of Hf-Ta-Cr-B, Hf-Ta-Cr-Al, and Hf-Ta-Ir-Al alloys, selected compositions exhibited up to 450 hr of protective oxidation resistance at 2500°F.

The principal application of Hf-20Ta alloys has been in the form of coatings or claddings for protection of short-life missile components at temperatures to 4000°F. Hf-Ta alloys roll-bonded to a Ta-10W substrate have been used for nozzle inserts in small, uncooled-liquid rocket engines. Several nozzles successfully endured cyclic firing tests simu-lating engine restarts for 800 sec, utilizing 5000°F flame-temperature propellants. Also, in air arc-jet tests generating leading edge temperatures to 3900°F, Hf-Ta alloys when used as coatings on Ta–10W and graphite, or a solid Hf-20Ta-2Mo alloy, have demonstrated a capability for the retention of sharp edges.

Oxides

The primary attractiveness of ceramic-type coatings is based on their thermal insulating ability and chemical stability in oxidizing atmospheres. However, current oxide or ceramic coatings are generally limited to plasma-sprayed oxides or castable ceramics utilizing honeycomb or woven-wire rein-forcements. Due to the thermal stress sensitivity of these materials, utilization is relegated to minimum-area, maximum-temperature zones, such as nose sections and possibly segmented leading edges of hypersonic vehicles. Reuse capability is usually limited to one and possibly two cycles. Temperature capability is excellent for certain types of hypersonic vehicles, since 4000° to 5000°F allowable temperatures may be sustained.

Coatings for Graphite

The principal attractiveness of graphites for hypersonic structures is an advantageous strength-to-weight ratio at elevated temperatures. However, because graphites are brittle, only limited utilization has been made of the mate-rial in hypersonic vehicle design. For this reason, as well as for technological reasons, development of oxidation-resistant coatings for graphite has been limited, and only a few coatings are available. Principally, these coating sys-tems are based upon silicon carbide, iridium, and flame-sprayed oxides. Flame-sprayed oxides are only marginally referred to as a coating, since bonding, compatibility, and life are extremely limited. Of the two remaining coatings, neither provides long life or equals the temperature resist-ance of graphite. Silicon carbide coatings are limited by the rate of SiO_2 vaporization at 3100°F. Iridium is more rate-limited; however, when emittance is taken into considera-tion, it is found that SiC/C can accommodate a higher heat flux for a given surface temperature. The oxidation rate of iridium coatings limits their use only a few mission cycles at very high temperatures. Silicon carbide coatings would be useful for longer periods of time; their reliability is fair.

FUTURE NEEDS

Intermetallics

Comparison of the previously discussed requirements and coating capabilities reveals that present coating systems do not fully provide the life or reliability sought. In some cases, the coatings do not exhibit the desired temperature capability. These disparities do not mean that available coatings are unacceptable for most of the foreseeable aerospace applications. Rather, coated refractory metal parts will require careful inspection, perhaps after each flight, and the parts may have to be repaired, refurbished, or replaced at more frequent intervals than is desirable from a practical or economic standpoint. Therefore, it is very desirable to increase the average life and reliability of coatings so as to reduce the overall operating costs of reusable aerospace vehicles.

The intrinsic (i.e., oxidation "wear out") coating life may in certain instances be improved by reducing the rate of formation of nonprotective intermetallic phases at the coating-substrate interface, or in other cases by retarding the rate of oxide formation and loss at the air-coating interface. The number and size of microcracks in the coatings due to thermal expansion differences might be reduced, and in some cases the rate of oxygen diffusion through the coating should be lowered. Such factors are discussed in detail in Chapter 2. Of greater immediate importance is an improvement in coating reliability—that is, reducing the spread between the present intrinsic "wear out" coating life and the average or minimum protective performance. Factors affecting coating reliability include:

Thickness uniformity
Probability of compositional defects
Probability of pinholes, cracks, etc.
Edge condition
Design of the part to be coated

It is evident that protective life is to some degree a function of the coating thickness. Therefore, thickness should be accurately controlled for reproducible coating life from batch to batch or for uniform protection over a finished part. Lack of thickness control is no doubt one of the major reasons for premature coating failures and can be particularly troublesome for aerospace structures, in view of their complexity and their fabrication from thin sheet metal. In particular, for fused-slurry silicide-coated columbium alloys, a good correlation between minimum nondestructively measured thickness and coating life has been established. In this example, the edges tend to be thinly coated in the event of inadequate control of slurry viscosity in dip processing or poor control of spray application. All coating systems have problems of thickness control to varying

degrees. The problem can be attacked by establishing better control of the coating process, and by extensive use of nondestructive thickness-measurement techniques. Applying an excessive coating thickness to avoid this problem usually is not practicable because the consumption of thin-gauge substrate must be controlled, weight increase may be unacceptable, and excessively thick coatings can cause other problems, notably edge growth or thermal stress cracking.

Certain coating systems (such as the Cr-Ti-Si system) are much more likely than others to encounter localized composition inhomogeneities. The probability and nature of compositional defects are thus important considerations in selecting or tailoring a coating system. In some coatings, small overall variations in composition are critical to oxidation performance, while other coatings accept fairly wide ranges of composition without significant decrease in protection. Coatings formed from fused (molten) slurries are least likely to suffer from local variability. Complex multilayer coatings, deposited in two or more distinct processing steps, such as the Cr-Ti-Si system, are most likely to exhibit either gross or localized compositional defects or variability.

Pinholes and macrocracks are less likely than the above factors to be a serious problem with diffusion coatings. Proper selection of the coating-metal system and careful control of cleaning, processing, and handling should virtually eliminate this type of defect. Nondestructive inspection methods are available that permit rejection of any parts where such physical defects occur.

Premature coating failures occur most frequently at edges. This can be due either to thinly coated areas or to the formation of edge fissures. To prevent, or at least minimize, edge fissures, the edges of the substrate generally are rounded or "radiused" prior to coating. Neglecting to prepare an acceptable edge condition on parts prior to coating can, therefore, be the cause of premature coating failures. This may be more crucial with vapor-deposited coatings than with fused slurry systems. Improvement of edge protection is desirable. However, this problem is much less critical than generally supposed, since sheet edge failures have minimal effect on the structural integrity of heat-shield panel structures. Much more importance must be attached to proper preparation, coating, and inspection of holes for fasteners, where coating failure could be structurally critical.

Aerospace components are usually relatively complex and are generally fabricated from thin-gauge material. This accentuates the problems of coating-thickness control and edge condition. It may also reduce the efficiency of inspection. The component design, manufacturing sequence, and coating system must be very carefully integrated for maximum coating performance. The coating cannot be expected to perform equally well for all designs. Some compromises between design and coating requirement may be necessary to achieve the most effective coated refractory metal structure. To provide additional data on design and coating relationships, more extensive work is needed on re-entry-simulation testing of representative designs and constructions. Such testing should include inducing realistic three-dimensional strains, such as are provided in bending, to provide correlation of actual flight and to test part behavior.

For single-flight hypersonic vehicles, coatings do not overly restrict the use of coated refractory metals for heat shields and structures because their effective life is more than satisfactory. The major shortcoming in this case is usually the lack of very high-temperature resistance. For multiple-flight vehicles, however, coatings impose one of the greatest single restrictions on the use of refractory metals. Inadequate reliability and a relatively short effective life are the major shortcomings. The reasons for these shortcomings have been previously discussed and one major factor is early coating failure due to defects. Premature coating failure is the factor which causes poor reliability; it is now generally accepted, however, that small, isolated, coating-defect failures will not cause structural failure of a part during the "flight" on which the coating failure occurs. Nevertheless, it is important to locate the early failures as soon as possible so that the part may be replaced or repaired. Thus, NDT techniques will be important. However, because of limited accessibility, complete inspection will be difficult.

Coated refractory-metal components are inherently expensive, and the coating contributes to the high cost. The actual coating-application process makes less of a contribution to total coating cost than the requirements for edge preparation, cleaning, inspection, and handling and joining restrictions. Although coating-related cost can be reduced, it is doubtful that a very large cost reduction will be possible. If the coating life were very long, factors such as high initial cost would not be significant, because the cost could be spread over a large number of flights. The actual reuse capability of state-of-the-art coatings is not accurately known and must be established before the situation can be properly assessed.

Another coating limitation for refractory metal components for reusable hypersonic vehicles is temperature capability for abort heating. If the flight of a lifting re-entry vehicle must be aborted just before orbital velocity is reached, re-entry temperatures will be higher than for a normal re-entry from orbit. Therefore, the coating must be able to withstand these higher temperatures, in addition to providing a long, effective life for multiple normal re-entries. It is difficult for a coating to be optimal for both conditions. Usually the coating is selected that is nearly optimal for the normal re-entry, yet capable of one abort entry. Thus, the coated refractory metal parts must be replaced after one abort entry. This is not overly expensive, because such aborts would be very infrequent. Sometimes the anticipated abort temperatures are higher than allowable for the

long-life coatings. In such cases, thin ablative overlays are proposed.

The major coating improvement required is an increase in the reliable effective life for reusable hypersonic vehicles. This improvement must be accomplished by (a) development of longer-life coating compositions and (b) increase in the effective life of present coatings when applied to typical hypersonic hardware. The problems involved in development of longer-life coatings are discussed in Chapter 2, and will not be considered further in this section. Emphasis should be placed on narrowing the gap between demonstrated potential and accomplished performance of available coating compositions.

The first step toward improving the effective life of coatings when applied to typical hardware is to define the coating life accurately by testing representative constructions under simulated flight conditions that include temperature, pressure, and stress as a function of time. Such testing, which is now within the capability of the state of the art, will permit the clear identification of specific coating–metal deficiencies, and these, hopefully, can lead to improving the effective coating life.

Improvement is likely to require better coating processes and process control to ensure more uniform coating thicknesses and the elimination of defects. For several coating systems, process and process control improvements appear feasible. Expanded use of NDT must be an important part of any overall improvement. The process improvement program should be broader than coating application *per se* and should include cleaning, edge preparation, part design, and manufacturing sequence. Careful integration of these factors will probably result in improved performance (life) of coated refractory metal parts for hypersonic vehicles and is also likely to provide some cost reduction.

The major improvements required in oxidation-protective coatings for superalloys are (a) longer effective life (5,000 hr) at temperatures of $1600°$ to $2200°F$ and (b) lower coating weight. These improvements are required for manned hypersonic aircraft that will become operational in the more distant future. For the limited life requirements (100 hr) associated with lifting re-entry vehicles, it appears that oxidation-protective coatings are not required. It does not appear likely that coatings of less than about 0.003-in. thickness can be developed to provide oxidation protection for 5,000 hr in the vicinity of $2000°F$. Therefore, the weight of coating will have to be traded off against more frequent replacement of uncoated heat shields and structures that operate at these high temperatures.

Noble Metals

Platinum-group metal coatings are not likely to be used for oxidation protection of metallic substrates to be used on hypersonic vehicles. An exception may be their use for very specialized application involving small areas and limited production. There does appear to be a role for the use of iridium-base coatings for high-temperature oxidation protection of graphite materials.

Oxidation-Resistant Alloys

Hf–Ta alloys currently are of interest principally for the protection of refractory metals against oxidation in the 3000°–4000°F range, where silicide or aluminide coatings are inadequate or useless. The protection provided by Hf–Ta alloys in this temperature range is of short (minutes to a few hours) duration, inasmuch as the HfO_2 protective oxide grows rapidly and eventually allows transport of oxygen to the substrate. Little is known of the mechanical properties of Hf–Ta alloys or their compatibility with refractory metal and graphite substrates. Considerable further work is necessary (a) to increase protective life, particularly at moderate temperatures, thereby broadening the field of potential applications; (b) to develop optimum alloy–substrate combinations and reliable processing operations; and (c) to develop design data. Improvement of protective capability by alloying additions of Al, Cr, Si, B, and Ir has been demonstrated in the 2000°–2800°F range in static air furnace tests. Increased life is therefore attainable, although the mechanism has not been studied in detail.

With regard to mechanical properties, hafnium–tantalum alloys have a potential for advantageous ductility. It may be difficult to obtain ductility at ambient temperatures in alloys with desired oxidation resistance, but ductility is likely at elevated temperatures. It should therefore be possible to develop refractory metals coated or clad with Hf–Ta with attractive properties compared with silicide- or aluminide-coated materials.

Oxides

The major limiting factors for oxide coatings are thermal shock and attachment means. A number of stabilized ceramic compositions are available that exhibit excellent thermal shock resistance and temperature capability; however, the methods of applying these materials to refractory metal or superalloy substrates are marginally useful and unreliable. The oxide coatings are considered primarily as thermal insulation rather than for oxidation protection. Some progress in adherence and thermal stress resistance has been made through the development of graded plasma-sprayed coatings and foamed ceramics. A promising approach is through the use of composite structures so that the reliability of metal reinforcements and the protection of oxide matrices may both be utilized.

Coatings for Graphite

For effective and realistic utilization of graphite structures in hypersonic vehicles, the major requirement is the development of coatings having a temperature capability equal to the graphite. Since the present coatings are temperature-limited, new compositions and techniques must be developed.

The most temperature-resistant coating is iridium, and new coating techniques are required to improve reliability and reproducibility. Fused-salt deposition is limited, both in size and coating efficiency. Plasma-spraying, due to the high cost of iridium, is extremely expensive, and the resultant coatings are not as protective as desired.

Considerable progress has been made in recent years in the development of various composite graphite structures having improved mechanical properties. A major limiting factor in the utilization of these materials is the lack of long life and reusable oxidation-protective coatings. Although iridium offers considerable promise for hypersonic structures, the extremely high cost ($126/ft^2/mil thick) precludes its use for all but minor or critical applications. The use of silicon-based coatings is restricted by their poor reliability and low temperature limits. To provide satisfactory coatings for graphite, new coating-material concepts must be developed. Present programs investigating complex silicides and borides show promise of an increased temperature capability compatible with graphite requirements. Additional studies are needed to integrate the new coating-material concepts with graphite and to develop adequate manufacturing techniques.

MANUFACTURING TECHNOLOGY

A thorough discussion of manufacturing technology is presented in Chapter 5. The following are some of the major considerations for the processing of hypersonic vehicle structures: (a) Very thin substrate metal gauges will be used; (b) many structures will be relatively large and complex; and (c) a significant amount of small hardware, such as washers and threaded fasteners, will be required. The use of thin substrate metal demands close control over the amount of base metal reacted to form the coating. Complex structures, such as single-faced corrugation-stiffened heat shields, will have faying surfaces and areas of limited access that must be coated. Not only is coating application on such assemblies restricted, but cleaning and inspection are made very difficult—in some cases impossible. Small hardware, such as fasteners and washers, presents a problem in handling and racking during the coating; also, a high ratio of edge to surface area exists for such items.

In most cases, the coating application process is as important to producing effective coatings as the coating chemistry itself. This is particularly true for some of the complex hypersonic vehicle structures. Improvements in reliability which will reduce the disparity between inherent coating oxidation life and minimum coating operative life is the major advancement required for hypersonic vehicles. Advances in manufacturing technology are an important part of achieving these improvements. There may be instances where the coating composition with maximum life should be compromised to permit use of a process that would improve the effective coating life, (i.e., the useful coating life) on hypersonic vehicle structures. Likewise, design compromises will be required in order to be certain of achieving the longest reliable life and thereby the most cost-effective coated structure. Data must be generated to show the difference between design approaches in terms of coatability and, consequently, part survivability.

Manufacturing technology considerations go beyond the process of coating application. Several manufacturing steps prior to actual coating application have a significant bearing on the performance of the coating. Those involved in coatings development and application must concern themselves with these related factors; otherwise, no concern for such aspects of processing will be manifested. These factors include contamination due to welding and metalworking, effect of welding parameters on faying surface condition, edge preparation and inspection, cleanliness, and marking control, cleaning prior to coating, and handling after coating. Many of these factors are particularly important for hypersonic structures because of the nature of their construction from sheet metal and because of their complexity. Considerable welding of various types may be performed. Mechanics must mark parts to build up an assembly. The parts will contact metal and nonmetal tooling. An assembly may be built up over a period of several months; when the assembly is complete, it cannot always be cleaned by the preferred methods because of inaccessibility—for example, the problem of acid retention in faying surfaces. If the substrate is not thoroughly clean, the desired coating may not form properly. Here again, the opportunity of a trade-off exists—minimum sensitivity of a coating system to contamination versus a coating which has maximum coating life, but only if applied under ideal processing conditions.

INTERMETALLICS

Pack cementation is the basic process used for the deposition of silicide on Mo-base alloys aluminides on superalloys. This process and its subsequent modifications are discussed in Chapter 5 of this report. Initial attempts to apply these processes to the coating of Cb- and Ta-base alloys were largely unsuccessful. Because simple silicide systems did not provide adequate protection, complex coatings had to be developed. This procedure necessitated the development of a variety of new coating processes and a more sophisticated manufacturing technology.

As previously mentioned, a two-step vacuum-pack process

was employed for the deposition of complex coatings on Cb alloys. The vacuum-pack Cr–Ti–Si coating process has been scaled up for coating large parts and has been used extensively to protect columbium test articles for several hardware prototype and fabrication development programs. Flight testing on the ASSET vehicles included a nose-cap-attachment bulkhead and aft leading edges fabricated of Cr–Ti–Si-coated D-14 Cb alloy. Many thousands of fasteners have been coated for the ASSET and BGRV vehicles by this process. The coating has been applied to virtually all available Cb alloys in sheet thicknesses as thin as 0.006 in. In the past, there has been variability of performance among coating batches. Usually, test specimens sustained over 100 hr at 2500°F in static oxidation; occasionally, however, a failure occurred prematurely as early as 30 hr. Some of the representative flight hardware that was coated exhibited very early coating failure. The Air Force has supported considerable research to improve the reliability and reduce the scatter in the performance of the Cr–Ti–Si coating. As a result, several of the process features contributing to irreproductibility have been isolated and remedied, so that the current process (1969) is greatly improved over that of two years ago. Also, nondestructive inspection techniques have been devised to permit identification of imperfect coatings. Limitations of size and configuration of parts that can be properly coated have been eased by a recently completed manufacturing-methods program. However, parts with fairly deep blind holes or faying surfaces cannot be coated satisfactorily by any pack process.

Several organizations have worked on methods of applying the Cr–Ti–Si coating, other than vacuum-pack cementation. A two-cycle slurry-pack technique can be used to produce equivalent coatings. This process is essentially a variation of the vacuum-pack process in which the pack is applied as a thin bisque to the parts to be coated; deposition is more uniform with less likelihood of the pack particles sintering to the part. The problems of preparing, blending, handling, and heating through massive packs are also avoided. Properties of Cr–Ti–Si coatings applied by this technique are basically the same as those of vacuum-pack coatings, but with somewhat better reproducibility.

Several other process approaches have been explored with varying degrees of success. Deposition of the Cr–Ti–Si coating system by a fused-salt process by both sequential and codeposition approaches has been pursued. However, reproducible formation of Cr–Ti–Si coatings comparable in properties to the vacuum-pack applied coating was not achieved by the fused-salt process. Electrophoresis was also investigated; the Cr–Ti coating was deposited electrophoretically and subsequently was isostatically pressed and sintered. Silicon was then generally applied in a second step by pack cementation. Cr–Ti–Si coatings comparable in protective properties to the vacuum-pack-applied Cr–Ti–Si coating were formed by this method on small specimens and on fasteners. Fluidized-bed methods for forming silicide coatings have been investigated extensively; the systems Si, V–Si, and Cr–Ti–Si received attention. Reproducible silicide coatings were formed on large heat-shield panels, but the Cr–Ti–Si composition could not be duplicated. Chemical vapor deposition has been explored as a means of forming the Cr–Ti–Si system on columbium. Both two-cycle and single-cycle laboratory procedures were developed; Cr–Ti–Si coatings comparable in properties to vacuum-pack-applied Cr–Ti–Si coatings can be deposited by the CVD process on a small scale.

Work on coatings for tantalum nozzle vane applications has brought forth a unique coating approach that also shows excellent potential for protecting columbium. The process consists of applying a particulate mixture of Mo–5Ti to the substrate surface, followed by vacuum-sintering to produce a porous product and subsequent pack-siliciding to eliminate interconnected porosity and produce $MoSi_2$. The surface is subsequently sealed with barium borosilicate glass, producing a multilayered system of approximately 6–8 mils thickness. The system appears promising for gas-turbine application but has not yet been evaluated for hypersonic flight or for re-entry vehicles.

The evolution of fused-slurry processes for the application of silicide coatings has been one of the most significant developments in recent years. The fused-silicide process is particularly adaptable to coating hardware; the process has been scaled up for large components, and numerous experimental parts have been coated and tested satisfactorily. Application of fused-slurry silicide coating ingredients is accomplished by dipping or spraying a slurry of elemental or alloy powders suspended in a nitrocellulose binder. The dried slurry-coat is then fired at 2500° to 2600°F in a vacuum furnace for 1 hr. During firing, the bisque melts, wets the surface, and flows into faying surfaces and other areas that may have been missed during application of the slurry. At this time, a small amount of the substrate refractory alloy (about 0.001 in./0.003 in. of coating) is dissolved into the molten coating, and upon solidification, a dense, tight, uniform coating with a high remelt temperature is formed. The processing method makes it possible to coat large complex components; however, the control of coating thickness is difficult on complex parts. Using a slurry composition of Si–20Cr–5Ti as the pilot coating, assembled full-scale re-entry vehicle heat-shield panels and various rocket-engine components have been coated and undergone test.

A wide variety of coating compositions can be deposited, and the desired chemistry is well-maintained because of the nature of application. Several compositions have been developed for various refractory alloys, and considerable latitude exists in tailoring of specific formulations for particular substrate alloys or test environments. Also significant is the fact that performance of the coatings is not grossly affected by small variations in composition—e.g., the Si–Cr–Fe system provides similar oxidation over the entire range of 15 to 40 percent Fe.

NOBLE METALS

Platinum-group metals can be deposited on refractory metals by using fused cyanide electrolytes.[28,29] They can also be applied to many substrates by cladding and pressure-bonding.[25,26] A slurry sinter technique has been used to apply Ir to graphite, and a fugitive-vehicle slurry technique to coat tungsten or tantalum with Ir. A relatively new process that shows promise as a means of depositing Pt-group metals involves pyrolytic deposition from organometallic vapors or solutions.[30] The potential of this technique will depend upon the development of suitable organometallic compounds of the Pt-group metals.

OXIDATION-RESISTANT ALLOYS

Hf–Ta coatings and claddings have received relatively little development study. A fused-slurry coating process (Sylvania 515) has been developed for applying an Hf–20Ta–0.25Si coating (the silicon addition reduces the fusion temperature from 4000°F to 3270°F). A 15-min post-treatment in argon or vacuum provides a protective coating system with a useful life of about 1 hr at 3000° to 4000°F; approximately 20 mils of coating are required for 1-hr protection at 4000°F. A duplex coating system has been developed using an initial coating of HfB_2 containing 10 percent $MoSi_2$ bonded to the substrate by a 15-min cycle at 3310°F in a vacuum. Pores in this first layer are sealed by a second step using the R515 (Hf–20Ta–2.5Si) slurry as an infiltrant and overlay. This procedure yields a 15–20-mil coating that is protective for 100 hr at 2500°F and 100 thermal cycles. In flame-testing at 3300°F (brightness temperature), a 3-hr life is obtained. Equivalent or enhanced life is possible with reduced-pressure conditions. This fact is in contrast to the degradation of silicide and aluminide coating where volatile products exist at 1 Torr.

Processes for cladding Ta–10W alloy with Hf–Ta alloys have been developed for use in manufacturing nozzle inserts for small uncooled liquid rocket engines. A press-forged Hf–20Ta sheet bar was rolled to heavy plate, assembled into a five-layer sandwich with a Ta–10W core and unalloyed Ta interleafs by edge-welding in argon, and subsequently hot-rolled into sheet at 2450°F. The roll-bonded sheet was successfully spun into a nozzle configuration in air at 1800° to 2100°F. New alloys and improved spinning techniques are under study, as well as the production of clad tubular spinning blanks by coextrusion. Satisfactory welding practices for clad sheet also have been developed.

OXIDES

For the plasma-sprayed and composite ceramics (mesh- or honeycomb-reinforced), current fabrication technology is moderately consistent with some requirements for propulsion-system thermal barriers. The major limiting factors, however, are reliable attachment and thermal shrinkage, particularly for multiple-reuse capability.

COATINGS FOR GRAPHITE

Current manufacturing capability for graphite coatings is limited to relatively small components. Silicon carbide coatings are limited primarily by furnace capabilities. Vapor-deposited silicon carbide coatings are even more limited (2–3-ft^2 surface areas). Siliconized (*in situ* SiC) coatings can be applied to approximately 8–10-ft^2 surface areas. Iridium coatings are formed by fused-salt deposition and plasma-spraying.

REFERENCES

1. Heldenfels, R. R. 1966. Structural prospects for hypersonic air vehicles. Presented at the 5th Congress of the International Council of the Aeronautical Sciences (ICAS), London, England. Aerospace Proceedings 1:561–583.

2. Jackson, L. Robert, John G. Davis, Jr., and Gregory R. Wichorek. 1966. Structural concepts for hydrogen-fueled hypersonic airplanes. NASA TN D-3162.

3. Pride, Richard A., Dick M. Royster, and Bobbie F. Helms. April 1964. Design, tests and analysis of a hot structure for lifting reentry vehicles. NASA TN D-2186.

4. Wichorek, Gregory R., and Bland A. Stein. Dec. 1964. Experimental investigation of insulating refractory-metal heat shield panels. NASA TN D-1861.

5. Mathauser, Eldon E. 1962. Research, design considerations, and technological problems for winged aerospace vehicles. Proceedings of the NASA-University Conference on the Science and Technology of Space Exploration 2:499–510. NASA SP-11. (also available as NASA AP-28.)

6. Stein, Bland A., and Gregory R. Wichorek. July, Sept. 1967. Results of current studies on coated tantalum alloy sheet at NASA Langley Research Center. Presented at the Thirteenth Meeting of the Refractory Composites Working Group, Seattle, Washington. Summary of paper in DMIC Memorandum 227, pp. 6–7.

7. Staff of NASA. 1967. Conference on aircraft aerodynamics. NASA SP-124, pp. 523–564. (Confidential) (Title unclassified)

8. Staff of NASA. 1967. Conference on hypersonic aircraft technology. NASA SP-148, pp. 471–584. (Confidential) (Title unclassified)

9. Norton, A. M. April 1968. (U) Hypersonic aerospace vehicle structures program. Interim Technical Documentary Report, Contract F33-615-67-C-1300. (S)

10. Plank, P. P., *et al.* Mar. 1968. (U) Preliminary design and experimental investigation of the FDL-5A unmanned high L/D spacecraft. AFFDL-TR-68-24, Part V (C).

11. Quest, R. G., *et al.* Oct. 1968. (U) Lifting reentry test vehicle preliminary designs for FDL-7MC and FDL-5MA configuration. AFFDL-TR-68-97 (S).

12. Roberts, W. E., E. J. Pearlman, *et al.* July 1967. Thermal protection for scramjet engines. AIAA Third Propulsion Joint Specialist Conference, Washington, D.C. (Confidential).

13. Manos, William P., *et al.* July 1960. Thermal protection of structural, propulsion, and temperature sensitive materials for hypersonic and space flight. WADC Technical Report 59-366, Part II (Unclassified).

14. Childers, Milford G. Oct./Nov. 1966. Structure and materials for hypersonic vehicles. SAMPE Journal 2:6 (Unclassified).

15. Laidlow, W. R., and E. W. Johnston. October 1966. Developing HST structural technology. Astronautics & Aeronautics (Unclassified).

16. Jamieson, Charles P., Duane M. Patterson, Charles C. Stewart, *et al.* April 1968. Passive thermal protection of scramjet engines, AVCO Corporation Technical Report AVSSD-0063-68-RR. Preliminary information under subcontract to The Marquardt Corporation under Prime Contract AF33(615)-3776, AFAPL-WPAFB, Ohio (Confidential).

17. Packer, C. M., and R. A. Perkins. 1969. Performance of coated refractory metals in low-pressure environments. Proceedings of the conference on Refractory Metals and Alloys, AIME, French Lick, Ind., 1965, Gordon and Breach, New York, pp. 1045–1053.

18. Gunderson, J. M., D. V. Lindh, and C. A. Krier. Oct. 3–5, 1965. An emittance improvement topcoat for disilicide-coated TZM molybdenum alloy. From: Fourth Symposium on Refractory Metals, AIME.

19. Allen, T. H., C. R. Johnson, and E. L. Rusert. April 19–21, 1967. High temperature emittance of coated refractory metals. From: SAMPE 11th National Symposium and Exhibit, Vol. II, The effects of the space environment on materials, p. 111.

20. Goetzel, C. G., P. S. Venkatesan, and R. F. Bunshah. Feb. 1960. Development of protective coatings for refractory metals, WADC Technical Report 59-405.

21. Pentecost, J. L. 1963. Coating materials and coating systems. High-temperature inorganic coatings (Editor: J. Huminick, Jr.), Reinhold Publishing Corporation, New York, pp. 45–47.

22. Jaffee, R. I., and D. J. Maykuth. February 26, 1960. Refractory materials. DMIC Memorandum 44.

23. Ware, G. C. 1965. Platinum-group metals. Minerals Yearbook, Vol. 1, Metals and Minerals, U.S. Government Printing Office, Washington, D.C., pp. 735–746.

24. Beamish, F. E., W. A. E. McBryde, and R. R. Barefoot. 1954. The platinum metals. Rare Metals Handbook (Editor: C. A. Hampel), Reinhold Publishing Corporation, New York, pp. 291–328.

25. Buchinski, J. J., and E. H. Girard. Oct. 1964. Study of ductile coatings for the oxidation protection of columbium and molybdenum alloys. Final Report Prepared Under Bureau of Naval Weapons Contract NOw 63-0706-C. (AD-608135).

26. Criscione, J. M., J. Rexer, and R. G. Fenish. High-temperature protective coatings for refractory metals. Yearly Summary Report Prepared Under Contract No. NASw-1030, not dated (NASA CR-74187).

27. Dickson, D. T., R. T. Wimber, and A. R. Stetson. Oct. 1966. Very high temperature coatings for tantalum alloys. Air Force Materials Laboratory Technical Report AFML-TR-66-317.

28. Withers, J. C., and P. E. Ritt. June 17–20, 1957. Iridium plating and its high-temperature oxidation resistance. Proc. Am. Electroplaters Soc., pp. 124–129.

29. Rhoda, R. N. 1962. Deposition of several platinum metals from molten cyanide electrolytes. J. Am. Electroplaters Soc. 49:69.

30. Srp, O. O. Private Communication, August 14, 1968.

ENERGY-CONVERSION SYSTEMS

INTRODUCTION

Energy-conversion systems convert heat or chemical energy into electrical energy. So long as the system is stationary and operates on earth, where weight is not a problem and economy is not a major consideration, energy-conversion systems operate at low temperature, and coatings are not needed or used in critical operations. Efficient energy-conversion systems capable of producing large quantities of electrical energy at high energy-to-weight ratios, or smaller quantities reliably for long periods of time without attention, are needed for orbital and space vehicles. These are called auxiliary power units (APU).

A wide variety of power units exists, classified by energy source or conversion principle. Where nuclear energy is the source, the systems are called by the acronym SNAP (Systems for Nuclear Auxiliary Power). These are of two types: Those that use radioactive isotopes and thermoelectric generators (RTG) are the odd-number SNAP systems, while those employing turbine alternators are the even-number SNAP systems, like SNAP-8. Generally, the RTG's are considered for power levels less than 1 kW, while the turbine generator-alternators are used where larger power levels are required, such as 3–300 kW.

Solar radiation is used as the energy source in many other systems. Some employ so-called solar concentrators in which mirrors focus solar energy on an absorbing device that stores the heat for transmission to the conversion device. Others use banks of solar cells to generate power.

Thermocouple elements such as lead telluride or silicon-germanium arrayed in thermopiles are used in conjunction with the radioisotope heat source and also with nuclear reactors to generate power. Higher power levels may be obtained with thermionic diodes, which operate at the highest temperatures, 1600°–1800°C, generally in conjunction with nuclear heat sources but also, in military mobile systems, with gas-fired heat sources.

In other systems, particularly nuclear-reactor heat sources, turbine alternators are used. At present, the highest-temperature nuclear reactor operates at about 700°C—rather a low temperature for considering coatings in the turbine-alternator-radiator system. Coatings are needed for the reactor fuel elements themselves to prevent corrosion and retain products of fission.

The applications of coatings in energy-conversion systems appear to be the following:

Nuclear reactors
Radioisotope thermal generators
Thermoelectric generator systems
Turbine-alternator systems
Thermionic diode systems

These will be discussed in relation to time and temperature of operation, stress, and environmental conditions to which the coatings are subject.

APPLICATIONS AND ENVIRONMENTS

NUCLEAR REACTORS

All high-temperature nuclear reactors that transfer thermal energy to gas, pressurized water, steam, or liquid metal have a coating or jacket to isolate the fuel elements from the heat transfer fluid. Generally, this coating is a relatively thick cladding of a corrosion-resistant metal applied by diffusion-bonding or other metallurgical bonding systems. The cladding is probably best considered as an integral part of the fuel element, which serves as a pressure vessel to retain gaseous fission products. The fuel-element cladding is subject to the highest temperature of the nuclear fuel systems, with internal stresses often high enough to cause plastic deformation over a period of time (creep). This may result in fuel swelling and ultimate failure of the fuel element. Cladding failures are intergranular and generally occur at lower creep strains than in the absence of radiation. Some fuel elements, such as in the S8ER (SNAP-8 Experimental Reactor), have coatings in addition to the metallic cladding. The U-Zr-H fuel is surrounded by a 10-mil Hastelloy cladding, chromized with 1–2 mils of chromium-enriched surface for the purpose of improving the adherence of a glass coating applied to retain hydrogen released from the uranium-zirconium hydride fuel. The fuel element operates at about 750°C with very long operational times up to 10,000 hr. The stresses in the cladding are high enough to initiate plastic deformation. Requirements for reliability and integrity are very great.

Lawrence Radiation Laboratory and General Electric NMPO are developing reactors for operation at much higher temperatures (1100°–1700°C). These reactors use a refractory metal cladding, typically W-Mo-Re alloys instead of superalloys. The problems are expected to be somewhat similar. The environment will certainly be more difficult. High-temperature fuel-element coatings are a critical and unsolved problem.

RADIOISOTOPE THERMAL GENERATORS

These devices are fueled by metals and oxides of plutonium, polonium, curium, cobalt, etc., which, if released, would create a major safety problem. The amount of isotope material used per capsule is selected to produce a suitable operating temperature for a given thermoelectric generator, say 700°C for PbTe and 1000°C for SiGe.

Another application for isotopes is as a heat source for a dynamic power system; here, the operating temperatures may reach approximately 1150°C. The radioisotope containment capsule is generally multilayered for high-temperature applications with tungsten inside, then a layer of high-strength tantalum alloy, and an outer layer of oxidation-protection material. The low-temperature radioisotope sources have been contained in a shell of high-strength, oxidation-resistant superalloy.

The chief safety hazards in the operational cycle of an RTG are the ground handling and launch operation and the re-entry from orbital flight. It is essential to provide oxidation protection to maintain fuel-container integrity during these times in order to maintain the RTG intact and not disperse fuel into the atmosphere. In addition to the function of oxidation protection, the coating must not react with the primary containment material for the time necessary to insure safety of the source. This time is very long, say, ten half-lives of plutonium, which approaches 900 years. Thus, metallurgical stability of the oxidation-protection coating must be considered over this long life. With these requirements in mind, research workers are presently studying a platinum-alloy coating. After a vehicle's re-entry, the environment of the capsule cannot be assured; it may be air, soil, or water. Therefore, the selected protection material must resist all of the above environments and not just an oxidizing environment.

The last requirement for heat-source coatings that must be described here is a coating to enhance thermal emission from the source to a dynamic system. The coating is required because testing, ground setup, and prelaunch checkouts of a system of this type require that personnel have access to the power system. This requires that the radioisotope heat source not be an integral part of the power system and thus dictates that the heat-transfer mode from the source to the power-conversion system be radiation, which requires a high-emissivity coating to achieve the highest efficiency.

THERMOELECTRIC SYSTEMS

There are only secondary applications for coatings in thermoelectric generators. Insulating coatings, such as boron nitride, separate the module cladding from an iron strap conductor. Typically, the BN coating is 10 mils thick and may be applied by chemical vapor deposition. Temperatures are about 700°C. Stresses are mainly thermal; service times may be up to 20,000 hr. Coatings are also used to provide a diffusion barrier against the iron second leg in thermoelectric generators to prevent deterioration and loss of efficiency by mechanisms as yet not understood. Bare thermoelectrical surfaces deteriorate more rapidly than coated ones.

TURBINE-ALTERNATOR SYSTEMS

High-temperature turbines intended for space power do not have to be protected against oxidation, since they are started up in space rather than on the ground. It would be risky to use coatings for ground testing, so it is best to ground-test the system in vacuum chambers.

Turbine power systems are of two varieties: Rankine and Brayton. The Rankine cycle employs a liquid metal, such as potassium, which is vaporized in a boiler, producing potassium vapor, which drives a turbine–alternator set. The working fluid is condensed, cooled in a radiator, and returned to the boiler. The radiator operates by radiation and must have a high emissivity, which is generally provided by a coating. The conditions under which the coating operates are not very severe, and failure would not be very detrimental, since it would only make the radiator somewhat less efficient.

The second type of turbine system employs the Brayton cycle, which is a closed gas cycle, and is somewhat simpler in operation, although the temperatures of operation are higher. The need for a high-emissivity coating in the radiator is greater, since larger radiators are required for Brayton cycle systems than with the Rankine cycle, and the influence of the coating is greater at higher temperatures, radiation being proportional to the fourth power of temperature.

THERMIONIC DIODE SYSTEMS

Thermionic diodes are intended for use in the highest power systems and employ very high-temperature heat sources—nuclear, solar concentrator, or chemical—operating at 1600°-1800°C. The emitting surface generally is made of refractory metals and is insulated from the collector surface. If the thermionic converter operates in vacuum, the gap between the emitter and the collector is very small. If ionized cesium vapor is added to the gap, the space is considerably larger, and the insulating and dimensional requirements are less rigorous. The collector is cooled by radiation or a liquid metal system employing a radiator. Thus, coatings on the radiator surface are used to increase emissivity. Development of thermionic converters is not advanced; hopefully, they might operate at high power levels for thousands of hours, but problems connected with the swelling and stability of the high-temperature reactors are far from solved.

ANALYSIS OF THE COATING PROBLEM

From the previous discussion, it may be seen that coatings for energy-conversion systems are used primarily for secondary purposes such as diffusion barriers and electrical insulation, for maintenance of efficiency of thermoelectric materials, or for enhancement of thermal emission from high-temperature surfaces such as radiators. The only current application for oxidation protection is in connection with high-temperature refractory-metal isotope-containment capsules.

Although the other coatings satisfy secondary purposes and are not within the scope of this report, it should be pointed out that some problems exist for such coatings. The major problem with the secondary-purpose coatings is one of increasing the reliability and maximum operating temperatures.

Since only oxidation-resistant coatings are covered in this report, and coatings for RTG's are the only ones used for this purpose in energy conversion, the balance of this discussion will be devoted to them. Metallic sheathings are used in RTG's to protect against oxidation during prelaunching operations, during re-entry, and for maintaining capsule integrity after return from orbit while the capsule must remain intact to prevent fuel dispersal that would be a safety hazard. The operational temperatures for these heat sources may be as high as $1150°C$. The materials selected differ according to application, with oxidation-resistant superalloys being used at present on low-temperature systems; approximately 20 mils of Pt–20Rh sheath is isolated from the refractory metal containment material by a refractory oxide to prevent interaction. A silicide and aluminide coating would not be reliable for long-term operation because of the very long reaction exposure with the substrate at a very high operating temperature.

The Pt–20Rh coating and the diffusion barrier present no materials compatibility problem that can be foreseen at this time. However, fabrication development will be required to apply the oxide barrier reliably and to seal the Pt–20Rh material to prevent any breach of capsule integrity.

Unless a reliable venting system for fission gases is developed, the maximum operating temperature of isotope heat sources will be limited by the ability, i.e., strength of the capsule, to retain fission gases. The fission gases can cause stress up to approximately 15,000 psi within the capsules at elevated temperatures.

INDUSTRIAL APPLICATIONS

INTRODUCTION

High-temperature coatings serve to protect surfaces from degradation caused by corrosion, erosion, wear, etc., or they control the flow of heat or electricity. Since aerospace and military applications frequently require that materials perform near their upper temperature limits to allow maximization of system performance, coatings are often needed. These coatings must survive hostile environments ranging from those associated with military jet engines to rocket nozzles. Service temperatures can extend from $1000°F$ to $4000°F$. Performance is all-important, and cost of the coated parts (the total number of which is usually small) is often of secondary concern.

Industry, however, deals with large numbers of parts and so is much more concerned about unit costs. Indeed, performance optimization is always tempered by cost trade-offs. Since costs are usually lower when industrial processes are operated at low temperatures, industry generally avoids high-temperature operation and thus avoids the consequent requirements for high-temperature coatings. Even if higher temperatures increase thermodynamic efficiencies, product reliability, or part life, they are usually avoided because higher-temperature operations frequently require changeover in processing equipment, increased maintenance costs, complex safety procedures, upgrading of process controls, and retraining of the labor force. Furthermore, process changeover must be supported by a strong technical backup in order to have a smooth transition. Most industries simply cannot afford to support such a technical staff.

In some industries, product requirements or competition demands that operations be conducted at high temperatures. These industries have either developed their own technology for coping with the high-temperature environments or they have taken advantage of aerospace–defense spin-off. Such technology transfer has been aided by efforts of the NASA Technology Utilization Program and by the publications of the Department of Defense Information Centers.[1-4] Such spin-off must first be converted to an economical procedure or process. This takes time—in some cases the lag may be five years or more. For example, plasma-arc spraying was introduced in the late fifties and was immediately put to use in coating rocket nozzles. It has been only relatively recently, however, that plasma-spray-coated industrial products such as knife blades, soldering-iron tips, and furnace refractories have begun to appear on the market.

Information on industrial uses of high-temperature coatings was solicited from industrial, government, and university personnel, too numerous to mention specifically. Many applications were discovered. The economic barriers have apparently been overcome in these cases. The following review presents specific examples of industrial applications of high-temperature coatings. No attempt has been made, however, to identify and document every real or potential industrial application. The proprietary nature of many industrial processes and the vastness of American industry make such a task beyond the scope of this review. Rather, the industry has been sampled for the purpose of illustrating the variety of high-temperature coating applications now in use or potentially needed in the future.

APPLICATIONS

It appears that the greatest use and the greatest potential for high-temperature coatings is with those organizations that produce materials—the melting, casting, refining, and primary metalworking industries. The hardware or manufacturing industries appear to have less need for coatings except in high-temperature shaping or for improving the corrosion resistance of their individual products in service.

The coatings listed in the following paragraphs do not always represent solutions that are completely satisfactory. Some coatings discussed are purely stop-gap measures awaiting better solutions, while others are strictly experimental approaches.

GLASS MELTING

In melting glass, great care must be taken to avoid melt contamination. Contamination, even in small amounts can produce unwanted coloration or alter other glass properties. For this reason, the glass processing industry uses large amounts of platinum to protect against contamination. Melt pots are lined with a thin layer of platinum foil, which does not contaminate the glass. Any utensils that are inserted into the molten glass, such as stirrers (Pt on Al_2O_3-coated Mo) and heater electrodes (Pt on Mo), are platinum-coated. Patents have been issue for a process that circumvents the problem of fragility of foil-gauge platinum melt-pot liners through the use of plasma-sprayed coatings of adherent, impervious platinum on bulk refractories. These coatings must be defect-free and no thicker than the original foil liners, or their economic advantage is rapidly lost.

Since the large use of platinum involves a substantial investment in the metal, alternative protection schemes have been sought for some time. For certain glass-melt compositions, crucibles have been lined with mullite, calcium zirconate, and magnesium zirconate. Instead of platinum coatings, silicide coatings have been tried for molybdenum melting electrodes. At best, however, coated molybdenum has a very short life in molten glass, and coating failure produces unacceptable molybdenum contamination of the melt. Presently, some of the more advanced multicomponent silicide coatings are being evaluated for this task. Also, there has been considerable activity directed toward finding materials to replace platinum used in die bushings for glass-fiber production.

Plasma-sprayed NiAl can be ground to a smooth finish, and this coating has been used to protect glass pressure molds from corrosion.

METAL MELTING AND CASTING

Crucible materials for melting metals also must not contaminate the melt. In some cases, crucibles are constructed from low-cost refractories and are coated with a lining that is inert to the metal being melted. This coating approach offers the potential of tailoring desirable features into the composite crucible through selective grading of thermal expansion and conductivity. Some crucible liners presently in use or being considered are listed below, along with some of the metals for which they are used.

Liner	Melt
ThO_2	Most metals
Mullite	Al
Magnesium zirconate	U
ZrO_2	Ru, Rh, Ri, Cr
Al_2O_3	Al, Ni, Co, Fe
Clay-based	Fe, Al
TiN, ZrN	Ce, Fe
AlN	Al
AlN and SiC	Al

In the production of aluminum, chlorine is sometimes bubbled through the melt to "clean it up." The steel tube that introduces the chlorine can reach $1200°F$ and ceramic coatings have served to prevent corrosive attack of the tube by either the chlorine or the aluminum.

Casting hardware usually includes feeder tubes and troughs, molds, and mold inserts. In aluminum casting work, troughs of the following materials have been used with varying degrees of success: Iron-based materials have been coated with Al_2O_3 and NiAl; tungsten has been used with silicide coatings; and Al_2O_3 wash coats have been used on fused silica. Mold inserts of alumina-coated iron alloys have also been evaluated for the casting of aluminum.

Die casting of aluminum involves the use of carefully machined high-temperature steel dies. The cost of these dies makes it mandatory that they serve for at least several hundred thousand injections. Dies have been coated with various oxide layers, either self-generated during "break-in" or by the addition of wash coats. Die cores of H-13 steel have been chromized prior to service in aluminum die-casting operations. Such chromized coatings extend the die life about three times, primarily by improving the thermal-fatigue resistance of the core materials.

There is considerable effort these days to develop procedures to die-cast steel. Generally, uncoated tungsten or molybdenum alloy dies have been considered, with only a mold wash lubricating spray prior to casting.

FURNACES

Coated resistance-heating elements are commercially available for high-temperature operation in air. These include aluminized superalloys, 18-8 stainless steel plasma-sprayed with NiAl, siliconized refractory metals, SiC-coated graphite, and SiO_2-coated SiC. The coated superalloy and stainless steel elements are protected by a natural alumina scale, and these operate in the same range as the nickel–chromium and iron–chromium–aluminum type elements. The more refractory materials operate up to $3000°F$ and depend on a silica film to resist oxidation. However, thermal cycling of the silicon carbide elements and slow heating of the others in the $1200°$ to $1600°F$ range where the protective silica layer cannot self-heal can cause premature failure.

Induction furnaces also employ coatings to good effect. Primarily, electrical and thermal insulative coatings of Al_2O_3, etc., are troweled around or thermal-sprayed on the induction coils to prevent their overheating by radiation from the hot susceptor. When metal or graphite susceptors are used in air, they are sometimes coated to prevent oxidation. Again, superalloys are aluminized, refractory metals are silicided, and graphite is coated with silicon carbide.

Furnace hardware, kiln furniture, etc., are frequently coated to extend their useful lives. Ferrous alloys coated with thermal-sprayed Al_2O_3 or "NiAl" are used in kilns for holding and positioning ceramic ware. These coatings must not only provide oxidation resistance for the substrate; they must also be nonreactive to the ceramic ware. Nonreactivity is especially important in firing electronic ceramics. ZrO_2-coated refractories are used in titanate firing, while Al_2O_3-coated SiC is used in ferrite processing.

In heat-treating furnaces, similar coating systems are used. Steel annealing mill covers have been successfully protected against hot corrosion by being plasma-spray-coated with nickel–chromium alloys. Aluminized superalloys serve as fused-salt-bath fixtures and stress-rupture grips in materials tests.

In general, the needs of coatings for heat-treatment facilities continue to exist, and better coatings could be used immediately.

Industry is constantly seeking to lengthen the lives of thermocouple protection systems in order to alleviate the need for frequent checking and replacement in process furnaces. A number of protection systems are in use. Chromel–alumel thermocouples have been successfully protected from oxidation to $2200°F$ with vitreous enamels. Protection tubes for thermocouples of silicon carbide-coated graphite, siliconized refractory alloys, aluminized superalloys, and NiAl plasma-sprayed iron alloys have been used. For a variety of applications, including molten metal probing, two-layer ceramic protection systems are finding application. The substrate consists of an impervious, chemically resistant structure, and this is coated with a porous, thermally insulative outer layer. The porous outer layer allows short-term probing of molten metal pools without the problems of overheating or thermal shocking of the substrate. Problems still exist in thermocouple protection. Besides longer lives for the more common thermocouples,

a requirement exists for protecting the newer W–Re thermocouples and others that operate above $3000°F$.

Vitreous coatings are used to provide temporary protection to alloys during heat treatment. It is desired that such coatings spall off during cooling. These coatings offer a unique means of low-cost processing of air-sensitive alloys. Exploitation of this process could, in some cases, eliminate the need for inert-atmosphere or vacuum-heat-treating processes. This approach is not widely employed, however, since the use of the coatings would require changes in existing process equipment.

METALWORKING

The yield strength of metals and alloys decreases with increasing temperature. This situation allows heated metals to be more extensively deformed than unheated metals in any given operation. Economically, then, hot-working is attractive. However, heated metals are more prone to react with their surroundings. Thus, oxidation and reactions with tooling become a problem. Also, as tooling temperatures rise, tool materials lose strength and deform during working. Coatings are being increasingly used in the metal-working industry to inhibit reactions and insulate the workpieces.

In the extrusion process, extrusion dies and chamber liners suffer severe erosion from the passage of heated billets. Segmented dies have been coated with thermally sprayed oxides applied over roughened surfaces or over metal undercoats to improve adherence. For example, die coatings of Al_2O_3 and ZrO_2 have been used with some success in the extrusion of tungsten around $3200°F$. The Air Force is seeking to improve on the performance of these materials and is evaluating ZrB_2 and HfB_2.

Coating life usually is short, and after one or two extrusions the die parts must be recoated. Extrusion pressures reaching 100 Torr/in.^{2} and, under similar laboratory conditions, billet temperatures as high as $5000°F$ compound this problem. The unfavorable combination of impact loading, mechanical stress, and thermal shock has not yielded to coating technology, and industry is studying massive, nonsegmented dies and liners made from refractory oxides, refractory metals, intermetallic compounds, and combinations of the three.

Extrusion lubricants are also desirable. Presently, glass coatings with chemically controlled viscosities serve as extrusion lubricants above $2000°F$. They wet the billet and provide adequate lubrication. After extrusion, however, these glasses are generally difficult to remove from the workpiece. Molten salts and metal coatings also can serve as high-temperature lubricants, as can WS_2, BN, and graphite-glass cloth.

Coatings or clads are also commonly used to encapsulate or "can" the extrusion billet. Such surface protection provides a means to apply controlled compressive stress on brittle materials. Thus, canning is frequently used in refractory material extrusion. As with glass lubricants, removal of these cans is sometimes difficult and tedious.

Coatings are also being employed in forging processes. To successfully forge such materials as superalloys and titanium at temperatures of 1000°F and above, part chilling and subsequent surface cracking must be avoided. Parts must either be clad with thick (up to 1-1½ in.) metal heat reservoirs or slurry-coated with thermal insulators such as aluminum oxide. (The only alternative is to forge oversized parts and then grind off a considerable quantity of surface material until all cracks have been removed—a very uneconomical approach.) Insulative coatings are especially important in cases where the part is in contact with the die for long times. Explosive forging is very rapid; steam-hammer forging, relatively so; mechanical press forging, moderately fast; and hydraulic press forging, relatively slow. The last-named process, normally used for large parts, benefits most, then, from insulative coatings. Also, since most insulative oxide coatings are brittle, they survive better under a slower forging rate.

Titanium parts are frequently glass-coated to provide protection from surface contamination as well as insulation during forging. Some titanium alloys develop an undesirable alpha-phase surface layer without such coatings. When exposed to air, titanium alloys also form very abrasive oxides that shorten die life considerably. Glass coatings, however, are difficult to remove, and, when hot, vaporization of coating constituents produces unwanted films of oxides on the forging equipment. Better glass coatings are needed to eliminate alpha-phase formation on titanium and to reduce glass constituent vaporization. These coatings would have to improve metal flow by approximately 20 percent to be reliable and to provide consistent protection for a wide number of similar alloys before they would be widely accepted in commercial forging shops.

High-temperature forging work frequently employs nickel- or cobalt-base hardfacing on the dies to minimize die wear. When exposed to 1100°F or above for any length of time, the die steels not so treated are tempered back from optimum microstructure, and die faces soften. Dies then easily deform, and tolerances can no longer be held to precision forging values. To alleviate this problem, die surfaces are frequently coated with superalloys by weld deposition. Superalloys have higher hot hardness and strength than the die steels, even though at room temperature, superalloys properties fall below the strength and hardness of die steels.

At present, the forging of aluminum and magnesium alloys at temperatures up to approximately 800°F does not require coatings. Die steels are sufficiently strong and hard at these temperatures to perform satisfactorily.

Rolling processes have not employed coatings to any great extent. Titanium has been thermal-spray-coated with "NiAl" to protect it against oxidation during hot-rolling.

Alumina and zirconia have been plasma-sprayed on the rolls of billet furnaces for contact with materials heated to above 2000°F. Instead of insulating the specimen or rolls, the material being rolled is generally heated between passes. On continuous rolling, flash reheat furnaces can be located between rolling stations. Of course, when reductions are greater than about 15 percent, some heat is generated in the material being rolled, which compensates partially for conductive heat losses to the rolls. Most rolls are not highly preheated. However, Sendzimir mills are being developed for 1000°F operation. Thus, precision rolling may soon involve the use of coatings similar to those employed in forging.

Of the government-funded activity on metalworking, the U.S. Air Force has supported a great deal of the research directed toward the use of coatings for metalworking applications. A good summary of much of that effort is presented in a report on the AFML Metalworking Technology Symposium, January 10–13, 1967, at Las Vegas, Nevada.

HOME APPLIANCES

High-temperature coatings are used in ovens, stoves, and domestic furnaces. Aluminized steel is used up to approximately 1250°F in combustion chambers of space heaters, barbecue grills, flues, liners and burners of hot water heaters, range broiler baffles and boxes, clothes dryers, roasters, griddles, and incinerators. Chromized steel is used as an oxidation-resistant diffuser plate in the burner assemblies of gas ranges. At temperatures below 900°F, glass coatings are also commonly used in hot water tanks and on range components. There is a need, however, for a higher-temperature porcelain enamel for self-cleaning ovens that could operate reliably at temperatures above the current limit of 900°F.

TRANSPORTATION

In the field of transportation by gasoline or diesel power, there are a few applications for coatings in the 1200° to 1600°F range. Presently, diesel engine pistons are being thermal-spray-coated with modified CeO to improve combustion efficiency. In marine service, diesel pistons are coated with Cr–Ni–B alloys to eliminate hot corrosion.

Auto parts being coated include manifolds and ceramic-coated mufflers. Automotive smog burners for exhaust-emission control rely on afterburners to clean up combustion products. Silicide-coated refractory-alloy ignition coils and aluminized superalloy structures are used in combustors. Resistance to thermal cycling and a variety of corrosive gases is a prerequisite for satisfactory operation of these devices. Potential exists for adding catalysts to ceramic coatings in mufflers, etc., to assist in the elimination of harmful exhaust constituents.

When small gas-turbine designs for trucks and cars are

perfected, coatings will play a major role. These engines will require long-lived oxidation- and hot-corrosion-resistant coatings for the inexpensive alloys required to make this concept economically competitive. Coating problems and use will be similar to those discussed earlier in this chapter in the section on gas turbines.

PETROCHEMICALS

At present, the petrochemical industry appears to have little use for high-temperature coatings. Most of the chemical processes are conducted below $1100°F$ because of inability to control reactions above this temperature and also because of the decomposition of some of the reactants and products. Thus, the prime concern seems to be with corrosion control at lower temperatures. Most solutions to corrosion problems involve changes in structural materials. Coatings are usually not even considered. Apparently, however, aluminized superalloys have been used in cracking units in the petrochemical industry. Also, some preliminary studies appear to have dealt with determining the feasibility of using tantalum-coated heat exchangers to provide protection at less cost than solid tantalum construction.

MISCELLANEOUS

Many wear-resistant coatings or hard facing systems are applied at high temperatures by welding or thermal spraying. Wear-resistant coatings currently find limited use at high temperatures. Ceramic-bonded calcium fluoride with a CaF overlay has been used as a solid-film lubricant for low-

friction-wear applications for short times at $1900°F$. The calcium fluoride–barium fluoride eutectic system has indicated good lubricating characteristics and quite reasonable wear life at $1000°F$. For $1000°$ to $1500°F$ service in gas-turbine engines, wear-resistant coatings of 75 percent chromium carbide–25 percent nickel–chromium alloy look promising.

Electrical insulation that will stand high temperature is becoming of greater interest for uncooled motors, a variety of control systems, and various electronic purposes. Motor windings can be coated with aluminum oxide. Ceramic coatings with dielectric top coats are being studied for insulative purposes. In vacuum tube construction, sintered Al_2O_3 coatings provide a means of electrically insulating and mechanically separating tungsten filaments. Alumina-coated asbestos insulators must be replaced frequently when used near an arc plasma-jet cathode because of thermal degradation from radiation. Such problems can only become more prevalent as power generation by MHD gradually evolves.

REFERENCES

1. Campbell, I. E., and E. M. Sherwood, Editor. 1967. High-temperature materials and technology. John Wiley & Sons, Inc., New York.
2. Huminik, John, Jr., Editor. 1963. High-temperature inorganic coatings. Reinhold, New York.
3. Plunkett, J. D. 1964. NASA contributions to the technology of inorganic coatings. NASA SP 5014.
4. Product Data Bulletins and Application Data Bulletins, METCO Corporation, Westbury, L.I.

5

Manufacturing Technology

INTRODUCTION

This section describes the availability of laboratory and manufacturing processes and facilities for the application of high-temperature protective coatings to refractory metals, superalloys, and graphite. A thorough survey was conducted of all known organizations in the country active in the application of high-temperature protective coatings. The organizations contacted are listed below.

Air Research Div., Garrett Corp.
Allison Div., General Motors Corp. (Now Detroit Diesel, Allison Div., GMC)
Alloy Surfaces Div., Reeves Industries
Arthur Tickle Engineering Works, Inc.
*Atomics International
*Battelle Memorial Institute
The Boeing Company
*Bay State Abrasives Div., AVCO
*Calorizing Corporation
Chromalloy Div., Chromalloy American Corp.
*Chromizing Div., Chromalloy American Corp.
*Composite Metal Products
E. I. du Pont de Nemours & Co., Inc.
Engelhard Minerals & Chemicals Corp., Div. Engelhard Industries
Ethyl Corp.
Fansteel Metallurgical Corp.
*General Dynamics, Inc.
General Electric Co., Evandale
General Technologies Corp.

Howmet Corp.
IIT Research Institute
Linde Div., Union Carbide Corp.
Lycoming Div., AVCO Corp.
*Marquardt Corp.
Martin-Marietta Corp.
McDonnell-Douglas Aircraft Corp.
*Metallizing Corp. of America
*Metals & Controls, Inc.
Metco, Inc.
National Bureau of Standards
National Lead Company
North American Rockwell Corp.
Olin Brass
Plasmadyne Div., Giannini Scientific Corp.
Pfaudler Co. (Ritter)
*Pratt & Whitney Aircraft Co., Florida
Pratt & Whitney Aircraft Co., Hartford
Norton Company
San Fernando Labs (Fansteel)
Solar Div. of International Harvester Corp.
Sylvania Electric Products, Inc. (High Temp Comp. Lab)
Stellite Div., Union Carbide Corp. (Now Stellite Div., Cabot Corp.)
*Texas Instruments, Inc.
TRW, Inc.
Union Carbide Corp.
*United Nuclear Corp.
Vac Hyd Processing Corp.
Vitro Corp. Labs
*Vought Div., Ling-Temco-Vought

166

Walbar
Wall Colmonoy Corp.
Westinghouse Electric Corp.
Whitfield Laboratories, Union Carbide Corp.
Wilbur B. Driver Company

Organizations denoted by an asterisk either failed to respond to our inquiry or responded negatively with regard to providing the solicited processing information. Each organization was requested to complete a table similar to those shown in Tables 38 through 46. All data in these tables are precisely as provided by the organizations. Nine process categories were defined, and these techniques are discussed individually in subsequent paragraphs. A brief description of each process is presented, along with a general discussion of the major coating systems available for each substrate. Process status is described briefly in each section.

CURRENT CAPABILITIES

CEMENTATION PROCESSES

PACK

By far the greatest quantity of coated high-temperature hardware is produced by pack application of the diffusion alloy coatings. Pack-coated materials have been predominately aluminides for the iron; nickel- and cobalt-based alloys, and simple and complex silicides for the refractory metals. Materials to be pack-coated are immersed in a granular pack medium that may include the metallic coating elements, inert filler material, and an energizer or activator such as a halide salt. In conventional pack-cementation processes, the thermal cycle experienced by the pack is conducted under a flowing or static inert gas atmosphere, such as hydrogen or argon, or under an environment of relatively low oxygen potential produced by volatization of a halide-pack additive. For the vacuum-pack process, the pack may be all metal or refractory-diluted, and the pack may or may not contain a halide activator. The thermal cycle for the vacuum-pack method may be conducted in a dynamic system (continuous evacuation) or under an inert gas atmosphere introduced into the evacuated retort.

For all pack-coating techniques, the coating elements may be provided from pure metal powder or from metallic alloys. Alloys or elemental powder mixtures are used both for codeposition of two or more elements and for control of the activity of the predominately depositing species. Inert diluents, such as Al_2O_3, ZrO_2, and MgO, are employed to lower the activity of the depositing elements and to prevent the solid-state sintering of metal powders to the hard-

ware surface. Transport of the metallic atoms from the source to the hardware is accomplished by chemical reaction with the halide additive or by physical vaporization, as in the case of the vacuum process. Diffusion growth of the coating occurs by a vapor–solid interaction at the gas–metal interface. The physical chemistry of the reaction occurring in pack-coating processes has not been explored in depth; consequently, many of the data and theoretical considerations required for implementing improvements in pack processing are not available.

One of the major limitations of pack processing with regard to hardware size is the limitation imposed by poor heat transfer in the granular pack. Severe thermal gradients exist in massive granular packs, causing hardware distortion and nonuniform coatings. Many small packs are generally used for the batch coating of large quantities of hardware; obviously, this approach is not feasible with large-size components. A second limitation of pack coatings is the inability of the process to adequately coat faying surfaces. Spot-welded, riveted, and other joint designs with faying surfaces are not amenable to coating by pack techniques, thus limiting the available fabrication methods for producing pack-coated hardware.

Table 32 presents the results of the survey conducted to ascertain the availability of sources for pack-coating high-temperature alloys. The refractory-metal and superalloy coating processes are discussed independently. All of the pack-applied diffusion coatings offered by the industry for refractory metals are silicides. Large pack-coating facilities are available at the Pfaudler Company, Chromalloy Corporation, TRW, Inc., and the Union Carbide Corp. for pack application of silicide coatings to refractory metals. Aluminide coatings may also be applied by pack deposition; however, the limited need for aluminides on refractory metals and the relative ease of slurry aluminide application have eliminated most activity on pack-aluminizing of these materials.

Simple and complex silicides have received wide use and acceptance for refractory metal aerospace hardware. The Chromalloy W-3 coating on molybdenum and the TRW Cr-Ti-Si coating on columbium alloys represent the most successful examples. The "pest" oxidation phenomena exhibited by most simple silicides have promoted the development of complex silicide systems. A major concern with the use of pack-applied coatings to refractory metals is repair. Although coating repair in the pack is feasible, the use of fused silicides for local repair has been very encouraging. The development of field repair techniques needs additional attention.

Hundreds of thousands of iron-, nickel-, and cobalt-based turbine components are pack-coated monthly in support of the aircraft gas-turbine business. Virtually all pack-applied coating systems employed on high-temperature materials are

TABLE 32 Pack-Coating Processes

Company[a]	Coating Process	Basic Substrates	Coating System (Designation and Composition)	Coating Facilities (Size and Type) Operating	Planned (Date)	Process Status[b]	Comments
			Refractory Metals				
TRW	Vacuum pack	Cb, Ta, Mo	Cr–Ti–Si	Vacuum 3 in. diam × 10 in. 7 in. diam × 20 in. 48 in. diam × 54 in.		C C C	
TRW	Vacuum pack	Cb, Ta, Mo, W	Simple silicides	48 in. diam × 54 in.		C	
Pfaudler	Pack	Molybdenum Mo–1/2 Ti TZM TZC	PFR–6 Alloy silicide	Globar (2 ft³ hot zone) Globar (27 ft³ hot zone)		C	
Pfaudler	Pack	B–66 FS–80 D–36	PFR–30 Alloy silicide	Globar (2 ft³ hot zone) Globar (27 ft³ hot zone)		AD	
Pfaudler	Pack	D–43 Cb–752 C–129 C–103	PFR–32 Alloy silicide	Globar (2 ft³ hot zone) Globar (27 ft³ hot zone)		AD	
Pfaudler	Pack	Ta Ta–10W Ta–111 Ta–222	PFR–50 Alloy silicide	Globar (2 ft³ hot zone) Globar (27 ft³ hot zone)		AD	
Chromalloy	Pack	All refractory metals	W–3 Disilicide	Production		C	
Union Carbide	Pack process	Columbium Molybdenum Tantalum and Tungsten–base alloys	WL–3 silicon	Production		PD	
Solar	Pack	Cb, Ta, W, Mo	Simple and modified silicides			PD, AD	
			Superalloys				
Lycoming	Vacuum pack	Nickel alloys	Vac Pac 701–Al	Ipsen vacuum furnaces		C	
Lycoming	Vacuum pack	Nickel alloys	Vac Pac 703–Cr–Al	Ipsen vacuum furnaces		AD	
Lycoming	Vacuum pack	Nickel alloys	Vac Pac 708–Modified aluminide	Ipsen vacuum furnaces		PD	
Chromalloy	Pack	WI–52 SM–302 HS–25 X–40 Inconel HS–21	UC–Complex aluminide	Production (proprietary)		C	

168

Company	Process	Substrate	Coating	Equipment / Status	Code	Notes
Chromalloy	Pack	B1900, U–700	DR–Complex aluminide	Production	AD	
Chromalloy	Pack	Kovar, Rodar, Molybdenum, Tungsten No. 52	GMS–Chromium	Production	C	
Chromalloy	Pack	4340, 1095, 52100, Cast iron	Chromacarb–chromium	Production	C	
Chromalloy	Pack	Amco iron, P. met iron, 1010 steels, 400 stainless	Chromalloy-G	Production	C	
Chromalloy	Pack	Inco 713, X–40	SAC–Complex aluminide	Production	C	
Chromalloy	Pack	Hastelloy	SUD–Complex aluminide	Production	C	
Chromalloy	Pack	(ThO$_2$ Disp) TD Ni, TDNichrome	BD–Complex aluminide	Production	C	
TRW	Pack cementation	Nickel and cobalt alloys	GE–CODEP series	Box furnaces	P	GE licensee
TRW	Vacuum pack	Ni alloys, Co alloys	Tapcoat–A-100 Al (Cr), Tapcoat–A-200 Al (Cr)	Vacuum 3 in. diam × 10 in., 7 in. diam × 20 in., 48 in. diam × 54 in.	C, C, AD	
TRW	Vacuum pack	DS alloys	Cr–Al	Vacuum 3 in. diam × 10 in., 7 in. diam × 20 in., 48 in. diam × 54 in.	C, C, PD	
TRW	Vacuum pack	Ni, Co, DS alloys	Cr, Ta, Mo, W, Hf, Mn, Mg, Fe modified aluminides	Vacuum 3 in. diam × 10 in., 7 in. diam × 20 in.	AD, AD	
Alloy Surfaces	Pack cementation	Nickel-base superalloys, Cobalt-base superalloys, Iron-base superalloys, Nickel	HI-15 Composition proprietary. Primarily chromium and aluminum.	Various furnaces of all types up to 36 in. diam × 60 in.	Production	
Alloy Surfaces	Pack cementation	Nickel- and cobalt-base superalloys	Tantalizing. Composition proprietary. Tantalum diffusion.		Development	Full production capability if required.

TABLE 32 (Continued)

Company[a]	Coating Process	Basic Substrates	Coating System (Designation and Composition)	Coating Facilities (Size and Type)		Process Status[b]	Comments
				Operating	Planned (Date)		
			Superalloys (continued)				
Alloy Surfaces	Pack cementation	Nickel-base super-alloys, cobalt- and iron-base superalloys	Tanchralizing. Composition proprietary. Primarily tantalum, chromium, and aluminum.			Development	Full production capability if required.
Alloy Surfaces	Granular pack	Ferrous metals low and high alloy steels, molybdenum, nickel, and their alloys, superalloys (nickel, cobalt, and iron)	HI-22 Chromium diffusion coating.			Production	
Alloy Surfaces	Gas convection or granular pack	Ferrous metals low and high alloy steels, molybdenum, nickel, and their alloys, superalloys (nickel, cobalt, and iron)	Alphatizing "CC" Alphatizing "MR" Alphatizing "OCR"	Various furnaces of all types up to 16 ft × 6 ft × 48 in.		Production	
Union Carbide	Pack process	Nickel alloys	C-9 Aluminum-rich	Maximum size part 30 in. high 26 in. diam		C	
Union Carbide	Pack process	Iron-base alloys	C-10 Aluminum-rich	Maximum size part 30 in. high 26 in. diam		C	
Union Carbide	Pack process	Cobalt-base alloys	C-12 Aluminum-rich	Maximum size part 30 in. high 26 in. diam		C	
Union Carbide	Pack process	High-carbon and mild steels	WL-5 Chromium	Maximum size part 30 in. high 26 in. diam		C	
Union Carbide	Pack process	Nickel and cobalt-base alloys	WL-6 Beryllium	Maximum size part 30 in. high 26 in. diam		PD	
Union Carbide	Pack process	Cobalt-base alloys	WL-8 Aluminum-rich	Maximum size part 30 in. high 26 in. diam		C	
Union Carbide	Pack process	Nickel alloys	WL-9 Aluminum-silicon	Maximum size part 30 in. high 26 in. diam		AD	
Union Carbide	Pack process	Nickel-base	WL-14 Aluminum-rich	Maximum size part 30 in. high 26 in. diam		C	

Company	Process	Material	Coating	Process/Size	Production	Status[b]	Remarks
Union Carbide	Pack process	Nickel, cobalt- and iron-base alloys	C-20 Aluminum-rich	Maximum size part 30 in. high 26 in. diam		C	
Union Carbide	Pack process	Nickel, and cobalt alloys	WL–1 Aluminum-rich	Maximum size part 30 in. high 26 in. diam		C	
Union Carbide	Pack process	Mild steels	WL–2 Chromium-rich	Maximum size part 30 in. high 26 in. diam		C	
Union Carbide	Pack process	Nickel-base alloys	WL–4 Aluminum–silicon–chromium	Maximum size part 30 in. high 26 in. diam		AD	
Sylvania	Pack process	Nickel- and cobalt-base alloys	NC-101 A	14 in. × 18 in. × 48 in.		C	
General Electric	Pack cementation	Nickel- and cobalt-base superalloys	CODEP A–Ti–Al; CODEP B–Ti–Al; CODEP C–Ti–Al	Everett, Mass. Bell Furnace 4 ft diam × 5 ft high		C	Process is also being used by Belgian, Italian, and Japanese licensees of J79 engine
Walbar	Pack	Nickel- and cobalt-base alloys	CODEP (GE) Series	Stoneham, Mass. box furnace		C	GE licensee
Allison	Pack	Nickel- and cobalt-base alloys	Alpak–nickel aluminide; Cobalt aluminide	50,000 pieces per month	70,000 (7/1/69)	C	
Howmet	Pack cementation	Cobalt alloys	MDC–7 Cr–Al–Si	H$_2$ atmo-Pit		C	Turbine components Oxidation and sulfidation
Howmet	Pack cementation	Nickel- and cobalt alloys	MDC–9 Duplex Cr–Al	H$_2$ atmo-Pit		AD	Turbine components Oxidation and sulfidation
Howmet	Vacuum pack	Nickel alloys	MDC–701 Al	Vacuum furnaces		C	Avco licensee
Howmet	Pack cementation	Nickel alloys	MDC–1 –Al	H$_2$ atmo-Pit		C	Turbine component Oxidation and sulfidation
Howmet	Pack cementation	Nickel alloys	MDC–1A Al + oxide particles	H$_2$ atmo-Pit		C	Turbine component Oxidation and sulfidation
Howmet	Pack cementation	Iron and nickel alloys	MDC–3 Chromizing	H$_2$ atmo-Pit		C	Corrosion, wear

[a]For full names of companies, see list on p. 166.
[b]PD, preliminary development; AD, advanced development; C, commercial.

aluminides. Chromium, tantalum, titanium, and other elements are predeposited or codeposited by pack techniques as aluminide modifiers. For a very limited number of applications, chromium is also pack-deposited as a primary coating. Each of the major gas-turbine engine manufacturers either applies its own proprietary pack aluminide or procures a proprietary aluminide coating from one of several sources available in the industry. Adequate pack-coating facilities are available in the industry or can be made available within several months for aluminizing the superalloy turbine components required by engine manufacturers. Automation in this industry is nil, however, and its absence is reflected in a relatively high cost for coating application.

The pack-aluminide coatings have met the immediate need for reducing superalloy oxidation and thereby have extended the lives of turbine blades and vanes. Where sulfidation or severe hot-corrosion conditions are prevalent, however, the pack aluminides are only an interim cure for the problem. Coatings with properties superior to those of existing aluminides are vitally needed to meet the challenge of hot corrosion (sulfidation) and higher engine operating temperatures. On the other hand, the need for pack aluminides is not likely to diminish over the next few years, since temperatures in other regions of the turbine will also increase with increasing turbine inlet temperatures, which means that coatings will be required where uncoated alloys currently suffice.

Two immediate problems related to pack aluminides for superalloys currently require attention. First, air-cooling of turbine blades and vanes is the major approach to coping with increasing turbine gas temperatures. Hot corrosion in these internal cooling passages is anticipated, and reliable methods for aluminizing cooling holes are not presently available. The influence of coatings on thin-wall superalloy sections is a corollary problem area. Second, much of the aluminized nickel alloy hardware currently operating in the engines of the commercial and military jet fleets has accumulated thousands of hours of service. Refurbishing this hardware will be required; a substantial portion of it is structurally sound but will require restoration of the coating. Although recoating of aluminized cobalt alloys (nonrotating parts) has been performed for several years, very little recoating of nickel alloy components has been conducted. Studies in this area are under way by the engine manufacturers and commercial coaters, but an accelerated effort will be required to meet the imminent high demand. Stripping and recoating versus overcoating and the influence of reprocessing on base metal properties are the primary problem areas.

A manufacturing capability for coating dispersion-strengthened materials also is needed. The Cr–Al coating system has demonstrated excellent protective properties on TD Nickel and TD Nichrome, but little manufacturing ca-

pability exists in the industry for applying this coating to these alloys.

FLUIDIZED BED

The fluidized-bed process is a variation of chemical vapor deposition that bears some similarity to pack coatings but eliminates the major disadvantage of poor heat transfer in a porous pack medium. The fluidized state of the bed of coating reactants gives rise to very high heat-transfer rates and rapid deposition. Any metal amenable to CVD or pack deposition can, in principle, be applied by fluidized-bed techniques. Two types of fluidized beds have been investigated: one in which the fixed bed consists of the coating metal and the fluidizing gases include a gaseous activator, such as iodine; and the other consisting of "inert" bed material through which externally generated reactant vapors are introduced.

Fluidized beds have been in commercial use for years for heat-treating and in chemical processes such as the reduction of ore. The use of fluidized beds to apply high-temperature diffusion coatings was pioneered by Boeing (Table 33) for the X-20 (Dynasoar) program in an effort to improve coating reliability and to reduce processing time for forming disilicide coatings on molybdenum and columbium. These objectives were achieved, but after cancellation of the X-20 program little further work was done. The production facilities formerly available have been dismantled. Some additional work was done showing feasibility for multiple-layer deposition—e.g., vanadium-modified silicide coating on columbium. However, the very high capital equipment costs and the large daily operating costs of a high-temperature fluidized-bed coating facility make it appropriate that the process be used only where no other processing approach appears feasible.

SLURRY

This process consists of applying a paint or slurry to the part and then firing it at an elevated temperature, usually in a protective atmosphere or vacuum. The major advantage of this process is the inherent ease of size scale-up for large sheet-metal components. There is no massive pack that requires major process modification due to poor heat transfer characteristics. The paint normally contains a fugitive vehicle—powders that form the coating in themselves or by reaction or diffusion with the substrate, or, in the case of a slip-pack, an additive that transfers the coating elements through the formation of a vapor phase. If the slurry melts and reacts vigorously with the substrate, coating coverage is enhanced. Slurry coating suppliers and processes are listed in Table 34.

There are several commercially available coatings and a

TABLE 33 Fluidized Bed Process Developed by the Boeing Co.

Basic Substrates	Coating System (Designation and Composition)	Coating Facilities (Size and Type)		Process Status[a]	Comments
		Operating	Planned (Date)		
Cb, Ta, Mo W alloys	Disilicide coating "Disil"	5-in. diam fluidized bed	None	C	The 5-inch fluidized bed is an Engineering Laboratory facility Production facility including 18-in. beds have been dismantled and are no longer in use
Cb, Ta, Mo W alloys	Vanadium-modified disilicide coating "V-Disil"	5-in. diam fluidized bed	None	AD	

[a]AD, advanced development; C, commercial.

number in the preliminary and advanced development stages. The R505 Sn–Al–Mo aluminide and R512 fused-silicide systems developed by Sylvania have received wide evaluation for Cb and Ta alloys in a variety of applications; slurry coatings on Mo and W have been studied to a lesser extent. The R505 aluminide has been used extensively for small attitude-control engine thrusters. Moderate size capability is available, with some future expansion planned. Vac-Hyd provides an aluminide system that has been used for Cb rocket-nozzle exhaust skirts and several silicides in the preliminary and advanced development stages. Fairly large components can be coated, and some size expansion is planned by Vac-Hyd. McDonnell-Douglas has slurry coating capabilities for large-size hardware but has not indicated that these are readily available to outsiders. Solar has several slurry coating systems for Ta and Cb alloys that are in various stages of development and that promise to become commercial in the near future. Size capabilities and expansion plans have not been reported.

The commercial Solar slurry coating listed for gas-turbine nickel-base alloys is the Pratt & Whitney Jo-Coat, which has been licensed to TRW. There is a large-volume capability at both Pratt & Whitney and TRW. Only Solar has a large line of ceramic-type slurries for gas-turbine applications that have been used extensively on a commercial scale. The size capabilities of several of these systems are reported.

A number of special-purpose coatings in the advanced development stages are available as a service or as starting materials at Engelhard Industries. It is not clear from the information available how they can be used in high-temperature oxidation, since they would have to be adapted or modified to meet particular problems. Apparently, a variety of sub-strates can be coated, and size capability is dependent primarily on furnace availability.

In summary, there is a fair commercial capability for coating refractory metal hardware with slurry-type coatings, although these processes are restricted to a few suppliers. The primary problem in obtaining expanded facilities is furnace size, and this problem can be overcome in 6–9 months for most requirements. However, the real problems lie in the hardware design—specifically, the complexity of the parts, including joined areas and the reliability with which they can be protected. Present programs aimed at these problems should provide increased reliability for future applications. The requirements for nonrefractory metal slurry coatings are neither widespread nor critical, and accordingly, little has been done or is planned for the future. With the increase in size of individual gas-turbine components, a need for a slurry process may evolve. However, it is not appropriate at present to recommend the specific development of slurry processes for this application.

HOT-DIPPING

The four survey replies received on hot-dipping processes are listed in Table 35. All are aluminizing processes and are commercial and available as a service or possibly for licensing. The process is simple, consisting essentially of cleaning the hardware suitably and dipping it into a molten aluminum or aluminum-rich bath. A precoat may be applied for alloying, such as in the AVCO 606B process. An aluminide is formed with the substrate during dipping, or is formed in a later treatment by diffusion of the aluminum picked up in dipping. Protective or cleaning fluxes may or may not be used

TABLE 34 Slurry Coating Processes

Company[a]	Coating Process	Basic Substrates	Coating System (Designation and Composition)	Coating Facilities (Size and Type) Operating	Planned (Date)	Process Status[b]	Comments
			Refractory Metals				
Sylvania	Slurry	Cb, Ta, Mo, W alloys	R505 Sn–Al–Mo	Vac. 36 in. diam × 36 in. long	1971	C	
Sylvania	Slurry	Cb, Ta, Mo, W alloys	R512 Si–Cr–Ti	Vac. 36 in. diam × 36 in. long	50 in. diam × 50 in. long	C	
Sylvania	Slurry	Cb, Ta, Mo, W alloys	R512 Si–Cr–Fe	Vac. 36 in. diam × 36 in. long	50 in. diam × 50 in. long	C	
Sylvania	Slurry	Cb, Ta, Mo, W alloys	R512 Si–Ti–Mo	Vac. 36 in. diam × 36 in. long	50 in. diam × 50 in. long	C	
Sylvania	Slurry	Cb, Ta, Mo, W alloys	R508 Ag–Si–Al–Mo	Vac.–Argon 12 in. diam × 18 in. long		C	
Sylvania	Slurry	Ta alloys	R516 Hf–Ta	Vac.–12 in. diam × 18 in. long		AD	
Vac Hyd	Slurry/diffusion	Columbium	Two highly modified silicide slurries have been extensively tested	25 in. × 36 in. Cold wall; 60 in. × 60 in. Cold wall	84 in. × 84 in. Cold wall 1970	AD	Coatings designated as Vac-Hyd 17 and 101.
Vac Hyd	Slurry/diffusion	Columbium Tantalum	Two highly modified silicide slurries in test	60 in. × 60 in. Cold wall	84 in. × 84 in. Cold wall 1970	PD	Coatings designated as Vac-Hyd 102 and 103.
Vac Hyd	Slurry/diffusion	Columbium	Highly modified aluminide	12 vacuum/argon furnaces from 20 to 60 in. diam	84 in. × 84 in. Cold wall 1970	C	Coating designated as Vac-Hyd 9.
McDonnell-Douglas	Halide-activated cold slurry (slip pack)	Molybdenum–Tungsten	L–7 straight silicide	3 ft diam × 5 ft long; Superalloy retort and furnace, atmosphere control		AD	Used for ASSET and BGRV flight hardware.
McDonnell-Douglas	Fused-slurry aluminide	Tantalum Columbium	LB–2 88 percent Al –10 percent Cr –2 percent Si	3 ft diam × 5 ft long; 600-gal slurry dip tank, superalloy retort, furnace, atmosphere control		C	Used for ASSET and BGRV flight hardware

Company	Process	Substrate	Coating	Size	Scale-up	Status	Remarks
McDonnell-Douglas	Fused-slurry silicide	Columbium	MD silicon eutectic series mixtures to give modified silicides	10 in. diam × 12 in. long	3 ft diam × 5 ft long vacuum furnace (1970)	PD	Plan scale-up to advanced development status during next 6 mo.
North American Rockwell	Slurry	Being used on Cb–103. Has been tested on Cb–752 and Ta–10W	NA–85 Aluminide	Vacuum furnace. Inconel retort holds part that is truncated cone 80 in. diam at base by 56 in. high. Furnace for heating retort at Certified Steel Treating Co., Vernon, Calif.	Any retort size possible. Limitation is available. Furnace size and temperature up to 1850°F	C	Used on Aerojet General AG10-137 engine on Apollo Service Module
Solar	Slurry-vacuum-sinter plus silicide pack	Ta and Cb alloys	TNV–7	14 in. × 14 in. × 8 in.	Could be scaled up to 3 ft × 3 ft × 4 ft in 6 mo.		Provide outstandingly reliable protection to Ta alloy for 600+ hr at 1600°F. Fewer data are available on protection of Cb alloy, but no failure in 200 hr at 1600° and 2400°F.
Solar	Trowel	Tungsten and tungsten-coated Ta alloys	Tungsten-reinforced thoria	Max. size to date 8 in. in diam	Could be scaled up to 6 ft × 6 ft × 8 ft	Lab.	Protection for up to 60 min.
Solar	Slurry-vacuum-sinter plus silicide pack	Ta alloys	TNV–13	14 in. × 14 in. × 8 in.—could be scaled up to 3 ft × 3 ft × 4 ft in 6 mo.		Lab. Pilot plant	Affords protection against oxidation to 3500°F. Has provided 600 hr protection at 2400° and 300 hr at 1600°F.
TRW	Slip pack	Cb, Ta, Mo, W alloys	Cr–Ti–Si, simple and modified silicides	Vacuum furnaces up to 48 in. diam × 54 in. high		AD	
TRW	Fused silicides	Cb, Ta, Mo, W alloys	Fused-complex silicides	Vacuum furnaces up to 48 in. diam × 54 in. high		PD	

TABLE 34 (Continued)

Company[a]	Coating Process	Basic Substrates	Coating System (Designation and Composition)	Coating Facilities (Size and Type)		Process Status[b]	Comments
				Operating	Planned (Date)		
			Superalloys				
Air Research	Slurry	Nickel- and cobalt-base alloys	Proprietary	Limited	None	PD	
Engelhard	Slurry	Quartz or other ceramics	Platinum (dimensional precision coating)	Custom service 18 in. × 18 in. × 8 in.		AD	Materials can be supplied to customer for his use. Size then dependent on customer furnace size.
Engelhard	Slurry	Ceramics	Palladium	Custom service		AD	U.S. Patent 3,344,586.
Engelhard	Slurry	Metal seals	Ceramic	Custom service		AD	U.S. Patent 3,367,696.
Engelhard	Slurry	Inconel Stainless steel	Diffusion barriers (precious or base metal)	Custom service		AD	U.S. Patent 3,176,678 3,176,679 3,363,090.
Sylvania	Slurry	TD Ni	NC–300–Ni–Cr–Al alloy	Vac-Hydrogen 10 in. diam × 18 in. long		PD	
TRW	Vacuum-slurry	Ni alloys	Jo–Coat-slurry aluminide (PWA-47)	3 ft × 4 ft × 4 ft (3) units		C	Approved PWA source
TRW	Slip-pack	Ni, Co, DS alloys	Modified aluminides	3 ft × 3 ft × 4 ft (3) 48 in. diam × 54 in. (1)		AD	
Wall Colmonoy	Slurry spray-fused	Alloys used for high-temp. service, high-stress, and those subjected to corrosive atmospheres	Cermet-type Nicrocoat Ni, Cr and ceramics	Detroit, Mich. San Antonio, Tex.	(1969) Dayton, Ohio Okla. City, Okla.	C	

Company	Process	Substrate	Coating	Location / Equipment	Size / Date	Status	Remarks
Wall Colmonoy	Slurry spray-fused	Alloys used for high-temp. service, high-stress, and those subjected to corrosive atmospheres	Nicrocoat alloys Ni, Cr, Si	Detroit, Mich. Morrisville, Pa. San Antonio, Tex. Okla. City, Okla. Dayton, Ohio Montebello, Calif. Montreal, Can. San Francisco, Calif. Milwaukee, Wisc.	(1970 or 1971)	C	
Vac-Hyd	Slurry/diffusion	Superalloys	Aluminide	12 Vacuum/argon furnaces from 20 to 60 in. diam	84 in. × 84 in. Cold wall 1970	C	Coatings designated as Vac-Hyd 11, 14, and 22.
Vac-Hyd	Slurry/diffusion	Titanium	Aluminide	12 Vacuum/argon furnaces from 20 to 60 in. diam	84 in. × 84 in. Cold wall 1970	PD	Being developed for fretting corrosion impact and oxidation.
Misco	Slurry spray	Ni alloys	Aluminide	Vac. furnace		AD	Oxidation–sulfidation.
Pratt & Whitney	Slurry	Ni alloys	Aluminide-PWA-47 Jo-Coat	Hydrogen 100,000 part/Mo		C	Blades and Vanes primarily.
Solar	Slurry	Nickel- and cobalt-based alloys; also some iron-based	S10-33A, S13-53C (Aluminizing)	5 ft × 5 ft × 8 ft		C	Processes require controlled atmosphere and are designed both for small components (for example, turbine blades and vanes and fuel vaporizer tubes) and for large components (e.g., turbine transition and combustion chamber liners). Masking is extremely easy. Protection against oxidation and sulfidation is effective to 2100°F.
Solar	Slurry	Low-alloy steels	S1177 (ALCERMET®)	6 ft × 6 ft × 8 ft		C	Prevention of oxidation and corrosion by the products of combustion to 1200°F.
Solar	Slurry	Titanium alloys	S1174 (Aluminizing)	4 ft × 4 ft × 6 ft		C	Protection of titanium from stress corrosion and oxidation to 1200°F.

TABLE 34 (Continued)

Company[a]	Coating Process	Basic Substrates	Coating System (Designation and Composition)	Coating Facilities (Size and Type)		Process Status[b]	Comments
				Operating	Planned (Date)		
			Superalloys (continued)				
Solar	Slurry	300 series stainless steel	S1166–107A (Aluminizing)	6 ft × 6 ft × 8 ft		C	Low-cost aluminizing process for the 300 series stainless steel to protect against oxidation and sulfidation to 2000°F.
Solar	Slurry	Ni-, Co- or Fe-base alloys containing greater than 5% Cr	S5210–2C (Solaramic®)	6 ft × 6 ft × 8 ft		C	Protective to 1800°F for hundreds of hours. Not recommended for turbine blades and vanes
Solar	Slurry	Ni-, Co- or Fe-base alloys containing 12% Cr	S6100M or S6100 (Solaramic®)	6 ft × 6 ft × 8 ft		C	Protective to 2000°F with maximum recommended continuous operating temp. to 1900°F. Not recommended for turbine blades and vanes

[a]For full names of companies, see list on p. 166.
[b]PD, Preliminary Development; AD, Advanced Development; C, Commercial

TABLE 35 Hot-Dipping Processes

Company[a]	Coating Process	Basic Substrates	Coating System (Designation and Composition)	Coating Facilities (Size and Type)		Process Status[b]	Comments
				Operating	Planned (Date)		
Arthur Tickle	Hot-dip pure aluminum	Ferrous alloys	Alumicoat	Plant of 100 ft × 200 ft. 6 furnaces and cleaning facilities; largest fabrication to 15 ft × 3½ ft diam (approx.)		C	Excellent atmospheric and high-temperature oxidation resistance
AVCO	Cr plate Al–Si dip	Ni-Base alloys	606B	Ajax Electrotherm Salt Furnace		C	
AVCO	Aluminum dip process	Ni-Base alloys	61A	Ajax Electrotherm Salt Furnace		C	
Union Carbide	Hot dipping	Nickel-, cobalt- and iron-base alloys	IAD Aluminum	Maximum-size part 12 in. × 12 in. × 12 in.		C	
Misco	Aluminum hot dip	Ni alloys	MDC 601	Hydrogen furnace		C	Sulfidation, oxidation
Misco	Aluminum hot dip	Ni alloys	MDC 606	Hydrogen furnace		C	Sulfidation, oxidation
Misco	Duplex Cr–Al	Ni alloys	MDC 14	Hydrogen furnace		PD	Sulfidation, oxidation

[a]For full names of companies, see list on p. 166.
[b]C, commercial; PD, preliminary development.

179

before or during dipping. Size capability is illustrated by the Arthur Tickle Engineering Works, Inc. facility, where parts up to 15 ft long can be handled. Usually, tubes and pipes are serviced in these sizes. Complexity of parts may be a problem in large components. However, small gas-turbine components are handled readily.

In general, hot-dipping has been successfully and widely used for ferrous metals and nickel-base alloys at many plants other than those listed. However, the more advanced and large-volume coatings for critical applications appear to be better applied by pack processes. Accordingly, there are no plans in the industry for expanding present hot-dipping facilities, nor is there any need for such increase. A similar lack of requirement for dipping facilities exists for refractory metals. A number of early aluminum-base coatings were applied by hot-dip processes, but these gradually lost favor because of the superior capabilities of silicides.

ELECTRODEPOSITION PROCESSES

Electrolytic deposition includes aqueous electroplating, electrodeposition from fused salts, and electrophoresis. Aqueous electroplating involves the deposition of metallic ions from an aqueous solution onto a negatively charged workpiece. Electrodeposition from fused salts is analogous to electroplating, where the electrolyte is generally a mixture of fused chloride and fluoride salts. Electrophoresis, on the other hand, involves the electrodeposition of charged particles from a liquid suspension of these particles in an appropriate electrolyte. A summary of current capabilities is given in Table 36.

For all three processes, proper shaping and location of the electrode and careful control of the solution characteristics are critical to effective electrodeposition. Dense pure metal and alloy coatings are deposited on cold substrates by aqueous plating; thus, a subsequent diffusion heat treatment or coating cycle is required to produce a bonded protective coating. Electroplating in a fused-salt bath is generally accomplished at a sufficiently high temperature to produce diffusion growth of the coating. Dense solid-solution alloys and intermetallic coatings such as borides, silicides, and chromides are formed by this method. Relatively porous coatings are formed by electrophoresis, although particles of virtually any metal, alloy, or metallic compound can be deposited by this method. Following deposition, a subsequent coating cycle, or isostatic pressing and heat-treating, is required to densify the porous deposit.

ELECTROLYTIC

In the survey conducted for this committee, little activity was indicated for the use of aqueous electroplating as a method of forming high-temperature protective coatings. AVCO employs electroplating to chromium-precoat turbine blades prior to hot-dip aluminizing. Certainly there are large electroplating facilities throughout the industry, and these facilities could be adapted to the coating of high-temperature hardware. At present, however, there are few high-temperature protective coating systems for which electroplating would function as a coating-application step.

Fused-salt electrodeposition has recently been promoted by General Electric, and several licensees of the GE Metalliding Process have been established. General Technologies Corporation indicated a current laboratory metalliding capability with a planned expansion to a 12 ft diam $\times$ 24 ft facility. The fused-salt method of coating deposition has also been investigated extensively by Solar and by Union Carbide. Fused-salt deposition permits the formation of a wide variety of very pure elemental and intermetallic coatings on high-temperature materials. Salt corrosion is a potential problem with this process, and some substrate–coating combinations are not possible by this method. The survey suggests that large production-scale facilities have not been established for fused-salt deposition of coatings, and considerable development work would be required to scale up this type of process from a laboratory-scale level.

ELECTROPHORESIS

Vitro Corporation has conducted the bulk of the research and development effort on electrophoretic deposition of high-temperature protective coatings. Large-scale facilities are indicated by Vitro for the application of coatings to both superalloys and refractory metals; however, the majority of these systems are only in the advanced development stage, thus requiring additional development work to render them commercial. The electrophoretic method is particularly amenable to the deposition of uniformly thick slurry coatings on complex shapes. At present, the electrophoretic method is being explored extensively for the application of low-temperature protective coatings to jet engine compressor components. The flexibility of this process for applying uniform deposits of complex alloys or intermetallics to irregular shapes makes it particularly attractive for applying those types of alloy coatings not amenable to formation by diffusion growth processes. The process has been used widely for the coating of refractory metal fasteners.

VAPOR DEPOSITION

CHEMICAL

In all versions of chemical vapor deposition (CVD)—also variously referred to as vapor plating or gas plating—a volatile compound of the material to be deposited is brought into contact with a heated substrate, and the desired coating is built-up in molecular steps via either displacement reduction, or thermal decomposition reactions of the plating compound at the surface of the workpiece. Both pack diffusion and fluidized-bed processes are special cases of chemical vapor deposition. In "pure" CVD the homogeneous vapor phase is allowed to impinge upon the heated substrate in a reaction chamber after external generation of the reactant vapors. Principal advantage of this process is that the deposition parameters can be independently controlled. Variations of the basic CVD process could involve codeposition of several elements, with possibilities for continuous variation of composition or multiple-layer depositions.

Producing a desired coating by CVD requires that the coating metal exist as a compound that can be readily vaporized at a relatively low temperature without thermal degradation and that it then be sufficiently unstable at some elevated temperature to be reduced or pyrolized ("cracked") to the desired metal with minimal side reactions. A wide variety of such compounds exist, ranging from several halides to various organometallics—including alkyls, aryls, arylalkyls, arenes, carbonyls, allyls, acetylacetonates, and cyclopentadienyls. A major limitation of many of the metalorganics is the tendency for carbon to be codeposited as an impurity. On the other hand, the pyrolysis of organometals can be conducted at significantly lower temperatures than the reduction or decomposition of more commonly used halides.

The CVD processes, with proper selection of starting materials, can be used to deposit a variety of coating materials under either reduced or atmospheric pressure (of inert or reducing gases) to produce dense coatings purer than deposits obtainable by most other methods. However, the prerequisite thermodynamics and kinetic studies have been done for only a small number of the pertinent reactions; hence, the technology of CVD coatings is at a very preliminary stage of development. There have been demonstrations of chromizing, aluminizing, and siliconizing by CVD, both with and without subsequent diffusion heat treatments. If deposition is carried out at high enough temperatures, diffusion of coating elements occurs simultaneously to form alloy or intermetallic coatings. Both silicon and silicon carbide coatings have been applied to graphite by many investigators using CVD. Resultant coatings are similar to pack-diffusion coatings. Commercially, however, most graphite parts are coated using pack methods. Similarly, CVD has been used in the laboratory to form silicide coatings for the refractory metals, but no commercial development has emerged.

There has been considerable interest in, and laboratory-scale investigation on, applying tungsten as a barrier layer on tantalum (or Mo on Cb) by CVD, to be followed by siliciding. Considerable difficulty has been encountered in ob-

taining uniform adherent deposits; problems include thickness variation, inadequate bonding, and edge buildup of the W or Mo coatings.

Reasonable success was met in a program to develop CVD techniques for applying the complex Cr–Ti–Si coating system to Cb alloys. Both two-step processing (i.e., codeposition of CrTi, followed by Si and codeposition of CrTi, followed by Si) and codeposition of all three elements in a single process produced coatings on laboratory samples that compared well with conventional two-step pack coatings. Investigation of the feasibility of coating small hardware items by this process is continuing.

As can be seen from Table 37, several organizations claim equipment and process capability to apply a variety of deposits to the substrates of interest, including Si, SiC, BN, NbC, Al_2O_3, ZrO_2, HfO_2 and refractory metal coatings for either graphite or refractory alloy substrates; aluminum coatings for superalloys; and W or Mo depositions on almost any substrate. However, most of these coatings have not been utilized to coat hardware, with the exception of the refractory metal coatings applied to rocket nozzles and tubing.

PHYSICAL

Vacuum vapor deposition processes encompass those methods of applying a coating that involve physical evaporation of a coating material source and line-of-sight condensation of this vapor phase on the surface to be coated. Vacuum metallizing, vacuum evaporation, and sputtering are common terminology for vacuum vapor deposition processes. In the case of vacuum metallizing and vacuum evaporation, chamber pressures of the order of 10^{-4}–10^{-7} Torr are required to achieve a clean environment and adequate metallic evaporation. The source of coating element(s) must be heated by resistance, induction, electron beam, or other means to promote evaporation. The rate of evaporation is controlled by the temperature and pressure conditions and is governed by the relative vapor pressures of the elements in the coating material source. The material to be coated may be heated during deposition to enhance diffusion bonding of the coating to the substrate.

Sputtering utilizes a principle different from that of evaporation. In sputtering, the source and material to be coated are mounted in parallel planes inside a chamber that can be evacuated to 10^{-4}–10^{-5} Torr. After evacuation, the chamber is backfilled with an appropriate pressure of inert gas, such as argon. The material to be deposited is charged negatively by a high-voltage dc power supply, thus becoming an electron emitter. Free electrons ionize the argon, and these high-energy positive ions in turn bombard the cathodic coating source. Atoms or molecules of coating material are ejected from the source and deposited on the substrate surface.

The methods described above have been used primarily

TABLE 36 Electrolytic Deposition Processes

Company	Coating Process	Basic Substrates	Coating System (Designation and Composition)	Coating Facilities (Size and Type)		Process Status	Comments
				Operating	Planned (Date)		
Vitro	Electrophoretic deposition	Stainless, low alloy steels, heat-resistant alloys	VEP-301 Sermetal® W (Inorganically bonded aluminum)	300 gal tank parts to 4 ft long × 2 ft diam		C	Salt spray corrosion and oxidation resistance
Vitro	Electrophoretic deposition	Steels	VEP-201 S-6100M, 558A, NBS 418 and other high-temp. frits	300 gal tank parts to 4 ft long × 2 ft diam		AD	Salt spray corrosion and oxidation resistance
Vitro	Electrophoretic deposition	Superalloys	VEP-302 **P**roprietary aluminide-diffusion coating	Turbine blades		AD	Sulfidation resistance
Vitro	Electrophoretic deposition	Superalloys	VEP-303 SermeTel®-J	Turbine blades		PD	Sulfidation resistance
Vitro	Electrophoretic deposition	Aluminum and steels	VEP-304 SermeTel®-385 Proprietary	Small parts		PD	Protection against galvanic corrosion
Vitro	Electrophoretic deposition	Columbium alloys	Cr/Ti-Si	Small parts		AD	Oxidation resistance to 2700°F
Vitro	Electrophoretic deposition	Tantalum alloys	Tungsten	Small parts		PD	Barrier layer for conversion to PWSi$_2$
Vitro	Electrophoretic deposition		Si/WSi$_2$	Small parts		PD	Binary coating—oxidation resistance to 2500°F

Company	Process	Coating	Substrate	Size	Status[b]	Remarks
Vitro	Electrophoretic deposition	MoSi$_2$–20TiSi$_2$/VSi$_2$		Small parts	PD	Binary coating—oxidation resistance to 2500°F
Vitro	Electrophoretic deposition	MoSi$_2$ and MoSi$_2$–Ni	Molybdenum alloys	Small parts	PD	Oxidation resistant to 2700°F
Vitro	Electrophoretic deposition	WSI$_2$	Tungsten alloys	Small parts	PD	Oxidation resistant to 3200°F
General Technologies	Metalliding (Fused-salt electrodiffusion process, GE licensee)	Berylliding Boriding Aluminiding Siliciding Titaniding Vanadiding Chromiding Nickeliding Zirconiding Columbiding Molybdeniding Hafniding Tantaliding Tungsteniding Rare-earthiding	Any metals with melting point above about 1200°F	All systems: Current 3 in. × 4 in.	AD, some C	Planned: 12 in. × 24 in. within 12 mo
Allison	Electrophoretic deposition	AEP-Aluminum plus additive elements	Nickel- and cobalt-base alloys		PD	
Lycoming	Electroplating	Chromium precoat for 606B coating	Ni alloys		C	Blades and vanes

[a] For full names of companies, see list on p. 166.
[b] PD, preliminary development; AD, advanced development; C, commercial.

183

TABLE 37 Chemical Vapor Deposition Processes

Company[a]	Coating Process	Basic Substrates	Coating System (Designation and Composition)	Coating Facilities (Size and Type)		Process Status[b]	Comments
				Operating	Planned (Date)		
Fansteel (San Fernando Labs)	CVD	Ta and Ta alloys	W–Mo, W–Re, Re, Ni, Al$_2$O$_3$ HfO$_2$, ZrO$_2$, BN, SiO$_2$, TaC, Ta$_2$C	18 units	Larger parts 24 in. diam × 50 in. length (10/68)	AD to C	Many of the combinations, e.g., those of steel and Ti, are used to construct free-standing items after removal of substrate.
Fansteel (San Fernando Labs)	CVD	Carbon/graphite	B, Si$_3$N$_4$, SiC, Ta, Re, W/Re, Al$_2$O$_3$, HfO$_2$, ZrO$_2$	*Tubing:* ID Coating Microns to tens of mils coating up to 4 in. diam or up to 15 in. length			
Fansteel (San Fernando Labs)	CVD	Mo and Mo alloys	BN, B, Ni, Ta, Cb, Re, W, Si, Al$_2$O$_3$, HfO$_2$, ZrO$_2$	OD Coating Microns to tens of mils coating up to 4 in. diam or up to 5 ft length			
Fansteel (San Fernando Labs)	CVD	Ni and Ni alloys	Fe, W, Ni, Ta, WC				
Fansteel (San Fernando Labs)	CVD	Pyrolytic graphite and alloys	Al$_2$O$_3$, HfO$_2$, B, ZrO$_2$	*Parts* Up to 8 in. diam by 24 in. length			
Fansteel (San Fernando Labs)	CVD	Re and Re alloys	W, Mo, Ta				
Fansteel (San Fernando Labs)	CVD	Steel–carbon stainless low alloy tool	W, Ni, Ta, Cb, Cb–Zr, TaC, TiC[c]	Up to 8 in. diam by 24 in. length			
Fansteel (San Fernando Labs)	CVD	U and compounds	W, W–Re, Mo	Up to 8 in. diam by 24 in. length			

Company	Process	Materials deposited	Substrate	Size	Status[b]
Fansteel (San Fernando Labs)	CVD	W, WC, Re, Si, SiC, Mo–Re, Cb, Ta	W and W alloys		
Fansteel (San Fernando Labs)	CVD	Mo, W, Re	Zr and Zr alloys		
Union Carbide	CVD	Pyrolytic graphite / Boron-pyrolytic Graphite	Graphite / Graphite	Up to 40 in. diam × 84 in. long	C
Union Carbide	CVD	Boron nitride / SiC	Graphite / Graphite		AD
Union Carbide	CVD	NbC	Graphite	Up to 1 ft diam × 4 ft long	AD
Sylvania	CVD	D–600 W, Mo	Ta, Cb, Mo, Ni, Fe alloys	12 in. diam × 16 in. long	AD
Sylvania	CVD	D–600 Ta	Cb, W, Mo, Ni, Fe alloys	12 in. diam × 16 in. long	AD
Sylvania	CVD	D–600 Re	Fe alloys (wire)	3 in. diam × 6 in. long	PD
Sylvania	CVD	D–600 TiC	Fe alloys	3 in. diam × 6 in. long	PD
Ethyl	CVD	Aluminum-form proprietary / Organoaluminum compound	Nickel- and cobalt-base superalloys	Pilot plant unit	AD
General Technologies	CVD	Refractory metals, Aluminum, Titanium, Chromium, Nickel, Platinum-group metals, Metal borides (TiB_2, B_4C, etc.), Carbides (TiC, SiC, etc.)	Metals and nonmetals	All systems: 8 in. × 15 in. depending on which metal or compound being deposited	Advanced development for aluminum, tungsten, and commercial SiC

[a] For full names of companies, see list on p. 166.
[b] PD, preliminary development; AD, advanced development; C, commercial.
[c] Not in stainless steel.

185

TABLE 38 Vacuum Vapor Deposition Processes Developed by Pratt & Whitney Aircraft Company

Basic Substrates	Coating System (Designation and Composition)	Coating Facilities (Size and Type)		Process Status[a]	Comments
		Operating	Planned (Date)		
Ni and Co alloys	FeCrAlY	6–8 parts batch		AD	Applicable to blades and vanes
Fe, Cr, Cb, W, Mo, Ta alloys	Various metals and alloys			AD	

[a]AD, advanced development.

for the deposition of thin film coatings on ornaments, mirrors, semiconductor components, camera lenses, and so on. Very little work has been done to employ these techniques for depositing 1–4-mil-thick protective coatings on high-temperature materials. In the survey conducted for this committee, only Pratt & Whitney Aircraft Company reported an active program directed to vacuum vapor deposition of high-temperature coatings (Table 38). Pratt & Whitney's principal effort is aimed at deposition of an FeCrAlY alloy on turbine blade and vane airfoils. The technique is in the advanced development stage, with facilities available for the coating of 6–8 parts per lot. Details of the evaporation techniques and deposition parameters are not presently available for publication. Organizations that manufacture electron beam equipment, such as Temescal and Electron Beam Corporation, are also currently exploring the deposition of high-temperature protective coatings by this method.

The utility of this method for depositing thick (1–4 mil) coatings on high-temperature materials has been well demonstrated by Pratt & Whitney. Relatively sophisticated vacuum facilities and evaporation techniques are required for this process, and a great deal of effort is needed to develop manufacturing methods applicable to coating blades, vanes, sheet panels, etc. The technique does, however, provide a means for depositing alloy coatings that developers have thus far been unable to deposit successfully by other means.

CLADDING PROCESSES

ROLL OR PRESSURE BONDING

The processes listed in Table 39 involve the straightforward sandwiching of a protective metal or alloy around similar or dissimilar substrates by the application of heat and/or pressure. Pressure can be applied explosively, as in Dupont's Detaclad process, or it can be applied by means of a rolling

mill or a gas envelope, as in isostatic pressing. The latter process lends itself more readily to producing finished hardware of moderately complex shapes but is limited in size as set by the combination of pressure and temperatures involved and the materials requirements of the pressure vessel.

The process of hot-roll cladding, on the other hand, is not applicable to finished hardware because of the problems involved in end and edge closures, as well as in bending and joining into structural shapes. The process is used in manufacturing sheet materials of sandwich construction in considerable amounts. Some fuel-element plates for nuclear reactors are normally made this way. The fuel core (substrate) can be successfully sealed in. Welding with appropriate filler rods can also be employed for edge sealing and other applications. There are few, if any, commercial roll-cladding facilities and processes for protecting refractory metals at high temperatures. None is reported here, but it is known, for example, that platinum-clad molybdenum rods are used as electrodes in glass-melting. Accordingly, for special limited applications there may be a need for the development of similar precious metal claddings for other refractory metals. Apparently, there is commercial capability to produce large quantities of materials in lengths and widths of practical importance in moderately high-temperature applications—i.e., the limits of superalloys. There are two processes under development for cladding tantalum and columbium alloys with Hf–Ta alloys. Applications of the resultant composites are at present limited to rocket nozzles.

In general it can be concluded that roll-cladding processes are controllable and reliable, as evidenced by nuclear applications, and materials are readily available, but most are not suitable for refractory metal applications.

PLASMA OR FLAME SPRAYING

A more practical approach to cladding with coating materials involves their physical deposition on the surface by

TABLE 39   Pressure Bonding Processes

Company[a]	Coating Process	Basic Substrates	Coating System (Designation and Composition)	Coating Facilities (Size and Type) Operating	Planned (Date)	Process Status[b]	Comments
Dupont	Detaclad®	Carbon steel, low-alloy steel, or stainless steel	300 series and 400 series stainless steels	Plates to 132 in. width × 480 in. length		C	Manufactured to ASTM and ASME specifications
Dupont	Detaclad®	Same as above	Nickel and nickel alloys (e.g., Monel, Inconel, Incoloy)	Same as above		C	Manufactured to ASTM and ASME specifications
Dupont	Detaclad®	Same as above	"Hastelloy" alloys B, C, C-276, X	Plates to 48 in. width × 140 in. length		C	Manufactured to ASTM and ASME specifications
Dupont	Detaclad®	Same as above	Commercially pure titanium	Plates to 105 in. width × 220 in. length		C	Manufactured to ASTM and ASME specifications
Dupont	Detaclad®	Carbon steel	Tantalum	Plates to 36 in. width × 96 in. length		AD	Manufactured to ASTM and ASME specifications
Dupont	Detaclad®	Carbon steel	Zirconium	Plates to 44 in. width × 120 in. length		AD	Manufactured to ASTM and ASME specifications
Olin Brass	Cladding POSIT-BOND (metallurgical bond)	Steel	5–50% Aluminum	0.005–0.125 × 12½ in. wide	(1/1/72) 0.005–0.150 in. × 26 in. wide	C	Coils or cut length
Olin Brass	Cladding POSIT-BOND (metallurgical bond)		5–50% Ni-Fe/Alloys	0.005–0.125 × 12½ in. wide	(1/1/72) 0.005–0.150 in. × 26 in. wide	C	Coils or cut length
Olin Brass	Cladding POSIT-BOND (metallurgical bond)	Steel	5–50% Nickel	0.005–0.125 × 12½ in. wide	(1/1/72) 0.005–0.150 in. × 26 in. wide	AD	Coils or cut length
Olin Brass	Cladding POSIT-BOND (metallurgical bond)	Steel	5–50% Stainless steel	0.005–0.125 × 12½ in. wide	(1/1/72) 0.005–0.150 in. × 26 in. wide	AD	Coils or cut length
Olin Brass	Cladding POSIT-BOND (metallurgical bond)	Steel	5–50% Titanium	0.005–0.125 × 12½ in. wide	(1/1/72) 0.005–0.150 in. × 26 in. wide	PD	Coils or cut length
Olin Brass	Cladding POSIT-BOND (metallurgical bond)	Steel	5–50% Copper and alloys	0.005–0.125 × 12½ in. wide	(1/1/72) 0.005–0.150 in. × 26 in. wide	C	Coils or cut length

187

TABLE 39 (Continued)

Company[a]	Coating Process	Basic Substrates	Coating System (Designation and Composition)	Coating Facilities (Size and Type)		Process Status[b]	Comments
				Operating	Planned (Date)		
Olin Brass	Cladding POSIT-BOND (metallurgical bond)	Stainless	5–50% Aluminum	0.005–0.125 × 12½ in. wide	(1/1/72) 0.005–0.150 in. × 26 in. wide	C	Coils or cut length
Olin Brass	Cladding POSIT-BOND (metallurgical bond)	Stainless	5–50% zinc	0.005–0.125 × 12½ in. wide	(1/1/71) 0.005–0.150 × 26 in. wide	C	Coils or cut length
Olin Brass	Cladding POSIT-BOND (metallurgical bond)	Stainless	5–50% copper and alloys	0.005–0.125 × 12½ in. wide	(1/1/72) 0.005–0.150 × 26 in. wide	C	Coils or cut length
IITRI	Cladding	Ta, Cb alloys	Hf–Ta Cladding	–	–	AD	–
Fansteel	Cladding	Ta	Hf–Ta	–	–	AD	–
W. B. Driver	Hot bonding by rolling	Bonding by hot rolling of clad materials with two or more metallic components that are compatible with each other in relation to temperature of hot-rolling, hot-shortness, etc.		Birdsboro two-high hot-rolling mill and associated welding preparation equipment		C	
W. B. Driver	Cold bonding by rolling	Bonding by cold rolling of clad materials with two or more metallic components		Pittsburgh four-high cold-rolling mill with associated wire-brushing equipment		C	
National Lead	Hot-roll cladding	Stainless Cermet	Stainless Stainless	Several rolling mills up to 24-in. rolls		C	
National Lead	Hot isostatic cladding	Stainless Cermet Zirc-U	Stainless Stainless Zircaloy	4 in. diam × 36 in. long Argon at high pressure--to 2200°F		C	
Engelhard Minerals & Chemicals	Swaging	Base metals	Precious metal	In production		C	
Engelhard Industries	Hot bonding Warm bonding	Ferrous and nonferrous	Ferrous, nickel, iron, precious metals, copper (can be complete or selective area overlay)	In production		C	

[a]For full names of companies, see list on p. 166.
[b]PD, preliminary development; AD, advanced development; C, commercial.

188

TABLE 40 Plasma and Flame Spray Processes

Company[a]	Coating Process	Basic Substrates	Coating System (Designation and Composition)	Coating Facilities (Size and Type) Operating	Planned (Date)	Process Status[b]	Comments
Metco	Plasma and flame spray	Almost all metals and alloys	CrC–Ni–Cr	Various contractors throughout U.S.		C	Turbine-fretting
			CrC–Ni–Cr–B–Si–NiAl			C	Wear-erosion
			WC–Ni–Cr–B–Si–NiAl			C	Wear-erosion
			Co–Cr–Ni–W			C	Wear-oxidation
			Ta, W			C	On graphite-erosion
			ZrO_2			C	Thermal barrier
			Mg ZrO_2			C	Over NiAl-flame erosions
			Ni alloy + BN			C	Abradable coating
			NiAl			C	Bonding coating-fretting
			Ni–Cr			C	Oxidation
Plasma-dyne	Plasma spray	Essentially unlimited	Metals, ceramics, combinations	Parts 4½ ft diam × 6 ft long and larger		C	
Wall Colmonoy	Flame spray	Various	Too numerous to specify—see catalog #80, *Colmonoy Hard-Surfacing Alloys*	Detroit; Sante Fe Springs, Calif.; Houston, Tex.; Hicksville, N.Y.; and others		C	"Fusewelder" "Sprayer" "Wirespray"
Wall Colmonoy	Plasma	Various	Almost any metallic element	Detroit; Santa Fe Springs, Calif.; and others	San Antonio, Tex. (1968) Hicksville, N.Y. (1969) Houston, Tex. (1969)	C	
Norton Company	Flame and plasma spray	Metals and metal alloys	NOROC 202 (chrome oxide) NOROC 214 (aluminum oxide) NOROC 226 (zirconium oxide) NOROC 244 (magnesium zirconate)	50,000 ft² All flame and plasma spray		C	Materials developed for special application, grinding and finishing of coating also provided

189

TABLE 40 (Continued)

Company[a]	Coating Process	Basic Substrates	Coating System (Designation and Composition)		Coating Facilities (Size and Type)			Comments
					Operating	Planned (Date)	Process Status[b]	
Norton Company	Flame and plasma spray	Metals and metal alloys	NOROC 252 (magnesium aluminate) NOROC 271 (chromia-alumina) NOROC 310 (tungsten carbide) NOROC 320 (chromium carbide) NOROC 327 (alumina-titania) All metals and alloys, various oxide mixtures, metal carbides and borides					
Union Carbide (Coatings Service)	Detonation gun	All metals plus Ni-, Fe-, and Co-base alloys	LA-2 LA-7 LC-1B LW-1 LW-5 LC-5	Al_2O_3 Al_2O_3 Cr_2C_3+Ni-Cr WC + Co WC + Ni Cr_2O_3+Al_2O_3	Indianapolis, Ind. Charlotte, N.C. North Haven, Conn. Miami, Fla. Kansas City, Mo. Los Angeles, Calif.		C	Union Carbide supplies all UCAR coatings on a service basis—applied in their own plants around the world. Coating processes are not for sale or lease
Union Carbide (Coatings Service)	Plasma plating	All metals plus Ni-, Fe-, and Co-base alloys	LA-6 LC-2 LC-4 LZ-1 LM-6 LT-1 LW-6 LS-31	Al_2O_3 Cr_2C_3+Ni Cr Cr_2O_3 ZrO_2 Molybdenum Tantalum Tungsten Co-base alloy	Houston, Tex. Malton, Ont. Glossop, England Warwick, England Tokyo, Japan Geneva, Switzerland			
Union Carbide Corp.	Flame spray	Nickel-, cobalt-, and iron-base alloys	IAS aluminum		Maximum size part 48 in. high 26 in. diam		C	
Solar	Plasma spray	Ta alloy	W plus HfO_2		Unlimited		Laboratory	Protection for up to 30 min at 5000°F

[a]For full names of companies, see list on p. 166.
[b]C, Commercial

190

plasma-arc or flame-spray techniques. Plasma- and flame-spray coatings are in wide use as hard, wear-resistant materials applied to contacting or rubbing surfaces of iron-, nickel-, and cobalt-base alloys. Flame-spray coatings are deposited using an oxyacetylene or oxypropane gun-torch into which the coating material is fed as a powder or wire. The material being sprayed is heated to a highly plastic or molten state such that wetting and/or deformation interlocking of the depositing material is accomplished as the particles strike the substrate surface. Flame spraying is limited to materials with melting points below 4600°F. Oxides can be sprayed with metallic binders to produce cermet-type deposits.

Plasma spraying of coatings is employed for applying high-melting-point materials not amenable to flame spraying and where a protective environment is required during deposition. Argon, argon–hydrogen, nitrogen, and nitrogen–hydrogen gases are used in plasma spraying, and a wide range of metal and oxide powders can be successfully deposited on a wide range of metal substrates. Only powders are sprayed by the plasma-gun method.

A third, unique method of coating application in this category is the Linde detonation-gun technique. The metal and oxide powders are fired at high velocity from a detonation chamber, and upon impact with the substrate surface the plasticized particles produce a dense, hard deposit. This process is proprietary, and not for license or lease.

Table 40 lists the plasma and flame-spray processes available for applying high temperature coatings, based on the survey responses. A wide range of carbides, borides, aluminides, and oxides are available for coatings resistant to high-temperature wear, fretting erosion, abrasion, and corrosion. In general, the plasma- and flame-spray coatings have not exhibited corrosion-resistant properties equivalent to those of diffusion coatings, and certainly this approach to coating application would be more expensive than pack coating. However, the need for complex-alloy coatings not easily formed by diffusion methods lends impetus to expanding these techniques for forming high-temperature protective coatings.

The responses from the organizations queried regarding flame and plasma spraying were extremely general, citing "unlimited" capabilities for the application of "all metals, alloys and oxides" to "various" and "all" metals and alloys. It is concluded that methods and facilities for plasma and flame spraying are currently adequate and that improved equipment is under independent development in the industry. Since the composite material combinations that can be produced by these materials are virtually unlimited, the primary development effort should be directed at utilizing these methods for producing coatings not amenable to formation by diffusion-growth processes. Complex iron, nickel, and cobalt alloys, intermetallics, borides, etc., have demonstrated excellent corrosion resistance, but formation of these coatings by conventional methods has not been successful. The achievement of very dense, bonded coatings is a major requirement in this area.

6

Testing and Inspection

INTRODUCTION

Testing and inspection procedures are integral to every phase of product development. From the initial evaluation of a material concept to in-service evaluation of finished products, tests of various types are employed to obtain information for decision-making.

Testing technology is broad, interdisciplinary, and quite complex. A comprehensive review of the entire testing field was considered both impractical and unnecessary. Therefore, this section is restricted to a review of testing and inspection procedures for coatings used principally to protect refractory metal, superalloy, and graphite substrates from corrosion at service temperatures exceeding 1800°F. The imposition of this lower limit of use temperature eliminates most industrial coating applications except those involved with the processing of molten metals and ceramic materials. Conversely, there are many aerospace and propulsion applications that require materials that can function satisfactorily at temperatures exceeding 1800°F. Gas-turbine engines, lifting re-entry vehicle panels and leading edges, and rocket engines are the principal systems that make extensive use of high-temperature coatings. Consequently, the review of testing methods was oriented towards these applications.

CLASSIFICATION OF TESTS

For discussion purposes, testing and inspection procedures can be classified into four broad areas:

Standard characterization (property measurement)
Nondestructive testing (NDT)
Screening
Environmental simulation

Tests that could be classified as standard characterization are those that quantitatively measure such properties as hardness, density, thickness, and melting point. Tests that measure the coating–substrate properties of tensile and creep strength, elastic modulus, and fatigue life also fall into the category of property measurements. These latter tests are usually performed to determine if the physical presence of the coating or its application process affected the properties of the substrate.

Micrographic evaluation of sectioned samples to obtain a visual evaluation of coating thickness, microstructure diffusion zone depth, and flaw density can be classified as a standard characterization. Similarly, other evaluations involving sectioned samples, such as election microprobe analysis and microhardness traverses, would also fall in this category.

Nondestructive testing (NDT) includes any inspection technique that does not disturb or alter the physical or chemical characteristics of a coating system. It is most frequently employed in process control to measure coating thickness and invisible flaws such as debonded regions, internal cracks, inclusions, and thin spots. Many techniques are employed, including dye penetrants, ultrasonic emission or transmission, radiography, infrared emission, eddy current emission, and thermoelectric response. Nondestructive testing is a new and evolving technology that offers the promise of quick and inexpensive coating characterization.

Many of the techniques being developed are suitable for total inspection of production-line parts and in-service components.

Screening tests are simple and inexpensive procedures for obtaining comparative performance data. There is usually no attempt to simulate a complicated service environment; rather, fixed levels of certain critical environmental parameters, such as pressure, temperature, or atmospheric composition, that represent either peak or average values encountered in service are selected. Intelligent use of screening tests greatly reduces the probability that unsatisfactory materials systems will survive preliminary evaluations only to be subsequently disqualified in time-consuming and expensive proof tests.

Screening tests must be used properly; there are risks, some not so obvious, in their indiscriminate use. For example, it is not uncommon for coating systems to become performance-optimized to the screening test employed during their development. This can have serious consequences if critical environmental parameters are absent from the screening test. Moreover, performance criteria in screening tests are often arbitrarily established and may be more or less severe than necessary.

Perhaps the most difficult to classify are the simulated environmental tests. No laboratory test routine, regardless of its sophistication, can accurately represent the total environment in which high-temperature coatings are used. Several factors militate against realistic environmental simulation. In some applications, such as rocket propulsion and re-entry, duplication of the combined environment of pressure, temperature, atmospheric chemistry, and mechanical loading through multivariant control of all significant environmental parameters simply exceeds current technical resources. In long-term applications such as aircraft gas turbines, the environmental experiences of one engine lifetime comprise statistical events. The number of hot starts and temperature overshoots and quantities of abrasive and corrosive materials ingested by an engine are indeterminant and cannot be standardized for laboratory evaluation purposes.

Consequently, most simulated environmental tests represent a compromise between the actual use environment and that which can be achieved in the laboratory at a reasonable expenditure of manpower, time, and materials. Generally, simulated environmental tests involve a greater number of test variables than screening tests, and the resultant data are more quantitative.

TEST STANDARDS

The comprehensive evaluation of a coating system can be an expensive and time-consuming process. To expedite the identification of a manageable number of potentially effective coating systems from a large number of candidates, most laboratories depend upon published evaluation data from other organizations. Because there is an almost complete absence of conformity among testing laboratories, proper assessment of published data requires a thorough knowledge of testing procedures employed, equipment configuration, specimen geometry, failure criteria, and evaluation objectives. Unfortunately, such procedural details are rarely published. This lack of uniform test procedures can be traced back to the early years of coating development.

Testing procedures for high-temperature coatings evolved in an atmosphere of limited communication between laboratories. In the case of coated superalloys, testing precedents had been established by years of experience with uncoated superalloys among organizations not inclined to exchange ideas with their competitors.

Refractory-alloy and graphite coatings were oriented towards aerospace environments that could not be simulated in the laboratory. The compromises required to accomplish laboratory testing capability led to a proliferation of evaluation techniques. Some were well formulated and are still retained in the inventory of test procedures; others were quite unorthodox. For example, in the late 1950s, the ductility and impact resistance of coated molybdenum was demonstrated by tying strip specimens in knots, by shooting coated panels with a shotgun, or by simply striking a specimen with a ball peen hammer. Fortunately, none of these crude tests has ever achieved widespread acceptance.

As late as 1961, confusion regarding test interpretation led Krier[1] to state:

Procedures and techniques for evaluating coating systems are about as plentiful as the number of organizations working on the problem of protecting refractory metals. There are no standardized tests to which coating systems are subjected; consequently, comparison of results from different organizations is sometimes impossible, always difficult, and frequently dangerous and misleading.

The situation has since improved somewhat. In 1964, the Materials Advisory Board published a collection of recommended testing procedures for coated refractory alloy sheet.[2] It was intended that this publication introduce some semblance of uniformity in the testing of refractory metal coatings, and, to a degree, this objective was achieved. With rare exceptions, recent programs in the development and evaluation of coated refractory metal sheet have incorporated MAB recommendations in their testing procedures.

In September 1965, a new subcommittee of ASTM Committee C22 on Porcelain Enamel and Related Ceramic-Metal Systems was formed. The scope of this subcommittee, designated Subcommittee VI–High Temperature Protective Coatings, is as follows:

Formulation of methods of tests and specifications needed for the evaluation of high temperature protective coatings other than porcelain enamels. This will be accomplished by the review of existing standards and data from any recognized source and by the initiation of new investigations as required. Close coordination should be maintained with all other standardizing agencies.

The ultimate objective of Subcommittee VI is the establishment of test standards to cover all aspects of high-temperature coating evaluation. While the subcommittee has been quite active since its formation, no standard tests have been published to date.

TESTING METHODS

A review of the literature makes it apparent that for a given coating application the goals of a comprehensive evaluation are similar with different organizations. However, the means for achieving these goals through various test procedures will vary significantly. Regardless of the methods employed, there are two fundamental considerations that must be resolved in any high-temperature test—the attainment of uniform specimen heating and the accurate measurement of specimen temperature.

TEMPERATURE MEASUREMENT

Temperature measurement presents certain problems, particularly above 2500°F. Incompatibility between the thermocouple alloys and coatings precludes the use of contact thermocouples. If isothermal conditions can be established in the hot test zone, thermocouples strategically located near the specimen will provide fairly accurate measurements of specimen temperature.

Quite frequently, specimen temperatures are measured with radiation pyrometers. Unless such measurements are made in a blackbody cavity (emittance = 1.0), the emittance of the specimen surface must be known so that appropriate corrections can be applied to the pyrometer reading. According to Perkins and Packer,[3] a temperature error of 2 percent of the scale (50°F at 2500 °F) can result in an order of magnitude difference in coating life. It is quite important, therefore, to know the emittance of the specimen at the time pyrometer measurements are taken. It is not sufficient merely to estimate emittance (a fairly common practice) if truly accurate measurements are desired.

Two-color pyrometers have been developed that are allegedly insensitive to emittance effects. These pyrometers sense the radiation output of a high-temperature source at two different wavelengths and combine the two signals elec-

tronically to yield a voltage signal proportional to true specimen temperature. It is necessary that the temperature source radiate as a "gray body" which implies an emittance independent of wavelength. Unfortunately, most materials do not radiate as gray bodies; the extent of deviation from gray-body behavior within the interval of the two measurement wavelengths determines the sense and magnitude of two-color pyrometry errors.

HEATING METHODS

There are a variety of techniques for attaining uniform and reproducible hot test zones up to 3000°F in air. Electric resistance furnaces employing platinum wire, modified SiC, or modified $MoSi_2$ heating elements are commercially available at moderate cost and will operate quite effectively to approximately 3000°F. Gas-fired muffle furnaces and inductively heated susceptor furnaces of platinum, SiC, and commercial refractory composites such as Boride Z and JTA graphite have also been successfully used to 3000°F in air.

Self-resistance heating is sometimes employed where space or equipment limitations prevent the use of other methods. This technique is very simple and works quite well with specimens of uniform cross section. Opponents of this technique argue that undetected hot spots due to substrate or coating thickness variations can result in localized failures. It is also claimed that in most applications, coated hardware is heated by convection or radiation and internal heating thus represents an unnatural situation that affects diffusion behavior and thermally induced stresses. It is also argued that specimen temperatures measured at the surface are lower than substrate temperatures because of the insulating effect of the coating. Stress-rupture experiments conducted by Davis et al.[4] using both self-resistance and radiant-heating methods tend to confirm these objections.

Induction heating, in which the specimen is coupled directly to the induction coil, has all the inherent problems of self-resistance heating and is also susceptible to shape effects. The geometry of the heating coil must match the shape of the specimen to avoid localized overheating, particularly at edges and corners.

Special equipment is required for testing in air at temperatures above 3000°F. Platinum, SiC, $MoSi_2$, and the refractory composites that perform quite well at lower temperatures vaporize or are oxidized too rapidly. Low-cost refractories such as mullite and alumina soften and melt in the temperature range of 3000°–4000°F and are therefore unacceptable.

Impervious zirconia muffle furnaces are used quite extensively above 3000°F. The muffle tube is heated by a susceptor or a resistance element that surrounds the outside of the tube. The heating element, usually a graphite tube, is protected from oxidation by an outer enclosure that is continuously purged with an inert gas. Other methods for attaining specimen temperatures above 3000°F in air include torch heating (arc plasma jet or combustion torch), self-resistance heating, and direct induction heating.

Coatings for oxidation protection below 3000°F range in thickness from 1.0 to 3.0 mils and have reasonably high thermal conductivities. The insulating effect of these coatings is moderate and often can be neglected. In contrast, the current series of coatings under development for service at temperatures above 3000°F are much thicker (5 to 20 mils). Many of these coatings are oxides, or form oxides when exposed to high-temperature air. The insulating effect of these systems can be appreciable. It is not surprising, therefore, that Hill and Rausch[5] and others have found that the heating method may have a profound effect upon test results.

At equilibrium, specimens heated in an isothermal cavity are at constant temperature throughout their cross section. Coated specimens that are torch-heated on one side only will equilibrate at a surface temperature dependent upon the heat flux through the specimen. The thermal gradient through a torch-heated coating can reach several hundred degrees. Therefore, at the same coating surface temperature, substrate–coating diffusional reactions can be much slower in torch tests than in isothermal furnace tests.

Some coating systems, such as the Hf-Ta alloys, are thought to experience an exothermic reaction at the substrate/coating interface. In isothermal furnace tests, the interface temperature may rise significantly as a result of the insulating effect of the coating. This exothermic effect may not be as evident in torch tests because of heat transfer to the cool side of the specimen.

The problems inherent in internal heating of the substrate by self-resistance or induction techniques are magnified by the presence of thick, insulative coatings. It is impossible to measure the cross-section temperature distributions using either heating technique; therefore, both methods are impractical for all but the most crude oxidation tests.

STANDARD CHARACTERIZATION

MICROGRAPHIC ANALYSIS

Micrographic examination is a very useful destructive technique for evaluating the physical characteristics of coated specimens in the "as received" condition and after various environmental exposures. To achieve full benefit of micrographic analysis, proper specimen preparation is essential. This is a slow and painstaking operation because coatings are easily damaged by rough grinding and cutting operations

and because the coating and substrate polish at different rates because of their difference in hardness.

It is particularly difficult to preserve the outer edge of a coating during polishing because of rounding and pullout. Guard rings, cushion mounting, and metal plating have been successfully used to minimize this problem.

Carefully prepared metallographic specimens can be employed to measure coating thickness and the quality of coverage (e.g., the presence of cracks, voids, and irregularities in thickness). Reactions between the coating and substrate, which result in the formation of diffusion zones, can be characterized. Substrate dimensional change and microstructural changes can also be determined.

By sectioning through a failure site, the extent of corrosive attack, including interstitial contamination (in the case of metals), can be readily evaluated by visual examination.

Frequently, visual analysis is supplemented with microhardness measurements of the coating, substrate, and at the interface region. Such measurements can provide information regarding diffusional migration of coating elements and corrosives into metallic substrates.

Microprobe analysis can be considered a sophisticated extension of micrographic analysis. This technique is often employed to identify the composition of complex coatings that form layered structures. It is also useful in establishing the kinetics of diffusion in a coated metal system.

BEND TESTS

Bend tests are employed to evaluate the transition temperature and ductility of coated refractory metals. Typically, such a test involves the 90° free-bend deformation of a rectangular specimen in a punch-and-die fixture.

Ductile–brittle transition temperature is generally defined as the lowest temperature at which a ductile 90° bend can be formed about a specified radius. Although this is a qualitative test, it is quite useful in assessing the effect of a coating, or coating process, on the ductile–brittle behavior of the substrate. Frequently, when the "as coated" alloy exhibits 90° bend ductility at room temperature, bend tests are employed in the evaluation of retained ductility after oxidation exposure. In this manner, loss of ductility due to inward diffusion of coating elements, "leakage" of atmosphere contaminants, and other subtle effects not readily detectable by visual inspection can be characterized as a function of the time and temperature of exposure.

Specifications for bend testing were included in the Materials Advisory Board Report MAB-201.[2] Since the publication of this report, every organization in the mainstream of coating development or evaluation has adopted the procedure. Of all the tests specified in MAB-201, the bend test has received the most universal acceptance. The only significant deviation from specifications noted in the review of current work was the ram loading rate. While MAB-201 calls for a ram loading rate of 1 in./min, some organizations use a slower rate—either because of equipment limitations or to achieve better operator control of the test.

TENSILE TESTING

Specifications for room-temperature tensile testing of coated alloys as outlined in MAB-201 and in appropriate ASTM standards are being followed by most laboratories. At temperatures above 2400°F, however, there are no standards. Consequently, test procedures vary widely among laboratories.

The problem of measuring strain at temperatures above 2400°F is especially acute. Conventional clamp-on extensometers of nickel- and cobalt-base alloys that perform well at lower temperatures lack sufficient creep resistance and hot strength. Mechanical extensometers constructed from ceramics have been used with limited success. Problems arise from the fragility of such devices and the loss of accuracy due to sag and distortion of the strain transmission mechanism.

Metals and ceramics in contact with a coating may produce undesirable high-temperature reactions that lead to premature failure. To alleviate this problem, optical tracking devices have been used for strain measurements. This equipment is expensive and requires frequent maintenance and calibration. Moreover, optical-path distortion through sight ports, sometimes due to target "shimmer," can result in significant measurement errors.

CREEP-RUPTURE TESTING

The problems with strain measurement in high-temperature tensile testing are compounded in creep testing of coated refractory alloys because of the longer duration of these tests.

Designers require creep data in the range of 0.1 to 1 percent extension. Higher levels of creep-strain and stress-rupture data are useful only in making gross comparisons among different coating systems. Mechanical extensometry with sufficient precision and accuracy to meet design data requirements presents a multitude of technical problems. Oxidation failures arising from reactions between the coating and extensometer grips are more likely to occur as exposure time increases. Sag or creep of hot-zone components is frequently a source of serious error with these devices. As previously noted, optical strain measurements, either automatic or manual, circumvent the problems of mechanical systems, but this method is not entirely free of error-producing problems. To date, no completely satisfactory methods for high-temperature strain measurement have been devised.

Most creep data are now derived from deflection of the load train measured outside the specimen hot zone. While this method may be convenient for estimating creep extension, it is prone to serious error and is not compatible with design-data accuracy requirements. The analysis of creep data may be complicated in long-term tests by premature failure resulting from coating failure and subsequent catastrophic oxidation of the substrate. Thus, short-duration tests will yield a clearer picture of creep behavior, since long-term tests tend to become a combined test of creep resistance, strain tolerance, and oxidation resistance.

Because of their inherent oxidation resistance and the lower test temperatures involved, coated superalloys do not pose as serious a problem. Nevertheless, the interacting influences of oxidation and diffusion must be considered in long term exposures of coated superalloys.

FATIGUE

Most fatigue testing of coated alloys is performed on sheet and foil-gauge material at room temperature with tension-tension loading. The number of specimens tested is rarely sufficient to establish the statistical aspects of fatigue behavior. Usually, the data are sufficient to determine whether the coating has had any effect upon the fatigue characteristics of the substrate; however, a quantitative determination of this effect is generally not possible. Some programs are carried a step farther, and rudimentary S-N diagrams are obtained, but not with sufficient data points to perform a thorough statistical analysis.

NONDESTRUCTIVE EVALUATION OF HIGH-TEMPERATURE COATINGS

BACKGROUND

Destructive test methods may have considerable value for understanding and characterizing a type of coating, but it is apparent that these methods cannot be applied directly to the actual components or systems that will be used in a specific mission. It is also apparent that sampling techniques or the use of test tabs will not accurately represent all of the important variables (or their extent) that may be present in coated hardware. Indeed, it has been demonstrated on numerous occasions that components or specimens may fail catastrophically, while destructively evaluated test tabs from the same coating batch did not give any reason for suspecting this type of behavior. Further, the coating processes now being used most extensively may yield components with local areas of variability that may lead to premature failure.

Therefore, nondestructive test (NDT) techniques are essential for evaluating high-temperature coatings because of the types of processes used in applying the coatings and the possibility of the catastrophic failure of a component or an entire system due to a localized imperfection.

In general, the reliability problems associated with high-temperature coatings are a result of the following inadequacies[26]:

Insufficient understanding of failure modes and mechanisms

Inadequate destructive test technology for properly and thoroughly characterizing coated systems under service-related conditions

Lack of nondestructive tests for evaluating materials and process variables known to influence service life and upon which to base reliable life predictions and quality acceptance decisions

To properly and reliably utilize any coating, these inadequacies must be eliminated. This may be done during the development phases by following a procedure such as that broadly outlined below:

1. Obtain specimens of the candidate materials that represent defined processes and raw materials.

2. Sort out or screen specimens for compositional or structural variability and flaws using economical, rapid, and exhaustive NDT methods.

3. Submit specimens representing extremes of variability to environmental simulation to determine which fail prematurely and which succeed.

4. Analyze results of environmental tests to prove the significance of NDT results and to characterize failure-triggering variables. Submit counterpart specimens to metallographic analysis to identify and characterize the relevant variables.

5. Revise and improve the NDT techniques to further characterize important variables, to become more specific, or to become more sensitive to the variables associated with failure.

6. Improve or adjust manufacturing processes to eliminate or reduce variability.

This procedure was the basis for an Air Force Materials Laboratory program that has successfully identified relevant variables for several coatings and has resulted in the development of effective nondestructive test methods.[27-29] It was concluded from this program that each type of coating and coating process had its own specific set of variables (Table 47). It was also found that in addition to significant within-batch variability, batch-to-batch variability also existed. The NDT was also of considerable value in predicting actual failure sites and in observing changes in the coating and substrate as a function of the exposure time in the high-temperature environment. From the latter, it was possible to

correlate NDT measurements with growth of the coating by diffusion and oxidation and to note the time at which coating growth changed from parabolic to linear, indicating impending failure.

A brief description of each of the successfully applied NDT techniques is given below, along with the relevant variables detected and monitored by each. It should be borne in mind that this list is not complete and, as stated above, that each type of coating has its own specific set of variables and that therefore, NDT techniques that are useful for evaluating one type of coating may have little value on another.

NONDESTRUCTIVE TEST TECHNIQUES

RADIOGRAPHY

An x-ray or gamma-ray source is used in conjunction with a sensitized film. Radiation passing through the component is absorbed as a function of the density or chemistry variations within the component. Local coating areas of varying composition and structure have been detected in some coatings with this method. Variations, such as pits and stringers, also have been seen in substrate materials. Two distinct disadvantages of this technique are low sensitivity to coating variations due to the very thin cross section of coating (0.002 in.–0.005 in.) and the superposition of the top coating, substrate, and bottom coating images. It is not possible, without extensive analysis, to separate the effects from each of these zones.

BACKSCATTER RADIOGRAPHY

An x-ray or gamma-ray source is used to stimulate secondary x-ray or electron emission.[30] In this case, the film is placed on the same side of the specimen as the radiation source. As the radiation passes through the film, it causes a slight exposure but also stimulates the emission of electrons from the specimen surface and secondary x-rays from both the surface and subsurface areas. These secondary emissions are low in energy but efficient in their interaction with the film emulsion, causing strong exposure of the film in accordance with the intensity of the emission.

Local areas of coating variability in chemistry and structure can be seen readily. No superposition of images takes place, and only a single coating side is evaluated at any time without interfering images from the substrate or lower coating side. This technique is easy to apply and very sensitive to the variables mentioned above. A special vacuum cassette is necessary to place the x-ray film in close proximity to the specimen surface. Failure to use a vacuum cassette results in fuzzy images of local coating variations. Local areas that are

TABLE 47   Use of NDT to Detect, Measure, and Monitor Coating Variables

Coating System	Failure-Influencing Variable	Sensitive NDT Technique
Cr–Ti–Si on Cb 752 alloy	Chromium-poor areas	X-ray backscatter, dye penetrant, thermoelectric
Batch 1	Chromium-poor areas	Above techniques
Batch 2	Porosity	Dye penetrant, thermoelectric
	Thin coating	Eddy current, micrometer
W-3	Delaminated alloy	Dye penetrant, microscope
TZM alloy	Coating thickness	Eddy current, micrometer
R512A B-66 alloy	Coating thickness on edge	Thermoelectric
R512E Cb 752 alloy	Coating thickness on edge	Thermoelectric
R512C Tantalum alloy	Coating thickness on edge and surface near edge	Thermoelectric
R512A D-43 alloy	Coating thickness	Eddy current, thermoelectric
	Brazed area structure or chemistry	No evaluation done on these problem areas on this program
Aluminide (701) Inco 713C	Thickness of coating (variations at edges)	Thermoelectric test
Aluminide on cobalt alloys	Thickness, chemistry variations	Thermoelectric test
Zinc–iron 5508 stainless steel	Thickness of coating	Thermoelectric test

chromium-poor can be detected readily in the Cr–Ti–Si-type coatings on columbium substrates.

EDDY CURRENT

An electrical field is induced into the coating using specially designed coil and electronic apparatus. The interaction of the electrical field and the coating material results in changes in phase or amplitude of the field. These changes are converted into appropriate electrical signals and are read out on a meter or oscilloscope screen. Changes in meter reading or oscilloscope pattern can be correlated with variations in coating thickness, composition (conductivity), or structure. The selection of a proper frequency is necessary to assure that, for maximum sensitivity, the field is projected into the coating and only slightly into the underlying substrate material. In general, for silicide coatings ranging from 0.002–0.008 in. thickness, a test frequency of 2.0–8.0 MHz is adequate. As coating thicknesses increase beyond this range, lower frequencies should be used. Excellent commercial equipment is available with numerous probe-coil shapes and sizes available. This method has been used to sensitively

measure the thickness of silicide coatings on refractory alloys as a function of thermal gradients in a coating pack and to indicate batch-to-batch variations. It is possible to apply this method to the evaluation of coatings as they become oxidized and diffused because of high-temperature oxidation and to observe incipient failure as the oxidation growth rate changes from parabolic to linear. A disadvantage of this method is its lack of adequate resolution on edges and at areas of severe curvature and irregular profile.

This technique is being applied at several facilities for inspecting finished components as well as for evaluating "green" coatings of the slurry type before they are sintered or fused.

THERMOELECTRIC

A heated tip of metal, such as copper, tungsten, or stainless steel, is placed in contact with the coating surface, and an electromotive force is generated as a function of the couple which is formed. The amplitude and sign (positive or negative emf) is related to the chemical composition of the coating at the local area of contact. The chemistry of the coating is directly related to processing variables and coating thickness. This technique has sensitively detected areas of thin coatings on many substrates and has been used successfully to predict failure sites. It has been of special value in the inspection of slurry-applied coatings where thin areas at edges are the most important variables that lead to premature failure. Local areas of compositional variation (chromium-poor areas) in Cr–Ti–Si pack-cementation coatings on refractory alloys have been detected by this technique. In addition, aluminide coatings on nickel- and cobalt-base superalloys have been successfully inspected for coating-thickness variations using this technique. It is especially useful for evaluating edge areas, interior holes, and other constricted areas where it is necessary to have a very small probe and where other techniques cannot be applied. A disadvantage of this technique is that the coating must be electrically conductive. Oxidized coatings cannot be evaluated in this manner.

DYE PENETRANT

A dye, carried in a penetrating oil, is applied to materials and allowed to seep into fine cracks or porosity where it is held by capillary action. When the surface is flushed with an appropriate solvent or detergent and is treated with a finely divided developing powder, the entrapped dye gradually bleeds out or is pulled out of cracks and pores by the blotting action of the developer. The dye, delineating the discontinuity, can then be observed against the developer background, and cracks, pores, and other such defects of sufficient size can be readily detected. Dye material can either be a bright color or a fluorescing material activated by an ultraviolet light. Areas of delamination in a molybdenum alloy have been detected with this technique, and porosity and cracking have been noted in some coatings. This technique is very simple to apply and very sensitive to coating discontinuities. However, it contaminates the surface with petroleum products and dyes, and it is very difficult to remove all traces from the specimens.

VISUAL AND OPTICAL

Use of the eyes to evaluate materials in a qualitative fashion should not be denigrated. Thorough documentation of visual observations by use of photographs and sketches can be of considerable value. Aids to visual inspection include magnifiers, microscopes, ultraviolet light, and gloss meters.

DEVELOPMENT OF SPECIFICATIONS FOR RELIABILITY AND QUALITY ASSURANCE

Nondestructive test techniques have been developed that have considerable value and should be applied in understanding the role of processing variables; in evaluating and controlling processes; in inspecting and insuring quality of hardware; in monitoring the effects of high-temperature exposure to detect conditions of incipient failure; and in evaluating refurbished and repaired hardware components.

During the development phase of any new coating system or process, large numbers of specimens are usually processed and destructively evaluated. The information gained during this phase is normally used in characterizing the coatings and processes and eventually leads to the preparation of specifications. It is during this early phase that nondestructive tests should also be developed and utilized for characterizing the coatings and for detecting variability normally associated with processing parameters. Nondestructive test techniques can often be utilized with considerable effectiveness at this stage for fast sorting and screening of large numbers of specimens so that appropriate units containing representative ranges of variability can be selected for the destructive analysis. This procedure can ultimately result in a considerable cost savings because it reduces the number of specimens needed to fully and accurately characterize the coatings and processes.

Such things in the processes as reused coating materials, thermal gradients in the pack, coating continuity, and composition at each stage in multistage coating processes should be monitored using NDT. In addition, processes developed for refurbishing and repairing materials—e.g., coating removal, recoating, heat treatment, weldments, or filler metal—should be checked by NDT.

To obtain total quality assurance, NDT inspection of hardware for local coating variations should be accomplished, with process specifications being written for each NDT process. These should outline all of the relevant variables of interest and should detail each step of the technique to be used for detecting these variables. Limits of acceptance can then be specified, based on realistic environmental and destructive test results.

Examples of NDT procedures used in specific systems for detecting various conditions are shown in Table 47.

SCREENING TESTS

FURNACE OXIDATION TESTING

Laboratory tests of oxidation resistance are integral to every development and evaluation program. The simplest of these is the so-called cyclic oxidation test for coated metals. In this test, groups of rectangular coated coupons (typically 1.0 in. × 0.5 in.) are furnace-heated in slowly moving air. The number of specimens in a test group will vary. Usually, at least 3 and sometimes as many as 50 specimens are involved. At periodic intervals, the specimens are removed from the furnace and inspected for failure. At the first evidence of substrate oxidation, the specimen is considered failed and retired from test. Testing is continued to a predetermined time limit or until the entire lot of specimens has failed.

In the evaluation of refractory alloy coatings, cyclic oxidation tests play a very important role, and in many programs this is the only procedure used to measure oxidation resistance. Cyclic oxidation tests of refractory-alloy coatings are usually of relatively short duration (less than 200 hr) with frequent specimen inspection, usually at 1- or 2-hour intervals. Performance in the temperature range of 2000° to 3000°F is emphasized. Exposures below 2000°F are usually conducted to check for low-temperature sensitivity, and tests are not normally run to coating failure.

Noncyclic or continuous furnace testing of coated refractory alloys is not a common practice. Generally, when this technique is employed, long-term exposures are involved. Tests are usually run for a predetermined time and may involve continuous weight-change measurements.

Furnace oxidation tests (continuous and cyclic) of coated superalloys and refractory alloys for gas-turbine applications are employed primarily for screening purposes in coating development programs. Test temperatures rarely exceed 2400°F; thus, problems with temperature measurement and control are minimal. Tests are frequently of long duration (600–1000 hr). Consequently, in cyclic exposures, the use of short-duration test cycles with frequent specimen inspection becomes impractical. Generally, in a long test, specimens are examined at 1- or 2-hr intervals for the first 10–20 hr of test and then at longer intervals, perhaps 24 hr, as the test progresses. Specimens are usually tested in lots of 2 to 5, and so the oxidation data are qualitative and not suitable for statistical analysis.

Cyclic oxidation test data usually exhibit a large amount of scatter, and it is not unusual for the minimum and maximum specimen lives to differ by a factor of 10 or more. Rarely is there good correlation of results between laboratories, even when specimens from the same coating batch are tested.

Undoubtedly, a number of factors contribute to the variability of these data. The test itself is a measure of the weakest site on the specimen, usually in the form of a thin spot, crack, or chip. Early failure could also be the result of a poorly formed coating over substrate imperfections, such as delamination, a surface contaminant, or an improperly finished edge. The density of life-shortening weak sites in a test specimen is a function of specimen size. It follows, therefore, that tests performed on different-sized specimens cannot be compared without normalizing to account for size difference. Another sensitive factor, which probably contributes to interlaboratory disparity of results, is the interpretation of specimen failure.

Failure detection of the molybdenum alloys is relatively

simple because the oxide (MoO_3) is volatile, and once the coating is penetrated, large substrate cavities form rapidly at test temperatures. In contrast, the oxides of columbium and tantalum are solid and frequently form various colored reaction products with coating constituents. Normal coating surface reactions can be confused with oxide growth through pinhole coating defects. Even when substrate oxidation is initiated and is recognized as such by the test operator, it is a matter of judgment as to when this condition becomes serious enough to constitute a failure. Quite often, suspected failure sites become evident many cycles before there is a positive visual identification of failure.

Superalloys and chromium-base alloys do not oxidize at catastrophic rates upon coating penetration; visual examination must, therefore, be supplemented with careful micrographic analysis to reveal more subtle forms of corrosion, such as intergranular and internal oxidation. In the case of chromium-base alloys, embrittlement due to nitrogen leakage through the coating may be the life-limiting factor. Bend ductility tests, micrographic examinations, and microhardness traverses of exposed specimens are employed to evaluate this aspect of coating performance.

The duration of the test cycle has been shown to affect coating life. This result is apparently due to the thermal expansion mismatch effects that crack the coating.

Another factor that can influence cyclic oxidation behavior is the choice of support media. At temperatures above 2500°F, a reaction between the coating and its support at the points of contact frequently cause problems. Materials commonly used to support specimens include ZrO_2, Al_2O_3, SiO_2, and $MoSi_2$. While most support-media reactions result in catastrophic destruction of the specimen, it is quite likely that there are instances where such reactions are more subtle and simply act to shorten coating life at the points of contact.

The importance of factors such as specimen size, test cycle duration, failure identification, and support media have not been clearly defined. It is reasonable to assume that these factors and others influence cyclic oxidation behavior. Studies to measure the relative importance of these variables individually and in combination are a necessary prerequisite for developing reasonable standards for this test.

In this respect, the specifications for cyclic oxidation testing outlined in MAB-201 probably allow too much latitude in the selection of specimen size. Furthermore, it is felt that more definitive specifications for failure identification and data analysis are necessary if interlaboratory consistency of results is to be achieved.

TORCH TESTING

Closely related to the furnace cyclic oxidation test is the so-called cyclic torch test. In this test, relatively large flat specimens (typically 1.5 in. × 1.5 in.) are heated by the combustion flame of an oxyacetylene, oxyhydrogen, or oxypropane torch. Occasionally, arc plasma jets and rocket engines are also employed in torch testing.

Specimens are usually sized such that the isothermal hot zone produced by the torch exceeds 0.5 in. in diameter but is contained within the boundaries of the specimen itself. In effect, edges are excluded in this test.

Proponents of this test argue that for applications where there are no exposed edges and where the hot air stream is moving (hot-gas ducting would be one example of this), the torch test is a more realistic screening procedure than the furnace cyclic-oxidation test. On the other hand, the test is frequently criticized because the number of edge-free applications is rather limited. Questions have been raised regarding lateral heat transfer effects due to variations in gripping techniques and regarding the difficulty of maintaining an oxidizing combustion environment. Another problem is temperature measurement and control, which is always accomplished by emittance-sensitive radiation pyrometry to avoid the difficulties of contact reactions with attached couples.

A sizable investment in torches, timing devices, and safety equipment is required for high-volume torch testing, whereas furnace testing requires a substantially lower initial investment. For this reason, the popularity of torch testing has declined measurably in the past several years. Screening tests to measure oxidation resistance are now almost exclusively some form of furnace test.

LOW-PRESSURE OXIDATION

Work by Perkins[6] and others in the early 1960's revealed that silicide coatings are susceptible to low-pressure failures at elevated temperature. The failure mechanism was identified as volatilization of silicon monoxide at low pressure that prevented the formation of a stable protective glassy oxide on the coating surface. Publication of these findings initiated a flurry of activity in low-pressure oxidation testing.

Testing procedures do not vary significantly among laboratories. Usually, a muffle-tube furnace is employed with one end of the tube connected to a vacuum pump and the other end to a control valve. Pressure is controlled in the muffle tube by operating the vacuum pump at maximum capacity and adjusting the control valve until air leakage through the tube equilibrates at the proper pressure level.

Low-pressure oxidation tests are frequently noncyclic and are conducted for specified periods of time. Analyses of test results are oriented toward the thermodynamic behavior of the coating system. This procedure is in contrast to cyclic oxidation tests at atmospheric pressure, which emphasize the defect-controlled protective behavior of coat-

ings. Results of low-pressure tests are often specified in terms of a "go, no go" boundary of pressure and temperature.

STEP-DOWN TESTS

Silicide coatings with good high-temperature oxidation resistance have been observed to fail prematurely when subjected to cycling at progressively lower temperatures. In the step-down test, specimens are heated by a torch or a furnace for 30 min at each of three decreasing temperatures and visually examined for failure at the end of each three-temperature cycle. Generally, each cycle is initiated by a 30-min exposure at the maximum use temperature of the coating. This stage is followed by 30 min at 2200°F and 30 min at 1400°F. Tests are usually terminated after five complete cycles.

The step-down oxidation test has not achieved widespread use among laboratories—probably because the results are difficult to relate, except in a very qualitative sense, to potential coating applications. Data that have been published[7] show this test to be a rather severe and damaging experience for some coatings that perform quite well in conventional cyclic oxidation exposures. In applications where thermal cycling is anticipated, the step-down oxidation test is a convenient means for identifying potential coating problems.

BURNER RIG TESTS

A test procedure frequently employed in the evaluation of the oxidation-erosion resistance of coated alloys for gas-turbine applications is the so-called burner rig test.[8–13] In this test, wedge- or airfoil-shaped specimens are mounted in clusters of 3 to 16 on a rotating fixture and heated by a high-velocity combustor nozzle (Mach 0.5 to 0.8). Most burner-rig facilities are fueled with JP-4 or JP-5 at air-fuel ratios of 18 to 30; air-kerosene, air–natural gas, and oxypropane are also used. Specimen design varies from simple wedge shapes machined from bar stock to rather complicated airfoil shapes representative of blade and vane configurations.

Since the specimens are larger than the hot gas stream, thermal gradients exist in the specimen. In the simplest burner rig tests, this temperature gradient is ignored, and the test is controlled as a function of highest pyrometer temperature measured at any location on the specimen. When more refined means of temperature measurement and control are desired, the specimen temperature gradient is characterized with calibration specimens containing internal hot hardness pins or thermocouples. Temperature-sensitive paints, radiation pyrometry, external spot-welded thermocouples, and infrared photography have also been employed for this purpose.

Temperature regulation during test is usually achieved by varying fuel-consumption rate as a function of specimen temperature, measured either by radiation pyrometry or an imbedded thermocouple. Neither method is completely satisfactory; coating-emittance changes and reflection from the hot gas stream can lead to significant errors in pyrometer readings. The slip ring configuration required to transfer thermocouple signals from the rotating specimen fixture to recording equipment can be troublesome and, unless properly designed and maintained, is subject to significant measurement errors. Of the two, the internal thermocouple technique is believed to be more likely to give satisfactory results.

The primary reason for specimen rotation is to achieve a uniform environmental exposure of the specimens. In practice, specimen rotational speed varies from 12 to 1,850 rpm, with the most common value being 1,725 rpm. The selection of specimen-fixture rotational speed is probably a function of electric drive motor availability and is not dictated by any other technical consideration.

Oxidation-erosion testing is normally conducted at specimen temperatures ranging from 1800° to 2400°F. Tests run from 100 to 600 hr in total duration, with periodic cooling for weighing and inspection of specimens at 20–30-hr intervals. Coating performance is judged by visual appearance, weight change, and micrographic analysis. The formation of visible substrate oxide, cracks that extend into the substrate, coating spall, excessive diffusion of the coating into the substrate, significant weight change, and general microstructural appearance are factors that are usually considered in identifying a coating failure.

HOT-CORROSION TESTING

Burner-rig testing is also employed in evaluating the resistance of superalloy coatings to corrosives other than oxygen. Trace elements of vanadium, lead, carbon, and sulfur in fuels are known to be potential sources of corrosion in gas-turbine engines. Corrosion by sulfur compounds has received particular attention in recent years, and almost all current hot-corrosion evaluations are directed towards this problem. Sulfidation is most serious with lower chromium-content superalloys that operate in marine environments where ingested sea salt reacts with the sulfur to form sodium sulfate.

Burner rigs have been adapted for sulfidation testing by providing for the injection of artificial seawater in concentrations on the order of 6–10 ppm.[9,14,15] Testing procedures are similar to those employed in oxidation-erosion evaluations. Visual appearance, weight change, and micrographic examination are the principal methods of analysis.

The duration of test cycles varies among laboratories from 10 min to 20 hr.

Allison Division of General Motors Corporation[16] employs a somewhat different procedure for sulfidation testing. Specimens consisting of actual turbine blades are mounted in clusters of 16 on a spindle and rotated at 1800 rpm in a natural-gas-fired furnace. At periodic intervals, the spindle is cooled with an aspirated spray of a 1 percent solution of sodium sulfate in deionized water. The test cycle consists of 1.5 min of heating and a 0.5-min cooling spray. The test is normally run for 500 cycles.

Phillips Petroleum Company[17] employs a fairly elaborate flame tunnel for hot-corrosion testing. The tunnel can accommodate six stationary specimens, 2.38 in. × 0.5 in. × 0.12 in., oriented at 45° to the direction of hot gas flow. The tunnel is operated on ASTM aviation type A turbine fuel at various air–fuel rations from 50 to 75. The fuel is essentially sulfur-free, sulfur content is adjusted to desired levels by additions to ditertiary butyl disulfide. Sea-salt simulation is achieved by additions of aqueous solutions of synthetic seawater (ASTM D665) to the gas stream in concentrations of 2 to 10 ppm. The facility can be operated over a pressure range from 5 to 15 atm; gas velocity and temperature at the specimen location can be varied from 50 to 664 fps and 1200° to 2200°F, respectively.

The work cited in Reference 16 dealt with uncoated superalloys, with the principal evaluation criteria being weight change and microstructural appearance. Presumably, coated superalloys would receive a similar evaluation, with emphasis on the preservation of coating continuity as evidenced by surface appearance.

THERMAL-FATIGUE TESTING

Burner rigs are also employed to evaluate the so-called thermal-fatigue characteristics of coatings for superalloys and refractory alloys. This test is quite similar to the oxidation-erosion test, except that rather short cyclic exposures, typically consisting of 1 min at peak temperature and 30 sec of air-blast cooling, are used.

Burner rig testing is just one of many ways that the thermal-fatigue characteristics of blade and vane materials are evaluated. At least four other techniques are employed, and widespread disagreement exists among laboratories regarding the merits of these various approaches.

Glenny and Taylor[18] employs a pair of fluidized beds for the cyclic heating and cooling of tapered discs. Cycle time varies from 1 to 10 min; the maximum and minimum temperatures can be varied, and either air or inert gases can be used for heating and cooling. Thermal-fatigue resistance is measured in terms of the number of cycles required to produce a crack of arbitrarily specified dimensions.

General Electric Company[19] uses a rotary table upon which wedge-shaped specimens are mounted. Four burners and two cooling stations are located around the periphery of the table. Specimens are indexed into the heating and cooling positions for a specified period of time, usually 10 sec. The burners are adjusted to produce a predetermined specimen temperature–time history. Thermal-fatigue resistance is rated in terms of the location, frequency, and length of cracks formed and by the number of thermal cycles required to induce this cracking.

Solar Division of International Harvester Co.[19] has developed a "hole-in-plate" thermal-fatigue test for coated superalloy sheet. As the name implies, this test employs a square plate specimen, usually 3 in. $\times$ 3 in. with a 0.5-in.-diam hole in the center. The specimen is locally heated around the periphery of the hole for 1.5 or 2 min with a torch and then cooled for 0.5 to 2 min with an air blast. Thermal-fatigue resistance is measured in terms of the number of cycles to crack initiation as a function of the peak specimen temperature. Solar claims good reproducibility for this test and cites its low cost of setup and operation as one major advantage.

Wall Colmonoy Corporation[19] is developing a simple, low-cost thermal-fatigue test in which up to 36 specimens (3 in. $\times$ 3 in. $\times$ 0.0625 in.) are mounted on a spindle and rotated past a burner. Test results are specified as the number of cycles to the first crack and/or as crack length as a function of the number of test cycles.

Glenny[20] has summarized the salient features of more than thirty other approaches to the measurement of thermal fatigue. In general, these tests are similar to those discussed in this report.

DAMAGE TOLERANCE

A performance criterion that is frequently evaluated is the protective character of a coated part that has been damaged in some manner, such as by mechanical deformation, ballistic impact, or abrasion. The tensile strain tolerance of coated refractory-alloy sheet can be evaluated, according to MAB-201, by prestraining tensile-type specimens at various temperatures up to 2000°F, followed by oxidation testing at either 1800° or 2600°F. Strain tolerance is specified in terms of the maximum allowable prestrain for 2-hr life at the test temperature. In similar tests, Pratt & Whitney Aircraft Company[21] has evaluated the strain tolerance of coated columbium bar stock. Solar[22] has used small coupons, prestrained by three-point bending and then furnace-tested, to evaluate coating strain tolerance.

Resistance to impact damage by ingested foreign objects is essential for gas-turbine engine materials. There are several techniques for evaluating the ballistic-impact resistance

of coating. The simplest of these is the so-called "Z" stamp test employed by General Electric Company and others to measure the adherence and ductility of a coating. A metal die is used to imprint a letter in the coating. It is claimed that if this can be accomplished without spalling damage, the coating will have good impact resistance. Test devices that employ a calibrated weight with a specific indentor geometry dropped from various heights are employed by Chromalloy Corporation[12] and Allison Division of General Motors Corporation[11] to obtain semiquantitative measurements of adherence impact resistance.

Adherence is also measured by grit-blasting the coating surface at room temperature with fine glass beads, silicon carbide, or alumina. This is followed by a high-temperature furnace exposure in air to develop indications of debonding such as blistering, spalling, or substrate oxidation.

A more direct measure of ballistic impact resistance is obtained by Pratt & Whitney[9] and others by firing metal pellets of specified size and weight at heated specimens. Tests are conducted at several levels of pellet velocity and are followed by oxidation tests to evaluate the extent of impact damage.

HOT-ABRASION TESTING

Abrasion damage to coatings can occur in gas turbines from three sources: the erosive action of carbon particles resulting from incomplete combustion, the ingestion of fine dust and sand, and fretting wear at the point of attachment of the turbine blades and wheel.

Erosion by fine particle abrasion can be evaluated in laboratory tests by heating coated specimens in a low-velocity flame tunnel or a furnace and impinging them with a high-velocity stream of fine Al_2O_3 or amorphous hard carbon particles until coating penetration occurs.

Fretting wear is usually evaluated at room or elevated temperatures in special test devices that produce a periodic displacement (vibratory or rotary) between two specimens preloaded against one another to some specified level. Fretting-wear resistance is related to the level of specimen damage produced in a specified period of time.

SIMULATED ENVIRONMENTAL TESTS

RE-ENTRY SIMULATION

A variety of techniques has been employed to simulate the re-entry environment. Generally, three test variables are involved: temperature, time, and pressure. Potentially important environmental inputs, such as high mass flow, and mechanical inputs, such as strain and vibration, are usually not considered.

Sylvania Electric Products, Inc.[23] employs a very simple technique in which test coupons are cycled from 800° to 2500°F over a 1-hr period. The temperature profile is a symmetrical bell-shaped curve. Pressure can be varied in the test chamber to simulate a typical re-entry. In the Solar facility,[24] specimens stressed by dead-weight loading are subjected to an asymmetrical temperature profile representative of a re-entry temperature–time history. Pressure is held constant during these tests at levels ranging from 0.5 to 760 Torr with a bleed-valve–vacuum-pump arrangement. The test chamber is constructed so that a flow of high-velocity air passes over the specimen during test.

McDonnell Aircraft[25] is evaluating coated refractory alloys for glide–re-entry applications using a very sophisticated facility that can simultaneously vary temperature, stress, and pressure according to predetermined time profiles.

GAS-TURBINE-ENGINE SIMULATION

Some of the more elaborate burner rig facilities are represented as turbine-engine simulators. In this report, however, these devices are classified as advance screening facilities.

There is very little reported information pertaining to the procedures employed by engine manufacturers for gas-turbine simulation tests. This is frequently considered proprietary information. At least one company uses a modular test engine into which experimental units can be mounted. Components can be tested in this manner without the expense of prototype engine fabrication.

REFERENCES

1. Krier, C. A. 1961. Coatings for the protection of refractory metals from oxidation. DMIC Report 161.
2. Procedures for evaluating coated refractory sheet. 1964. Materials Advisory Board, National Academy of Sciences, MAB-201M.
3. Perkins, R. A., and C. M. Packer. September 1965. Coatings for refractory metals in aerospace environments. Lockheed, AFML TR-65-351.
4. Davis, S. O., D. C. Watson, and C. R. Haulman. April 1965. Evaluation of resistance versus radiant heating method on the stress rupture properties of coated C103 columbium alloy in air at 2400°F. AFML TR-64-339.
5. Hill, V. L., and J. J. Rausch. September 1966. Protective coatings for tantalum-base alloys. IITRI, AFML TR-64-354 Part III.
6. Perkins, R. A. June 1962. Oxidation protection of structures for hypersonic re-entry. Lockheed, Summary of the Sixth Refractory Composites Working Group Meeting, ASD-TDR-63-610.
7. Wurst, J. C., J. A. Cherry, D. A. Gerdeman, and N. L. Hecht. July 1966. The evaluation of materials systems for high-temperature aerospace applications. Univ. of Dayton Research Inst., AFML TR-65-339 Part I.
8. Negrin, M. May and November 1966. Protective coatings for chromium alloys, Semi-Annual Reports 1 and 2. Chromalloy, NASA Contract NAS 3-7273.
9. Talboom, F. P., and J. A. Petrusha. Feb. 1966. Superalloy coatings for components for gas turbine engine applications. Pratt & Whitney, AFML TR-66-15.
10. Monson, L. A., and W. I. Pollock. March 1966. Development of coatings for protection of dispersion-strengthened nickel from oxidation, Part I Oxidation Studies, Coating Development, and Coating Analysis. DuPont, AFML TR-66-47.
11. James, W. A. January 1965. The evaluation and properties of coatings for superalloys and refractory metals. AGARD Working Paper, M35.
12. Epner, M. Aug. 1965. Activities of chromalloy division in the development of refractory composites (ductile aluminum-based coatings). Chromalloy Corp., Paper Presented at the Tenth Refractory Composites Working Group Meeting, AFML TR-65-207.
13. Moore, V. S., et al. September 1967. Evaluation of coatings for cobalt and nickel-base superalloys. Solar, Task I Report, NASA, NAS 3-9401.
14. Baker, F. R., et al. Aug. 1967. Further development of coatings for protection of dispersion-strengthened nickel from oxidation. DuPont, AFML TR-67-320.
15. Wheaton, H. L. July 1966–April 1967. Study of the hot corrosion of superalloys. Lycoming, Quarterly Progress Reports 1, 2 and 3, AF 33(615)-5212.
16. Ryan, K. H. Oct. 1966–April 1967. Investigation of hot corrosion of nickel-base superalloys used in gas turbine engines. Allison, Quarterly Progress Reports 2 and 3, AF 33(615)-5211.
17. Quigg, H. T., and R. M. Schermer. December 1964. Effect of JP-5 sulfur content on hot corrosion of superalloys in marine environment. Phillips Petroleum, Progress Report No. 2, NAVY BuWeps Contract 64-0443-d.
18. Glenny, E., and T. A. Taylor. 1959–60. A study of the thermal fatigue behavior of metals. Journal of the Institute of Metals, 88.
19. Private Communication with A. R. Stetson, Solar Research Laboratories, San Diego, California.
20. Glenny, E. 1961. Metallurgical reviews, 6:387–465.
21. Hauser, H. A., and J. F. Holloway. July 1966. Evaluation and improvement of coatings for columbium alloy gas turbine engine components. Pratt & Whitney, AFML TR-66-186 Part I.
22. Moore, V. S., and A. R. Stetson. Sept. 1963. Evaluation of coated refractory metal foils, Solar, RTD-TDR-63-4006 Part I.
23. Priceman, S., and L. Sama. Sept. 1965. Development of slurry coatings for tantalum, columbium, and molybdenum alloys. Sylcor, AFML TR-65-204.
24. Stetson, A. R., H. A. Cook, and V. S. Moore. June 1965. Development of protective coatings for tantalum base alloys. Solar, AFML TR-65-205 Part I.
25. Culp, J. D., and B. G. Fitzgerald. Oct. 1967. Evaluation of the fused slurry silicide coating considering component design and reuse. Report F904 AF 33(615)67-C-1574.
26. Stinebring, R. C. Jan. 1967. Understanding failure required for meaningful test selection. Proceedings of 1967 Annual Reliability Symposium, Washington, D.C.
27. Stinebring, R. C., and T. Sturiale. Nov. 1966. Development of nondestructive test methods for evaluating diffusion-formed coatings on metallic substrates. AFML TR-66-221.

28. Stinebring, R. C., and R. Cannon. June 1967. Development of nondestructive test methods for evaluating intrinsic variables in diffusion-formed coatings on metallic substrates, Part I. AFML TR-67-178.

29. Stinebring, R. C. Oct. 1968. Development of NDT techniques for detecting intrinsic variables in coated metals, Part II. AFML TR-67-178.

30. Shelton, W. L. Sept. 30, 1966. Two electron techniques for nondestructive materials evaluation. Proceedings of Symposium on Physics and Nondestructive Testing, Chicago, Illinois.

Bibliography

1. Ortner, M. H., and S. J. Kloch. Dec. 1966. Development of protective coatings for tantalum T-222 alloy. Vitro Laboratories, Final Summary Report, Contract NAS 3-7613, NASA Cr 54578.
2. Ortner, M. H., and K. A. Gebler. June 1965–June 1967. Electrophoretic deposition of refractory metals coatings. Vitro Laboratories, Interim Engineering Progress Reports 1–8, Contract AF 33(615)3090.
3. Criscione, J. M., J. Reper, and R. G. Fenish. July 1965. High temperature protective coatings for refractory metals. Union Carbide/Carbon Products Division, Annual Summary Report for NASA Contract NA Sw-1030.
4. Batiuk, W., et al. April 1967. Fluidized bed techniques for coating refractory metals. Boeing Aircraft, AFML TR-67-127.
5. Stacy, J. T. December 1963. Ductility of silicide coated TZM molybdenum alloy. Boeing Aircraft, AF 33(657)7132.
6. Engdahl, R. E., et al. 1966. Discussion on protective barrier systems requirements and three unique barrier systems for the protection of refractory metals at temperatures of 1300°–1450° and up to 2000°C. Consolidated Controls, Presented at the 12th Refractory Composites Working Group Meeting.
7. Phillips, W. M. January 1967. Evaluation of oxidation resistant coatings for inert-atmosphere application. Jet Propulsion Lab, TR 321014.
8. Lavendel, H. W. Aug. 1965. Investigation of modified silicide coatings for refractory metal alloys with improved low-pressure oxidation behavior. Lockheed, AFML TR-65-344.
9. Chao, P. J., G. J. Dormer, and B. S. Payne. May 1964. Advanced development of PFR-6. Pfaudler, ML-TDR-64-84.
10. Jefferys, R. A., and J. D. Gadd. 1961. Development and evaluation of high-temperature protective coatings for columbium alloys. TRW, Inc., ASD TR-61-66, Part II.
11. Gadd, J. D., and R. A. Jefferys. November 1962. Advancement of high temperature protective coating for columbium alloys. TRW, Inc., ASD-TDR-62-934.
12. Nolting, H. J., and R. A. Jefferys. May 1963. Oxidation resistant high temperature-protective coatings for tungsten. TRW, Inc., ASD-TDR-63-459.
13. Warmuth, D. B., J. D. Gadd, and R. A. Jefferys. April 1964. Advancement of high temperature protective coatings for columbium alloys (II). TRW, Inc., ASD TDR-62-934, Part II.
14. Gadd, J. D. April 1965. Advancement of protective coating systems for columbium and tantalum alloys. TRW, Inc., AFML TR-65-203.
15. Nejedlik, J. F. July 1966–April 1967. Coatings for long term intermediate protection of columbium alloys. TRW, Inc., Progress Reports 1, 2, & 3, Contract AF 33(615)5121.
16. Ebihorn, W. T. June 1966–June 1967. Development and characterization of high temperature coatings for tantalum alloys. TRW, Inc., Progress Reports 1, 2, 3, & 4, Contract AF 33(615)5011.
17. Hallowel, J. B., et al. April 1963. Coatings for tantalum base alloys. Battelle, ASD TDR-63-232.
18. Bartlett, R. W., and P. R. Gage. July 1964. Investigation of mechanisms for oxidation protection and failure of intermetallic coatings for refractory metals. Philco/Ford, ASD-TDR-63-733. Part II.
19. Lawthers, D. D., and L. Sama. May 1967. Development of coating for protection of high strength tantalum alloys in severe high-temperature environments. Sylcor, Progress Report on Contract AF 33(615)5086.
20. Priceman, S., and L. Sama. July 1967. Development of fused slurry silicide coatings for the elevated temperature oxidation protection of columbium and tantalum alloys. Sylcor, Progress Report on Contract AF 33(615)3272.
21. Stetson, A. R., and A. G. Metcalfe. September 1967. Development of coatings for columbium base alloys Part I: Basic property measurements and coating system development. Solar AFML TR-67-139.
22. Ohnysty, B., et al. January 1965. Task II—Development of technology applicable to coatings used in the 3000° to 4000°F temperature range. Solar, ML-TDR-64-294.
23. Wimber, T. M. November 1965. Development of protective coatings for tantalum base alloys. Solar, MLTDR-64-924, Part II.

207

24. Stetson, A. R., and V. S. Moore. June 1965. Development and evaluation of coating and joining methods for refractory metal foils. Solar, RTD-TDR-63-4006, Part III.

25. Wimber, R. T., and A. R. Stetson. July 1967. Development of coating for tantalum alloy nozzle vanes. Solar, NASA Cr-54529.

26. Dickson, D. T., R. T. Wimber, and A. R. Stetson. October 1966. Very high temperature coatings for tantalum alloys. Solar, AFML TR-66-317.

27. Aves, W. L., and G. W. Bourland. Aug. 1965. Investigation and development of techniques to extend the utility of pack processes and compositions for coating molybdenum and columbium alloys. LTV Aerospace Corp., AFML TR-65-272.

28. Wurst, J. C., and J. A. Cherry. September 1964. The evaluation of high temperature materials Section I, Evaluation of coatings for refractory alloys. Univ. of Dayton Research Inst., MLTDR-64-62.

29. Wurst, J. C., J. A. Cherry, D. A. Gerdeman, and N. L. Hecht. October 1966. The high temperature evaluation of aerospace materials. Univ. of Dayton Research Inst., AFML TR-66-308.

30. Wurst, J. C., W. E. Berner, J. A. Cherry, and D. A. Gerdeman. June 1967. The evaluation of materials for aerospace applications. Univ. of Dayton Research Inst., AFML TR-67-165.

31. Friedrich, L. A., E. C. Hirakis, and N. S. Bornstein. October 1965. Modified tin-aluminum slurry coatings for multi-thousand hour applications. Pratt & Whitney, TIM-945.

32. Bartsch, K. O., and A. Huebner. August 1966. The role of emittance in refractory metal coating performance. North American Rockwell, AFML TR-66-65, Part I Review and Analysis.

33. Leggett, H., J. L. Cook, D. E. Schwab, and C. T. Power. August 1965. Mechanical and physical properties of superalloy and coated refractory metal foils. Douglas, AFML TR-65-147.

34. Beck, E. J., and F. F. Schwartzberg. July 1965. Determination of mechanical and thermophysical properties of refractory metals. Martin, AFML TR-65-247.

35. Kerr, J. R., and J. D. Cox. February 1965. Effect of environmental exposure on the mechanical properties of several foil gage refractory alloys and superalloys. General Dynamics, AFML TR-65-82.

36. Brentnal, W. D., V. S. Moore, A. G. Metcalfe, and A. R. Stetson. August 1966. Protective coatings for chromium alloys. Solar, NASA CR 54535.

37. English, J. J., et al. January 1966–July 1966. Development of protective coatings for chromium base alloys. Battelle, First and Second Semi-Annual Reports NASA Contract NAS 3-7612.

38. Gadd, J. D. March 1966. Development of coatings for protection of dispersion strengthened nickel from oxidation. Part II: Development of Cr-Al coatings by vacuum pack techniques. TRW, Inc., AFML TR-66-47.

39. Wright, T. R., J. D. Weyand, and D. E. Kizer. July 1966. The fabrication of iridium and iridium alloy coatings on graphite by plasma arc deposition and gas pressure bonding. Battelle, First Quarterly Progress Report on Contract AF 33(615)3706.

40. Schulz, D. A., P. H. Higgs, and J. D. Cannon. July 1964. Research and development on advanced graphitic materials, Vol. XXXIV; Oxidation resistant coatings for graphite. WADD TR-61-72, National Carbon Company.

41. Krier, C. A., and J. M. Gunderson. February 1965. Oxidation resistant coatings, their applications and capabilities. Boeing Aircraft, Presented at the 1965 Western Metal Congress. ASM Technical Report W 11-4-65.

Members of Task Groups

AD HOC TASK GROUP ON CONCEPTS

Professor L. L. Seigle, *Chairman*
Department of Materials Sciences
College of Engineering
State University of New York
Stony Brook, L.I., New York 11790

Dr. Joan Berkowitz
Arthur D. Little Company, Inc.
35 Acorn Park
Cambridge, Massachusetts 02140

Dr. Robert I. Jaffee
Senior Fellow
Department of Physics
Battelle Memorial Institute
505 King Avenue
Columbus, Ohio 43201

Mr. Roger A. Perkins
Department 52-30
Palo Alto Research Laboratories
Lockheed Missiles & Space Company
3251 Hanover Street
Palo Alto, California 94304

Professor G. M. Pound
Department of Materials Science
Stanford University
Palo Alto, California 94305

Mr. Lawrence Sama, Manager
High Temperature Composites Laboratory
Chemical & Metallurgical Division
Sylvania Electric Products, Inc.
Hicksville, New York 11802

AD HOC TASK GROUP ON GAS-TURBINE APPLICATIONS

Dr. James R. Myers, *Chairman*
Civil Engineering School
Air Force Institute of Technology
Wright-Patterson AFB, Ohio 45433

Dr. John Gadd
Technological Development and Control
Jet Ordnance Division
TRW, Inc.
Harrisburg Works
1400 North Cameron Street
Harrisburg, Pennsylvania 17105

Mr. Frank Talboom
Project Metallurgist
Materials Development Laboratory
Pratt & Whitney Aircraft Company
East Hartford, Connecticut 06108

209

AD HOC TASK GROUP ON ROCKET PROPULSION APPLICATIONS

Dr. Herbert F. Volk, *Chairman*
Parma Technical Center
Carbon Products Division
Union Carbide Corporation
P.O. Box 6116
Cleveland, Ohio 44101

Mr. S. J. Grisaffe
Materials & Structure Division
Lewis Research Center, NASA
21000 Brookpark Road
Cleveland, Ohio 44135

Dr. James Stanley, Manager
Applied Metallurgy
Material Sciences Laboratory
The Aerospace Corporation
P.O. Box 95085
Los Angeles, California 90045

AD HOC TASK GROUP ON HYPERSONIC VEHICLE APPLICATIONS

Mr. Donald L. Kummer, *Chairman*
Chief Ceramics Engineer
Material & Process Department
McDonnell Douglas Aircraft Corporation
P.O. Box 516
St. Louis, Missouri 63166

Mr. Norman Geyer
Code MAMP
Air Force Materials Laboratory
Wright-Patterson AFB, Ohio 45433

Mr. Paul Lane, Jr.
Code FDMS
Wright-Patterson AFB, Ohio 45433

Mr. I. Machlin
AIR-520312
Materials Division
Naval Air Systems Command
Department of the Navy
Washington, D.C. 20360

Mr. Ken Marnoch
Mail Stop 20-01
The Marquardt Corporation
16555 Satiqoy Street
Van Nuys, California 91409

Mr. Eldon E. Mathauser, Head
Structural Materials Branch
Structures Research Division
Langley Research Center, NASA
Hampton, Virginia 23365

Dr. James R. Myers
Civil Engineering School
Air Force Institute of Technology
Wright-Patterson AFB, Ohio 45433

AD HOC TASK GROUP ON ENERGY CONVERSION APPLICATIONS

Dr. Robert I. Jaffee, *Chairman*
Senior Fellow
Department of Physics
Battelle Memorial Institute
505 King Avenue
Columbus, Ohio 43201

Mr. Robert L. Davies
Materials Section
Space Power Systems Division
Lewis Research Center, NASA
21000 Brookpark Road
Cleveland, Ohio 44135

Mr. James M. Taub
Group Leader—CMB-6
Los Alamos Scientific Laboratory
University of California
P.O. Box 1663
Los Alamos, New Mexico 87544

Dr. Herbert F. Volk
Parma Technical Center
Carbon Products Division
Union Carbide Corporation
P.O. Box 6116
Cleveland, Ohio 44101

AD HOC TASK GROUP ON INDUSTRIAL APPLICATIONS

Mr. S. J. Grisaffe, *Chairman*
Materials & Structures Division
Lewis Research Center, NASA
21000 Brookpark Road
Cleveland, Ohio 44135

Mr. John Wurst
Department of Ceramic Engineering
University of Illinois
Urbana, Illinois 61801

AD HOC TASK GROUP ON MANUFAC-TURING TECHNIQUES

Dr. John Gadd, *Chairman*
Technological Development and Control
Jet Ordnance Division
TRW, Inc.
Harrisburg Works
1400 North Cameron Street
Harrisburg, Pennsylvania 17105

Mr. Norman Geyer
Code MAMP
Air Force Materials Laboratory
Wright-Patterson AFB, Ohio 45433

Mr. Lawrence Sama, Manager
High Temperature Composites Laboratory
Chemical & Metallurgical Division
Sylvania Electric Products, Inc.
Hicksville, New York 11802

AD HOC TASK GROUP ON TESTING AND INSPECTION

Mr. John C. Wurst, *Chairman*
Department of Ceramic Engineering
University of Illinois
Urbana, Illinois 61801

Mr. Norman Geyer
Code MAMP
Air Force Materials Laboratory
Wright-Patterson AFB, Ohio 45433

Mr. Russell C. Stinebring
Supervising Engineer, Nondestructive Test Development
General Electric Company
3198 Chestnut Street, Room 5000 MB
Philadelphia, Pennsylvania 19101

Mr. Frank Talboom
Project Metallurgist
Materials Development Laboratory
Pratt & Whitney Aircraft Company
East Hartford, Connecticut 06108

Index

213